RAPTORS

RAPTORS

THE CURIOUS NATURE OF DIURNAL BIRDS OF PREY

Keith L. Bildstein

Comstock Publishing Associates
a division of
Cornell University Press
Ithaca and London

First published 2017 by Cornell University Press

Printed in the United States of America

Library of Congress Cataloging-in-Publication Data

Names: Bildstein, Keith L., author
Title: Raptors : the curious nature of diurnal birds of prey / Keith L. Bildstein.
Description: Ithaca ; London : Comstock Publishing Associates, a division of Cornell University Press, 2017. | Includes bibliographical references and index.
Identifiers: LCCN 2016044626 (print) | LCCN 2016045611 (ebook) | ISBN 9781501705793 (cloth : alk. paper) | ISBN 9781501707858 (epub/mobi) | ISBN 9781501707865 (pdf)
Subjects: LCSH: Birds of prey.
Classification: LCC QL677.78. B556 2017 (print) | LCC QL677.78 (ebook) | DDC 598.9—dc23
LC record available at https://lccn.loc.gov/2016044626

Cornell University Press strives to use environmentally responsible suppliers and materials to the fullest extent possible in the publishing of its books. Such materials include vegetable-based, low-VOC inks and acid-free papers that are recycled, totally chlorine-free, or partly composed of nonwood fibers. For further information, visit our website at www.cornellpress.cornell.edu.

To my father, Joe,
who opened my eyes to the world of birds
when I was nine

And to the many diurnal birds of prey
who have locked my vision
on them since

CONTENTS

Color photographs follow page 180.

PREFACE

SCIENTISTS STUDY what intrigues them most, and diurnal birds of prey—those predatory birds active during daylight hours—also known as raptors, have attracted my interest for as long as I can remember. Some of my colleagues have been drawn to study raptors—the hawks, eagles, and falcons of the world—by falconry, others by the more spectacular attributes of the birds themselves, including their ability to successfully capture, subdue, and kill large prey; to migrate long distances; to soar seemingly effortlessly for hours; to see movement much better than we; and to see colors that we can only imagine. As a scientist, my own interest in diurnal birds of prey has focused on their behavioral development and their movement ecology, along with their predatory abilities and social behavior.

Over the years, many other scientists have been equally intrigued by raptors. As a result of the work of my contemporaries and predecessors, we know more about raptor biology than we do about that of many other types of birds. And this, of course, is good, particularly when we are trying to protect them in our modern world. What is not good is that much of our knowledge about them lies hidden in a largely impenetrable morass of writings that we scientists call the technical literature. Consequently, raptors, although fascinating to many, remain an enigma to many more. The sizes of raptors and their ability to carry off large prey, for example, are frequently misjudged. Their predatory lifestyles are frequently misconstrued by people in ways that often make them more likely to be targeted for human persecution.

This book is an attempt to reveal highlights of the closely held technical literature about raptors so that lay naturalists, birders, hawk-watchers, science educators, school children, and the general public, along with new students in the field of raptor biology, can begin to understand and appreciate these birds. Written for those who want to know more about raptors as ecological entities, the book is intended to create a biological overview of this captivating group.

My journey through the technical literature has been a personal one, and in this book I have covered best what I know the most about. I have attempted to be rich in detail but not overwhelmingly so, as well as authoritative and instructive. I must admit, however, that I have not been encyclopedic. The growing raptor literature is simply far too vast and expansive for that. Many of the details described in these pages are answers to questions that I have been asked in my work as Sarkis Acopian Director of Conservation Science at Hawk Mountain Sanctuary, a place at which more than seventy thousand people a year come to see to see raptors in the wild and to learn more about them. Others are answers I've found in the literature to questions I myself have raised. Still others are answers to questions my own fieldwork helped me discover.

I have spent much of my scientific career studying raptors to better understand how these birds "work" biologically, how they affect the ecosystems in which they live, and, in turn, how humans and other aspects of their environments affect them. In some instances what I have learned has helped better protect birds of prey. Watching and studying raptors on five continents has instilled in me a firm belief in their importance, not only as ecological entities, but also as inspirational tokens of our natural and human-dominated landscapes. My hope is that this book and the answers it provides will infect readers with a similar enthusiasm for raptors, and that some will pursue careers that will help increase our understanding of these ecological actors.

What follows summarizes the institutional knowledge about raptors resulting from the careful studies of hundreds of scientists and raptor aficionados, without whose help I could not have written this book. Dean Amadon, David Barber, Marc Bechard, Jim Bednarz, David Bird, Gill Bohrer, Andre Botha, David Brandes, Leslie Brown, Tom Cade, Bill Clark, Michael Collopy, Miguel Ferrer, Laurie Goodrich, Frances Hamerstrom, Todd Katzner, Roland Kays, Robert Kenward, Lloyd Kiff, Sergio Lambertucci, Yossi Leshem, Mike McGrady, Bernd Meyburg, Ara Monadjem,

Peter Mundy, Juan José Negro, Ian Newton, Rob Simmons, Jean François Therrien, Jean-Marc Thiollay, Simon Thomsett, Munir Virani, Rick Watson, Martin Wikelski, and Reuven Yosef, esteemed respected colleagues and mentors all, immediately come to mind. There are many others as well, including many former students that I have had the opportunity to teach and learn from, and I thank them all. Numerous classes of international trainees at Hawk Mountain Sanctuary have heard or read most if not all of the book, and several have offered ideas for additional content. The board of directors of Hawk Mountain Sanctuary Association, together with my education and conservation science colleagues at the sanctuary, provided the intellectual environment necessary for me to complete this book, and I thank them for their support. I also thank all of the many fine people I have worked with at Cornell University Press, including my editor, Kitty Liu, and my production editor, Susan Specter.

I end with three disclaimers. First, *Raptors* focuses on diurnal birds of prey. Their nocturnal ecological equivalents, the owls, although mentioned throughout, are introduced largely in comparisons to demonstrate significant similarities and differences between these two wonderful groups of birds; they are sometimes noted but are not treated in equal detail. I have done so because my knowledge of the technical literature covering diurnal birds of prey is far more comprehensive than is my knowledge of the technical literature covering owls, and I have decided to go with my strength. It would be unfair to the owls to do otherwise. Second, this book does not offer "in one work all of the available knowledge on diurnal birds of prey of the world," a claim that Leslie Brown and Dean Amadon—the two great mid-twentieth-century pillars of raptor biology—justifiably made in their monumental *Eagles, Hawks and Falcons of the World* in 1968. The world of raptor biology was considerably smaller then than it is today, and what follows is best viewed as an introduction that briefly summarizes much of the current knowledge in the field, and not as a vast compendium of all that is now known. Third, I have written previously on the migration biology and conservation ecology of diurnal birds of prey, particularly in *Migrating Raptors of the World* (Cornell University Press, 2006). Chapters 7 and 8 of *Raptors* draw heavily on those writings, functioning as essential updates of those now-outdated efforts.

1 INTRODUCING RAPTORS

Birds evolved into the main groups or orders we know today at an early period in their history. The order Falconiformes, the subject of this book, is typical in this respect.
Leslie Brown and Dean Amadon, 1968

IF ONLY IT WERE AS SIMPLE to classify birds of prey now as it was in 1968. Although raptors are relatively easy to characterize, defining them biologically is another matter entirely. Let's start with their characteristics.

Raptors are relatively large and long-lived diurnal (active during daylight hours) birds of prey with keen eyesight. Highly successful, raptors first appeared in the fossil record some 50 million years ago. Today, they include approximately 330 species of hawks, eagles, falcons, vultures, and their allies globally. Found on every continent except Antarctica, as well as on many oceanic islands, raptors are one of the most cosmopolitan of all groups of birds. Many species migrate long distances seasonally, sometimes between continents. Others, including many that breed in the tropics, do not migrate at all, at least not as adults.

Like all birds, raptors have feathers and lay eggs. Unlike most other birds, however, raptors exhibit reversed size dimorphism, a phenomenon in which females are larger than their male counterparts. What follows is my attempt to detail the biology of these birds, to point out how they are both similar in many ways to other species of birds and, at the same time, distinctive. Although I sometimes mention owls, the nocturnal ecological equivalent of raptors, I do so mainly to point out similarities and differences between them and diurnal birds of prey. In this chapter, I define what makes a raptor a raptor, describe the different types of raptors and how they are related to one another, and describe their history in the fossil record. I also describe the common and scientific names of raptors. In sum, this chapter introduces the "players." The remaining chapters discuss their biology and ecology and their conservation status. (Technical terms are defined in the Glossary. Scientific names, or binomials, of all organisms mentioned in the book are listed in the Appendix.)

To reiterate, characterizing diurnal birds of prey—which is what I have done so far—is far easier than actually defining them. Unlike water, whose definition, if you'll pardon the pun, can be distilled into its chemical elements as simply H_2O, trying to define raptors is a bit more difficult.

Although there is a tendency to define the birds we group as raptors on the basis of their predatory habits, diurnal birds of prey are not defined by predation alone. Indeed, vultures, which most biologists (including me) call raptors, are largely nonpredatory, obligate scavenging birds. Furthermore, any bird that feeds on living animals is, by definition, predatory. But insect-eating warblers, worm-eating robins, and fish-eating herons are not considered raptors. So if predation does not distinguish raptors from other birds, what does?

Anatomy certainly plays a role. Unlike other "predatory" birds, raptors possess large, sometimes hooked beaks, and powerful, needle-sharp claws called talons. Most diurnal birds of prey use their talons to grasp, subdue, kill, and transport their prey. The English word "raptor" itself comes from the Latin combining form "rapt" meaning to seize or plunder. (That said, some raptors, including many falcons, use their beaks to kill their prey. But more on that later.) The oversized beaks and talons of raptors allow them to seize, rapidly kill, and consume many types of prey, including some that are relatively large compared with the raptor's own body mass and, as such, are potentially dangerous. Peregrine Falcons, for example, which are known to feed on more than 500 different species of birds globally, sometimes catch and kill waterfowl that are twice the mass that they are.

But in the end, ecology and anatomy alone do not define raptors. After all, several raptors, including most vultures, are not predatory at all, and a few eat fruit. Others have chicken-like beaks and modest talons. A third distinguishing characteristic, at least in part, is their evolutionary history. The ancestral relationship, or what biologists call their phylogeny, of diurnal birds of prey also helps define raptors. But unfortunately it, too, is not as simple as it appears. In fact, the quote at the head of this chapter is technically out-of-date.

All currently recognized groups of diurnal birds of prey trace their origins to several ancient avian lineages that evolved tens of millions of years ago. Some of the most recently published evidence, which are molecular phylogenies based on similarities in DNA sequencing to determine the lineages (i.e., birds that share more similar DNA are believed to be more

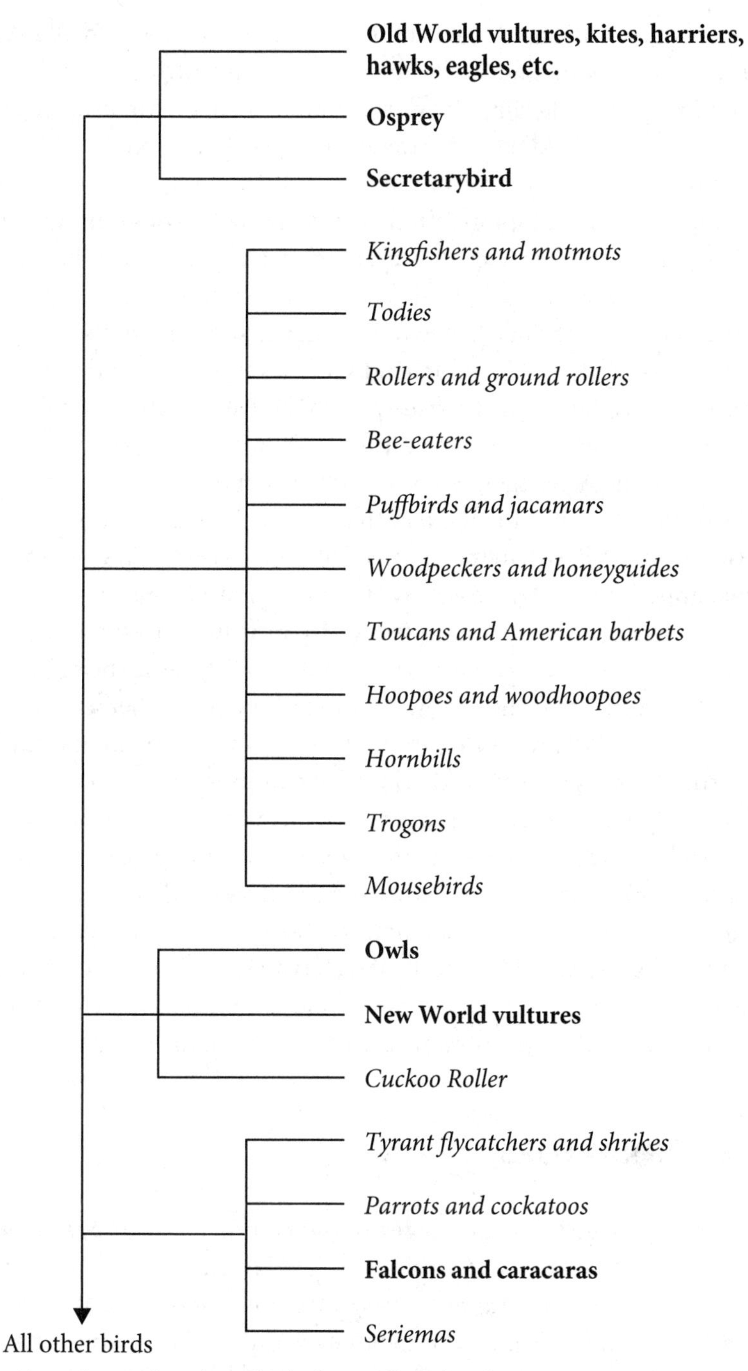

Family relationships involving diurnal birds of prey and owls based on a recent molecular analysis. Note the close relationships of most diurnal raptors and the more distant relationships of owls, New World vultures, and, particularly, falcons and caracaras. (After Ericson et al. 2006.)

closely related), indicate that hawks, eagles, kites, harriers, and Old World vultures (vultures that occur in Europe, Asia, and Africa) cluster tightly in the phylogenetic family Accipitridae and, thereafter, cluster together with the Secretarybird, the sole member of the family Sagittaridae, and the Osprey, the sole member of the family Pandionidae. These are what some might call "core raptors," in that they share a single common ancestor. These groups of species, however, are separated from other groups that most consider raptors by more than a dozen families of terrestrial and arboreal birds that include bee-eaters, trogons, and woodpeckers.

New World vultures (condors and vultures that are found only in the Western Hemisphere), in the family Cathartidae, which are phylogenetically distinct from Old World vultures, are next most closely related to the core raptors. After them come the falcons and caracaras, in the family Falconidae. Add to this the fact that nocturnal birds of prey, or owls, appear to be the closet living relatives of New World vultures, and that parrots appear to be the closest living relatives of falcons and caracaras, and you have a rather complicated and confusing ancestral lineage in which the birds we now call raptors are not monophyletic (belonging to a group derived from a single common ancestor), but rather are polyphyletic (belonging to a group with multiple origins whose biological similarities result from convergent evolution, not common ancestry).

Although the final verdict is not yet in, it is fair to say that the ancestral lineages of raptors are several and that whereas the group I call core raptors are each other's closet relatives, New World vultures and, to an even greater extent, falcons and caracaras, are not. All, however, are grouped as raptors because of their similar biological characteristics, not because of their evolutionary histories. So there you have it, about as complicated a definition as one can imagine. But even now, we are not yet out of the woods.

TYPES OF RAPTORS

Like other biologists, ornithologists define species as groups of interbreeding birds that are reproductively isolated from other groups of birds. Overall, members of the same species share suites of traits that other species lack. Such traits, which can include overall body size and shape and other anatomical features, as well as song, habitat use, diet, and behavioral traits, are said to be defining characteristics of a species. Traditionally, avian biologists used such features to classify raptors into species, as well as to determine the ancestral relationships among them. More recently, the DNA genetic analyses mentioned above have been used to determine which species fall into the different categories of raptors.

Old and New World Vultures

Ornithologists currently recognize twenty-two species of largely carrion-eating birds of prey as vultures, including seven species of New World vultures and condors and fifteen species of Old World vultures. (Africa's Palm-nut Vultures, despite their name, are not included here, as their ancestry remains unclear.) The most recently recognized species of vultures is southern Asia's Slender-billed Vulture, which was recognized as a distinct species around 2005 based on both molecular and anatomical evidence and separated from the Indian Vulture with whom it was formerly lumped. Africa, with eleven species of Old World vultures, is the center of vulture diversity in the Old World. South America, with six species of New World vultures, is the center of vulture diversity in the New World. There are no vultures in Australia or Antarctica and, with the exception of Egyptian Vultures and Turkey Vultures, vultures typically do not occur on oceanic islands.

The phylogeny of New World vultures, a distinct group of seven Western Hemisphere birds of prey that is distributed throughout most of the Americas, and that includes North America's California Condor and South America's Andean Condor, has long attracted the attention of avian taxonomists. Although they closely resemble Old World vultures in their feet, beaks, and largely featherless heads, New World vultures differ substantially from other diurnal birds of prey, including Old World vultures, in many anatomical and behavioral ways, which over the years have led many ornithologists to question their ancestral relationships.

In the early 1800s, taxonomists placed the New World vultures in a separate family from all other raptors, and in 1873, the influential English vertebrate zoologist A.H. Garrod removed the New World vultures from other raptors entirely, placing them next to storks in the family Ciconiidae, mainly because of similarities in the arrangements of their thigh muscles. The taxonomic confusion continued well into the twentieth century. In 1948, Washington State University's George Hudson claimed that New World vultures "have no more natural affinity with hawks and falcons than do owls" (something that more recent DNA analyses have confirmed). In the 1960s, two different biologists using similarities in egg-white proteins to determine bird relationships reached opposite conclusions regarding the New World vultures and other raptors. One concluded the two groups were closely related, the other that they were not.

In 1967, the University of Michigan's J. David Ligon published a paper that linked New World vultures to storks rather than to other raptors on the basis of "striking anatomical similarities" including perforated nostrils, nestling plumages, and the lack of syringeal, or "song-box," muscles. In 1983, the California museum worker Amadeo Rea reached essentially the same conclusion, pointing out several behavioral similarities between the two groups, including disgorging the contents of their stomachs when frightened, and urohidrosis, the habit of urinating on one's legs, presumably for thermoregulation.

In 1990, a then seminal molecular analysis by Charles Sibley and John Ahlquist at Yale University appeared to close the book on the subject when it concluded that the then popular DNA-DNA hybridization technique also linked New World vultures to storks. Shortly thereafter, the American Ornithologists' Union, the final arbitrator in North American ornithological taxonomy, removed New World vultures from the avian order Falconiformes and placed them next to storks in the wading-bird order Ciconiiformes. (Note that avian orders such as Falconiformes are used by taxonomists to group similar families of birds.)

The taxonomic tide, however, began to shift in 1994, when the American Museum of Natural History researcher Carole Griffiths published a detailed examination of the syringeal anatomy of raptors and several other groups of birds including both owls and storks. After examining more than 180 song boxes, Griffiths concluded that New World vultures were not at all closely related to storks and indeed belonged in the order Falconiformes. More recent DNA analyses, including that mentioned above, support her assessment and in 2007, the American Ornithologists' Union reversed itself and removed New World vultures from the order Ciconiiformes and placed them back in the order Falconiformes, making them, once again, raptors. Although the meandering history of the official taxonomic relationships of New World vultures offers something of an extreme case of shifting evidence and ideas, the more recent placement of falcons and their allies next to parrots confirms that a lot more remains to be learned regarding the ancestral relationship of the birds we call raptors.

Hawks, Eagles, and Their Allies

Estimates of the number of species of hawks, eagles, kites, and harriers alive today range from about 235 to about 240. There are approximately 35 species of kites and kite-like birds, 13 harriers, 60 eagles, and 120 or so hawks and hawk- and eagle-like birds. Representatives of these groups exist on many oceanic islands and on all continents except Antarctica. Several species within these groups are particularly widely distributed. The Northern Harrier, for example, breeds in North America, Europe, and Asia, and winters in all three continents, as well as in Central America, northern South America, and North Africa.

The English, or common, names of birds of prey can be confusing. A perfect example of this are the words "hawk" and "eagle," which are featured in the common names of many birds of prey, including several owls. Although one might expect that these two types of birds of prey differ in their ancestry, in fact the words "hawk" and "eagle" have little phylogenetic meaning.

Eagles include approximately sixty species from four decidedly different groups of raptors that have little in common except their last name.

Hawks, on the other hand, make up a rather diverse collection of about 120 species of raptors that are more readily characterized by what they are not (e.g., they are not ospreys, vultures, kites, harriers, eagles, caracaras, kestrels, falcons, or falconets) than by what they are. Adding to this confusion, many eagles are more closely related to hawks than to other eagles, and vice versa. And as if this weren't confusing enough, some raptors that North American ornithologists call hawks are known as buzzards in Europe, where the word "buzzard" (*buse* in French and *busardo* in Spanish) identifies raptors in the genus *Buteo*, a group of twenty-eight species that in North America includes the Red-tailed Hawk and Broad-winged Hawk, and in Europe includes the Common Buzzard and Long-legged Buzzard. But if hawks and eagles are not taxonomically distinct, what does make a hawk a "hawk" and an eagle an "eagle"?

The English word "eagle" is from the French, *aigle*, which is from the Latin, *aquila*, meaning eagle. *Aquila* is derived from the word-root *aquil*, which means dark. Etymologists suggest that the name *aquila* was used to describe the largest raptors, either because their expansive silhouettes "darkened" the sky, or because many had dark plumages. Whether the birds the Romans called eagles actually were eagles or large vultures

Box 1.1. Types of eagles

Sea eagles. Sea eagles, or fish eagles, are a group of ten species of largely aquatic birds of prey in the genera *Haliaeetus* and *Ichthyophaga*. North America's Bald Eagle, Eurasia's White-tailed Eagle, and the African Fish Eagle, are all sea eagles. The group, which is found worldwide, excepting Central and South America, appears to be closely related to some kites.

Snake eagles. Snake, or serpent, eagles are a group of fourteen species of short-toed Old World eagles in the genera *Circaetus, Spilornis, Dryotriorchis*, and *Eutriochis*. Most snake eagles, which feed largely on snakes and other reptiles, occur in Africa. Snake eagles appear to be closely related to the kites, as well as to another confusingly named group of raptors, the harrier-hawks.

Harpy or buteonine eagles. Buteonine, or buteo-like, eagles, make up a group of six species of Pacific Island and South American birds of prey. As their name suggests, buteo-like eagles are most closely related to the buteos, a group of broad-winged, short-tailed diurnal raptors including North and Central America's Red-tailed Hawk and Europe's Common Buzzard.

Booted eagles. Booted, or true, eagles make up a group of about thirty species of birds of prey whose legs and feet—unlike those of all other eagles—are fully feathered, or "booted," to the toes. Booted eagles, including the circumboreal Golden Eagle, are closely related to buteos.

remains yet another unresolved issue. *Nesher*, the Hebrew word for eagle, for example, is used in the Old Testament to describe both large vultures and eagles. Today we use the word eagle to describe especially large birds of prey, many of which weigh more than 2 kg (about 5 lb), and several of which weigh more than 6 kg (about 15 lb).

"Hawk" is a Middle English word derived from the Old English *hafoc*, as in "to have," in the sense of to grasp or to seize. Unlike the word eagle, there is nothing in the etymology of the word hawk that suggests size. And, indeed, many hawks are quite small. Male Sharp-shinned Hawks, for example, can weigh fewer than 100 g (4 oz). And about half of all hawks weigh fewer than 450 g (1 lb).

In the end, eagles are "eagles" and hawks are "hawks" mainly because of their relative sizes. Eagles are scaled-up versions of hawks: fierce, diurnal birds of prey with particularly large wingspans, beaks, and talons. North America's two species of eagles, the Bald Eagle and the Golden Eagle, each have wingspans that exceed 2 m (or greater than 6 ft), whereas the wingspans of North America's seventeen species of hawks range in size from about 0.5 to 1.5 m (or 20 in to about 5 ft). The size differential between hawks and eagles also holds in most other parts of the world.

That said, the sizes of hawks and eagles do overlap, and several large hawks are larger than several small eagles. North America's Red-tailed Hawk, for example, averages considerably larger than Australia's Little Eagle (1100 g, or 2.4 lb, compared with 820 g, or 1.8 lb). The fact that more than a dozen medium-size eagles are called "hawk-eagles," and that one South American eagle is called a "buzzard-eagle," only adds to the confusion.

Table 1.1 Raptors with potentially confusing common names

Common name	Mean mass (kg)	Continental distribution
Hawk-eagles		
African Hawk-Eagle	2.0	Africa
Ayres's Hawk-Eagle	0.8	Africa
Black-and-white Hawk-Eagle	0.9	Tropical America
Black Hawk-Eagle	1.0	Tropical America
Blyth's Hawk-Eagle	0.8	Asia and Pacific Islands
Cassin's Hawk-Eagle	1.0	Africa
Sulawesi Hawk-Eagle		Sulawesi
Crested Hawk-Eagle	1.5	Asia and Pacific Islands
Changeable Hawk-Eagle	1.5	Asia and Pacific Islands
Javan Hawk-Eagle		Java
Mountain Hawk-Eagle	3.0	Asia
Ornate Hawk-Eagle	1.2	Tropical America
Philippine Hawk-Eagle	1.1	Philippines
Wallace's Hawk-Eagle	0.6	Asia and Pacific Islands
Buzzard-eagles		
Black-chested Buzzard-Eagle	2.2	South America
Harrier-hawks		
African Harrier-Hawk	0.6	Africa
Madagascar Harrier-Hawk	No data	Madagascar

And it does not end with diurnal birds of prey. Owls, too, can have confusing hawk and eagle names. "Hawk-owls," and there are more than ten of them, tend to have longer and narrower wings and longer tails than most owls, along with less conspicuous facial discs, all features that make them appear hawk-like. On the other hand, "Eagle-owls" are some of the largest and most powerful owls.

Falcons, Caracaras, and Their Allies

Estimates of the number of species of falcons and their close relatives range from sixty to sixty-two, including thirty-eight to forty species of true falcons, at least seven forest falcons, seven pygmy falcons and falconets, nine caracaras, and the Laughing Falcon. The group exists on numerous oceanic islands and on all continents except Antarctica. The Peregrine Falcon, which breeds on six continents, is one of the most cosmopolitan of all birds.

Other Raptors

Ornithologists currently recognize one species of Osprey, which, except for the fact that it does not occur in Antarctica and does not breed in South America and southern Africa, is cosmopolitan, and a single species of Secretarybird, which occurs in sub-Saharan Africa.

By comparison, recent estimates of the number of species of owls, or nocturnal birds of prey, range from 162 to 178; including 149 to 161 species of what some ornithologists call true owls and their kin, and 13 to 17 Barn Owls. Like diurnal birds of prey, owls are found on many oceanic islands and on all continents except Antarctica. The Barn Owl is a nearly cosmopolitan species found throughout the warm and temperate regions of the world, as well as on many islands. The Short-eared Owl occurs in both North and South America and in Eurasia, as well as on many islands, including the West Indies, the Caroline Islands, the Galapagos Islands, the Falkland Islands, and Hawaii.

In sum, and depending on one's taxonomic proclivities, there are about 480 to 505 species of diurnal and nocturnal birds of prey, including about 310 species of raptors that feed principally on living prey, 22 species of raptors that feed principally on carrion, and 162 to 178 species of owls, all of which feed principally on living prey.

There are no nocturnal "vultures" most likely because low-cost soaring flight is difficult if not impossible over land at night, and many species of mammals, some of which would pose a threat to nocturnal avian scavengers, feed at night. Owls are thought to be less migratory than are diurnal birds of prey, and when the numbers of migratory species are subtracted out of the two groups, the remaining numbers of species—about 130 to 140 for each group—are similar. Viewed in this light, migration appears to have played a key role in the greater diversity of diurnal birds of prey.

Fossil and Extinct Raptors

Fossils are the remains or traces of organisms, including birds that are preserved in rocks. Because birds tend to be small and have delicate bones, their fossil record tends to be sparse compared with animals with sturdier bones. Unlike modern birds, most fossil birds do not have common names. The most widely known, and one of the oldest, fossil birds is *Archaeopteryx lithographica*, an ancient reptile-like but feathered creature that lived in Europe during the Jurassic Period 150 million years ago (MYA). The fossil record for diurnal birds of prey dates from the early Eocene of England, 50 MYA. A reliable fossil owl dates from the Paleocene of Colorado, 60 MYA.

Sixty-two palaeospecies of diurnal birds of prey were recognized by 1964. Many more have been described since. Some of the oldest known raptor fossils include well-preserved skulls of a small, presumably forest-dwelling raptor found in German oil shales, and a small falcon-like raptor whose fossilized remains have been collected in both France and the United States.

The New Zealand Eagle, the largest eagle that ever lived, is known only from fossils. This eagle, which had a wingspan of up to 3 m (9.5 ft) and tiger-sized talons, is said to have fed on extinct, flightless, emu-like birds called moas. Prior to the arrival of Polynesian settlers in New Zealand, New Zealand Eagles were the principal predators of moas, all of which disappeared shortly after the first human colonists started to feed on them. The largest fossil raptors of all were Ice Age teratorns. Believed to be related to New World vultures, some of these ancient soaring birds had wingspans of approximately 5 m (more than 15 ft). Extinct fossilized forms of these scavengers include *Teratornis merriami*, a 15-kg (33-lb) mega-vulture with a 4-m (>12-ft) wingspan, and the even larger

Teratornis incredibilis of Nevada and California, whose wingspan is said to have exceeded 5 m (17 ft). Although changing climates may have precipitated widespread extinctions in the Pleistocene, some paleontologists believe that increasingly sophisticated hunting by expanding populations of primitive humans also contributed to the extinction of many large mammals. If this is true, then humans may have been responsible, albeit indirectly, for the demise of the teratorns.

The most recent raptor to become extinct was Mexico's Guadalupe Caracara, a close relative of the Northern Crested Caracara. This somewhat vulture-like aggressive scavenger had the misfortune of occurring in low numbers on a small island off the west coast of Baja California, where it was exterminated by goat herders who believed the birds were killing young goats. The species was last seen alive, and the last specimens collected, in December 1900. Although no other raptors have since become extinct, at the beginning of the twenty-first century, fifty-five species of hawks, eagles, and falcons were globally threatened or endangered. The plights that these and other raptors now face from human threats are the subject of Chapter 8.

The fossil record of raptors provides many insights into the ancient distributions and geographic origins of several living species. Intriguingly, the fossils of so-called Old World vultures occur in the fossil record of North America, whereas those of New World vultures occur in the fossil record of Europe. Ancestral Ospreys are known to have occurred at least as far back as the Oligocene Epoch 30 MYA. Broad-winged Hawks, together with several other North American migrants, including Cooper's Hawks, Red-shouldered Hawks, and Merlins, are represented in the late Pleistocene almost as far back as 400,000 years ago.

Unfortunately, most present-day living species of birds of prey have little, if any, fossil record, suggesting that many extinct species of raptors likely also left little trace in the fossil record. Even so, the ones mentioned above that do exist tell interesting stories. I conclude this section with some background information about "extinction."

Extinction, the complete elimination of all members of a species, is a natural evolutionary process. And although human actions are increasing the rates at which species, including birds of prey, are threatened by extinction (see Chapter 8 for details), extinction has been under way almost since the first birds of prey evolved. Indeed, it is likely that there are many more species of extinct raptors than of living ones.

Raptors can become extinct in two very different ways. The first occurs when a species evolves to the point that it no longer resembles and—more importantly—no longer could reproduce with its distant ancestors. When this happens, the ancestral form is said to be extinct, and the new, or descended, form is given a new species name. Biologists refer to this type of extinction as pseudo- or phyletic extinction, and the creation of the new species is called successional speciation. In this type of extinction, raptor diversity does not change. The second type, and the one most people are familiar with and many are concerned about, occurs when all members of a species die off or are killed, leaving no descendants, and their phylogenetic lineage is terminated. In this type of phylogenetic, or terminal, extinction raptor diversity is reduced.

In both types of loss, rapidly changing environments substantially increase the occurrence of extinction. At the end of Pleistocene Ice Ages, some 10,000 years ago, for example, climates changed rapidly and many of the world's largest mammals became extinct. This, in turn, led to the terminal extinction of many species of vultures that fed on, and whose survival depended on, the carcasses of large mammals including the teratorns mentioned above.

SCIENTIFIC NAMES OF RAPTORS

As with all birds, each species of raptor has two types of names: its common name, and its official, scientific, name. Scientists often refer to scientific names as binomials, because these names consist of two parts. In most cases scientific names consist of words taken directly from classical Latin or so-called Latinate or Latinized words, which are taken from ancient Greek or other, mainly European, languages. In rare cases binomials are onomatopoeic. For the Southern Crested Caracara, for example, a crowlike South American raptor closely related to falcons, both its common name and its scientific name, *Caracara plancus*, are based on an onomatopoeic South American Indian word, which is based on the species' gutteral kik-kik-kik-kik-kerr call.

The first part of a scientific name, which is always capitalized, is the genus to which the species belongs and is referred to as the "generic" name. The second part of a scientific name, which is not capitalized, is the species, or "specific," name. In most publications, including this one, scientific names are italicized. Genera (the plural of "genus") consist of groups of two or more closely related species, or, in a few instances that I will note below, single species that are not closely related to any other species. The three largest genera of raptors are *Accipiter,* a group of fifty species of hawks that includes the Northern Goshawk; *Buteo,* a group twenty-eight species of hawks that includes the Red-tailed Hawk; and *Falco,* a group of thirty-eight kestrels and falcons that includes the Peregrine Falcon. And literally, true to form, these three genera can be described using only a few "generic" characteristics: members of the genus *Accipiter* are largely small to medium-sized forest-dwelling raptors with relatively short, rounded wings and long tails; members of the genus *Buteo* are largely medium- to large-sized open-habitat raptors with relatively broad wings and short tails; and members of the genus *Falco* are largely small to medium-sized open-habitat raptors with pointed wings and longish tails. If I tried to do this at the species level, I would need a far longer and decidedly more "specific" list of characteristics.

One significant difference between common and scientific names is that scientific names are guided by a universal set of rules, whereas common names are not. The scientific names of raptors—along with those of all other animals—are determined by a strict set of universal rules laid out in the International Code of Zoological Nomenclature. The rules are designed to increase the stability and universality of scientific names, which helps to improve communication among scientists that are working in different countries and speaking different languages.

Consider for example the Red-tailed Hawk, a common raptor throughout much of its North and Central American range. The scientific name of this largely open-habitat, medium-sized raptor with relatively broad wings and a short tail, which originally was collected on the island of Jamaica, is *Buteo jamaicensis*, or "Buteo of Jamaica," with the Latinized species name. In addition to the binomial, each species' official scientific name includes two additional parts, the name of the author who first named the species, and the year that the name was first published and officially recognized (e.g., *Buteo jamaicensis*, Gmelin, 1788). Although these two additional parts at times can be important, they are rarely used except in the technical taxonomic literature.

Some species of raptors, especially those that have large ranges, are divided further into "subspecies," or geographic races. Subspecies are groups within a species that, although anatomically distinct, can still interbreed with one another. The resulting scientific names of such groups are called "trinomials" because they contain a third, "subspecific," name, which follows the species name. More than a dozen subspecies of Red-tailed Hawks, for example, have been recognized by scientists, including eastern North America's *Buteo jamaicensis borealis* (*Buteo* of Jamaica found in the Boreal Forest).

Three principles guide the assignment of scientific names: (1) priority of names, (2) uniqueness of genus names, and (3) the conservation of traditional names. The principle of priority means that the first properly applied name for an organism is viewed as the correct name. Thus, if two or more individual organisms are originally described in the taxonomic record, and are named as representatives of separate species when, in fact, they represent a single species, when they are joined into a single species, the older name prevails. The principle of uniqueness of genus names means that each genus name can be used only once in the animal kingdom and only once in the plant kingdom. Species names, by contrast, can be used multiple times in each kingdom, but only once within each genus. This principle assures that all species of plants and animals bear unique scientific names. The principle of the conservation of traditional names means that well-established names are not replaced if a long-forgotten and unused prior name for a species exists and resurfaces.

The value of a global system of unique, all-inclusive names becomes obvious when one considers the difficulties associated with the use of common or vernacular names. The common names of the tundra-dwelling raptor with the scientific name of *Buteo lagopus* provide a good example. This raptor, which is known by a single scientific name throughout its North American, European, and Asian range, is called the Buse Pattue in France, the Rauhfußbussard in Germany, the Busardo Calzado in Spain, the Zimnyak in Russia, the Mao Jiao Kuang in China, the Rough-legged Buzzard in European English and the Rough-legged Hawk in North American English! Given the two latter English names, a recent list of "recommended English names" simply calls the bird the Roughleg, which is what I use here.

Many bird-watchers find it difficult to remember and understand the scientific names of the raptors they see in the field. Because of this, many bird-watchers refuse to use them despite their usefulness. One of the easiest ways of learning the scientific names of raptors is to develop an understanding of their component parts—their Latin, Greek, or other language roots—and then practice pronouncing them. Box 1.2 provides explanations for the binomials of eastern North American diurnal birds of prey, as well as information on how to pronounce them. The Appendix includes the binomials of all raptors, globally.

Box 1.2 Derivations of the scientific names of diurnal birds of prey: Examples from North America

Notes: Accents indicate the syllable to be stressed. The ascending accent (as in *Corágyps*) indicates the shortened sound of a vowel, the descending accent (as in *atràtus*) indicates a long English sound of the vowel. L = classical Latin; ML = middle Latin; G = ancient Greek.

Black Vulture (*Corágyps atràtus*) G. Cora + gyps: a crow or raven + a vulture. L. atrat: clothed in black as for mourning. The genus name is ancient Greek for crow-vulture. The species name refers to the species generally black plumage.

Turkey Vulture (*Cathártes áura*) G. Cathart: cleansing. American Indian (Mexican) aura (auroua): Turkey Vulture. The genus name refers to the species' scavenging habits. The species name is American Indian (Mexican) for Turkey Vulture.

Osprey (*Pandíon haliaètus*) G. Pandion: a mythological king of Athens. G. haliaet: a sea-eagle or osprey. The genus name is that of the mythological king whose daughters, Philomela and Procne, were metamorphosed into a nightingale and swallow, respectively. The species name is Greek for sea-eagle or osprey.

Bald Eagle (*Haliaèetus leucocéphalus*) G. haliaet: a sea-eagle or osprey. G. leuc + cephal: white + the head. The genus name is a misspelling of two Greek words "halos," the sea, and "aetos," the eagle. The species name refers to the species' white head.

Northern Harrier (*Círcus cyáneus*) G. circ: a hawk that wheels or circles. G. cyan: dark blue. The genus name is from a mythological hawk that circles. The species name is from the male's gray-blue back.

Sharp-shinned Hawk (*Accípiter striàtus*) L. accipere: to grasp or to take. L. stria: streaked, furrowed. The genus name is from classical Latin "accipere," which means to understand, grasp, or take. The species name is for the streaks on the underside feathers of juveniles.

Cooper's Hawk (*A. coòperii*) L. accipere: to grasp or to take. [English] Cooper: after William Cooper. The genus name is originally derived from the classical Latin "accipere," which means to understand, grasp, or take. The species was named for William Cooper (1798–1864), one of the founders of the New York Lyceum of Natural History, and the father of California ornithologist J. C. Cooper.

Northern Goshawk (*A. gentílis*) L. accipere: to grasp or to take. L. gentilis: belonging to the clan or tribe. The genus name is originally derived from the classical Latin "accipere," which means to understand, grasp, or take. The species name is from the classical Latin "gens" meaning tribe or clan. "Gentilis" has come to mean belonging to the wellborn, i.e., with class.

Red-shouldered Hawk (*Bùteo lineàtus*) L. buteo: a kind of hawk. L. lin: a line. The genus name is from the classical Latin for hawk or buzzard. The species name probably refers to the species' black-and-white banded wings and tail as seen from above.

Broad-winged Hawk (*B. platýpterus*) L. buteo: a kind of hawk. G. plat + pter: flat or broad + a wing. The genus name is from the classical Latin for hawk or buzzard. The species name refers to its broad wings.

Swainson's Hawk (*B. swàinsoni*) L. buteo: a kind of hawk. [English] swainson: after William Swainson. The genus name is from the classical Latin for hawk or buzzard. The species was named by Charles Lucien Bonaparte for William Swainson (1789–1855). Swainson was a well-traveled and highly regarded naturalist, who met Audubon in England and favorably reviewed his work.

Red-tailed Hawk (*B. jamaicénsis*) L. buteo: a kind of hawk. ML. jamaicensis: of Jamaica. The genus name is classical Latin for hawk or buzzard. The species name refers to the place where the first specimen was collected.

Roughleg (*B. lagòpus*) L. buteo: a kind of hawk. G. lago + -pus: rabbit or hare + footed. The genus name is classical Latin for hawk or buzzard. The species name refers to the species' feather-covered tarsometatarsi (shanks).

Golden Eagle (*Aquila chrysàetos*) L. aquil =a; an eagle. G. chrys + aetos: gold + eagle. The genus name is classical Latin for eagle. The species name is Greek for golden eagle, referring to the species' golden-feathered nape.

American Kestrel (*Fálco sparvérius*) L. falco: a falcon. L. sparveri: of a sparrow. The genus name is classical Latin for falcon. The species name refers to the bird's habit of preying heavily on sparrows, based on its initial misidentification as the Eurasian Sparrowhawk by European settlers. As a result of this misidentification, the species' common name into the middle of the twentieth century was Sparrow Hawk.

Merlin (*F. columbárius*) L. falco: a falcon. L. columb: of a dove or pigeon. The genus name is classical Latin for falcon. The species name refers to the species' reputation for taking pigeons.

Peregrine Falcon (*F. peregrìnus*) L. falco: a falcon. L. peregrin: a wanderer, traveler. The genus name is classical Latin for falcon. The species name refers either to the species' wandering tendencies, or to the fact that falconers usually captured the species in passage (i.e., during migration) rather than at the nest.

Gyrfalcon (*F. rustícolus*) L. falco: a falcon. L. rustic: of the country, a rustic. The genus name is classical Latin for falcon. The species name refers to the species' distribution in remote tundra regions.

SYNTHESIS AND CONCLUSIONS

1. Raptors are relatively large, long-lived diurnal birds of prey with keen eyesight and sharp talons.
2. Like all birds, raptors have feathers and lay eggs.
3. Unlike most birds, raptors exhibit reversed size dimorphism, a phenomenon in which females are larger than their male counterparts.

4. Aside from their characteristic ecology and anatomy, raptors are defined, but only in part, by their ancestry.
5. Although not all each other's closest living relatives, raptors belong to five groups of birds: New World vultures; hawks, eagles, and their allies; falcons, caracaras, and their allies; the Osprey; and the Secretarybird.
6. Altogether there are about 480 to 505 species of diurnal and nocturnal birds of prey. These include about 310 species of raptors that feed principally on living prey, 22 species of raptors that feed principally on carrion, and 162 to 178 species of owls, all of which feed principally on living prey.
7. Diurnal birds of prey first appear in the fossil record in the Eocene Epoch 50 MYA.
8. All raptors have common and scientific species names. Scientific species names, which are universal, are called binomials, and include a Latinized genus and species name. Common names, which are descriptive, vary geographically.

2 FORM AND FUNCTION

> Science owes to hawking the first scientific study of bird structure . . . Frederick the II's thirteenth century manuscript *The art of falconry.*
> **Nick Fox, 1995**

TO APPRECIATE AND CARE for the world's diurnal birds of prey, you need to know something about their biology. Falconers have known this for thousands of years. Today, raptor biologists and conservationists realize it as well. Understanding how raptors work as ecological units in natural- and human-dominated landscapes requires an understanding of their sizes, shapes, constructions, and colorations, as well as how they grow, develop, maintain their internal functions, and die. I address each of these areas of their biology in this chapter. Chapter 3 addresses another important aspect of raptor biology, sensory abilities, or how raptors see, hear, smell, feel, and comprehend the world in which they live. These chapters lay the biological foundation for understanding the global distribution and abundance of the world's diurnal birds of prey (discussed in detail in Chapter 4); their breeding, feeding, and movement ecology (Chapters 5, 6, and 7); and the risks that they face in our modern world (Chapter 8).

I begin this chapter by describing the feathers and the featherless body surfaces of raptors, their wings and tails, and their body masses. I then detail their growth and development, discuss their warm-bloodedness, and conclude with the factors that influence their health, the things that kill them, and how long they live.

FEATHERS

Feathers, not flight, distinguish living birds from all other living animals. If an animal has feathers, it is a bird; if it does not have feathers, it is not a bird. It is that simple. The telltale impressions of feathers helped nineteenth-century scientists identify the more than 150-million-year-old fossilized remains of *Archaeopteryx lithographica* as those of an ancient bird. The fossil remains of extinct birds found more recently, both older and newer species, also left traces of their feathers. Most recently, paleontologists discovered traces of feathers on fossilized theropod dinosaurs that clearly were not ancient birds. As a result, paleontologists now recognize that ancient reptiles, as well as ancient birds, had feathers. Nevertheless, the first three sentences of this paragraph still hold true for all living organisms. Nowadays, feathers do define living birds, and they, along with long-distance migration, best characterize modern birds anatomically and behaviorally.

But feathers are far more important to raptors than simply as signifiers of their avian heritage. Feathers provide a rigid airfoil for flight and act to protect and insulate the bodies of raptors from solar radiation and low temperatures. Together with body oils, feathers help shield their skin from dust, debris, and, to a limited extent, water. Tightly packed specialized head feathers and feathers arranged in facial disks help owls, as well as some diurnal birds of prey, hear better (see Chapter 3 for details). The various colors and patterns of feathers help to both camouflage and advertise the presence of a raptor.

Not surprisingly, raptors have lots of feathers, especially around heat-sensitive areas such as the head and neck, where about 40% of all feathers are found. Smaller birds tend to have fewer feathers than do larger birds. A 4- to 5-g (0.14-oz) Ruby-throated Hummingbird, for example, has about 950 feathers, compared with the 25,000 feathers on an 8-kg (18-lb) Tundra Swan. But what about diurnal birds of prey? Despite their tremendous importance, the total number of feathers for most species of raptors remains unstudied for most species. Size-for-size, owls, which are active during the coolest parts of the 24-hour cycle, have more feathers than daytime-active birds of prey. An 800-g (1.7-lb) Barred Owl, for example, has approximately 9200 feathers overall, whereas a 4500-g (10-lb) Bald Eagle has about 7100 feathers, 30% of which are white head feathers. (The common name for the Bald Eagle derives from a seldom-used Old English definition of the word *bald* as being "streaked or marked with white," and not from the more widely used definition of *bald* as being "hairless or featherless.") Totaling almost 700 g (1.5 lb), the plumage of a Bald Eagle makes up approximately one-sixth of the eagle's total weight, almost three times that of its skeleton.

Almost all raptors hatch with a thin covering of down feathers and develop a second, thicker, covering several days later. Depending on the species involved, fully shafted feathers with vanes begin to appear at between 10 and 40 days of age. Smaller raptors "feather out" faster than larger ones. By the time nestlings are ready to fledge (fly from their nest) young raptors have acquired most of their first year's plumage.

Typically, feathers cover all but a few portions of the faces, feet, and toes of diurnal birds of prey. There are exceptions, however. One of the more notable is the largely or completely featherless heads of many vultures, and the featherless and often colorful ceres, lores, and eye-rings of many diurnal birds of prey, all of which are detailed below under "Bare Parts." Owls, by comparison, are typically covered with feathers: head, face, feet, and toes included. In fact in many instances, only the owl's eyes, beak, and talons remain uncovered. The fully feathered face, feet, and toes of these nocturnal birds of prey help muffle flight noise, allowing owls to fly almost silently in the still nighttime air while searching for prey. The owl's feathered feet and toes also protect them against bites from struggling prey. Snowy Owls have particularly well feathered feet and toes that offer useful insulation against the Arctic cold, an adaptation that the aptly named tundra-dwelling Roughleg shares. On the other hand, Old World fishing owls, which snatch their prey from bodies of water, have unfeathered feet and toes, as does the Osprey.

A complex extension of the skin of raptors, feathers consist mainly of keratin, the fibrous protein that makes up human hair and fingernails. Although avian anatomists have organized feathers into dozens of types depending upon their construction, the typical, or "classic," form is the contour feather, the outermost feathering that covers most of a raptor's body, and the feather type that includes the wing and tail flight feathers. Contour feathers consist of a hollow, cylindrical shaft, called the quill, from which projects a long series of side branches, or barbs, that form the vane of the feather. Barbs, in turn, have their own series of interlocking side branches called barbules that serve to align and hold the barbs in place along the vane. Simple looking and seemingly lighter than air, feathers are structurally complex. Each of the twelve tail feathers of the Northern Harrier, for example, has about 1.25 million barbules. Viewed in this light, it is not surprising that raptors spend so much time preening their feathers.

Raptors have two types of stiff-quilled, long flight feathers on their wings. The outermost flight feathers, which emerge from their carpal, or "hand," bones, are called primaries; the inner flight feathers, which emerge from their "forearms," are called secondaries. Like most birds, raptors have ten primaries on each wing, along with fourteen to twenty-five secondaries, and twelve or, in a few large raptors, fourteen tail feathers. Larger species tend to have greater numbers of secondaries. Northern Harriers, for example, have fourteen, Cooper's Hawks fifteen, Black Vultures nineteen, Ospreys twenty, King Vultures twenty-one, California Condors twenty-two, and Andean Condors twenty-five. Falcons typically possess sixteen secondaries. In comparison, the relatively short-winged Northern Bobwhite has ten secondaries and the extremely long-winged Wandering Albatross has thirty-two.

Many species of raptors have specialized down feathers called powder down. Scattered throughout their body, these somewhat matted feathers grow continually, with parts breaking off as a fine powder that helps absorb the debris that raptors frequently come into contact with.

Old World Bearded Vultures use an external cosmetic, iron oxide, to stain white feathers on their heads, necks, and underparts a rustlike rufous. The staining, which occurs across the species range except on the Mediterranean islands of Corsica and Crete, happens when individuals purposefully bathe in iron-rich water, mud, or soil. Adult Bearded Vultures tend to be more intensely stained than juveniles, and adult females—which are larger than their male counterparts—tend to be stained more heavily than adult males, suggesting that the staining signals dominance. Some scientists believe that the topical application of iron oxide may also help mobilize vitamin A, as well as protect a bird against bacterial infections. If this is true, then staining creates what animal behaviorists refer to as an "honest signal" of an individual's overall health.

Molt

Fully grown feathers are dead tissues that cannot be repaired if broken and, as such, must be replaced as they wear out. The complex process of feather replacement is called molt. In most birds, the molt cycle, or ordered replacement of all feathers, follows a genetically enforced sequence, in which old feathers are pushed out of their follicles by new feathers that grow and develop on schedules that in large raptors can take years to complete. The production of new feathers is costly, not only because it compromises insulation and impairs flight ability by creating gaps in the airframe, but also because the construction of the new feathers is energetically expensive. One consequence of the latter is that smaller raptors, which have higher rates of metabolism than larger raptors, molt more rapidly. Another is that molt tends to occur at times when birds are not busy breeding or migrating long distances.

In the Temperate Zone, raptors almost always molt in late summer after they have finished raising their young. Molt typically extends beyond late summer, with different species having different schedules, and with most migratory species suspending molt, should it still be under way, during their migrations.

Most raptors undergo a single annual molt occurring over a series of at least several months. Newly-hatched young grow all of their "juvenal" feathers at one time, with feather growth beginning when they are still in the nest, and finishing shortly after they have fledged. This means that the trailing edges of the wings and tails of juvenile birds of prey appear "cleaner" and more even than those of older birds, often to the point of making it a useful field mark for distinguishing the silhouettes of first-year birds from those of adults and subadults.

Feathers that are damaged between molts, including those that break off, typically are not replaced unless the quill is completely removed from its base or follicle. Like other large birds, many species of raptors undergo a series of molts between juvenal and adult, or "definitive basic," plumages that result in one or more recognizable "subadult" plumages, a phenomenon that allows both birdwatchers and scientists to assign individuals to yearly ages classes until they are 3, 4, 5, and, in some species, even 6 or 7 years old. This progression of age-specific plumages is more common in larger birds of prey. Bald Eagles, for example, can be aged by their plumage until they are about 5 or 6 years of age. On the other hand, smaller raptors such as American Kestrels, which typically breed in their second calendar year, lack subadult plumages and simply molt from juvenal to adult plumage in the spring of their second calendar year.

Molt sequences also are known to change when conditions merit. Large raptors with delayed maturation that live in small, isolated island populations are more likely to skip subadult plumages than are their continental counterparts, and also are more likely to reproduce in subadult plumages. Both of these phenomena reduce the overall generation time, or the age at first breeding of females in the populations involved. Simulation models indicate that shorter generation times reduce the likelihood of a small population being eliminated because of a series of negative chance events on an island. It seems that life in the generational fast lane has its advantages on small islands, and that natural selection has acted to change species plumage sequences in response to this selection pressure. Of course, life in the generational slow lane also works, as the continental counterparts of island populations demonstrate.

Several long-distance migratory species, including first-year Ospreys, Egyptian Vultures, and European Honey Buzzards, exhibit delayed return migration as juveniles, a phenomenon in which non-breeding-age juveniles and subadults remain on their wintering grounds for a year and half, and, in some instances, two-and-a-half years, before returning to their potential breeding grounds. Delaying return migration to the breeding grounds has several advantages for young raptors, not the least of which is growing up in an area that for more than half of the year is unpopulated by older and more experienced competitors at a time when young raptors are still developing the social and hunting skills they will need to initiate successful breeding attempts. A second advantage of delaying their return to the breeding grounds is that it gives young raptors additional time to undergo a more protracted molt and, presumably, less energetically challenging annual molt.

In most raptors, body-feather molt begins before flight-feather molt, with molt in general progressing from the front to the rear of the bird. Although most species replace all of their body feathers annually, larger species can take up to several years to do so. In species where adult males provide most if not all of the prey for developing young, adult females typically begin molting flight feathers while still incubating their young. Some particularly long-distance migrants, including the high-latitude breeding Steppe Buzzard race of the Common Buzzard, take fully two years to complete their flight-feather molt.

Flight feathers molt symmetrically, with pairs of feathers molting and being replaced in sequence. In falcons, primary-feather molt begins with the fourth innermost primary and works simultaneously inward and outward. Other raptors molt their innermost primary first, with molt progressing outward. Secondary-feather molt is far more complicated. Tail molt usually begins with third and fourth feathers on each side of the bird's central tail feathers and proceeds simultaneously outward and inward. Overall, primary, secondary, and tail feather molts occur during the same time of year. The growth of individual flight feathers takes from 2–3 weeks in smaller species to 2–3 months in larger species. In many large raptors, including vultures and eagles, flight-feather molt takes more than a year to complete. In species with large geographic ranges, regional populations tend to molt at different times of the year. This suggests that molt is under strong selective pressure linked to regional variations in the timing of breeding and movement ecology. In the Temperate Zone, many species molt insulating body feathers in early summer but replace them in autumn just in time for winter. Even so, despite the details outlined above, we still know little about feather molt in the overwhelming majority of diurnal birds of prey, including that of many abundant and widespread species.

Patterns in Plumage

Most species of raptors are darker and less patterned above than below. Presumably such "countershading" makes them less conspicuous from below when in flight and from above when perched. Typical plumage colors include subtle grays and browns, and numerous blends of the two, as well as white, off-white, cream, tan, rust, rufous, black, and most of the possibilities in between. Bright reds and yellows are rare in birds of prey, and greens and blues are all but entirely lacking. In most cases the head is more intricately and distinctively patterned than the rest of the body, and tails are often banded or barred. Wings, which sometimes have diagnostic lighter or darker patches, lack the wing bars found in many songbirds. Juvenal plumages and, in some species, those of adult females as well, tend to be more cryptic than those of adult males.

Most species have decidedly different juvenal and adult plumages, and species with delayed maturation almost always have one or more subadult plumages. Juvenal and subadult plumages tend to be paler overall and more subtle than those of adults and, when patterned, are streaked vertically rather than barred horizontally, as they are in adults. In many species, including some kestrels and all harriers, adults are sexually dimorphic in plumage, with males being decidedly more colorful than their mates. A few species, including the Ornate Hawk-Eagle of Central and South America, the Long-crested Eagle of Africa, and the Black Baza of southeastern Asia, have stiff or floppy long crests. Other species have less pronounced crests.

Conspicuous white upper-wing patches and bright secondaries in Black Bazas almost certainly serve as social signals in this largely crepuscular bird of prey that routinely feeds during low-light periods of the day. Similarly, the white rumps of harriers, and the bold white bars found on tips and bases of the tails of many species of birds of prey, most likely serve to signal an individual's presence to conspecifics. That said, there are few thorough studies of plumage patterns in diurnal birds of prey, and this facet of their biology remains very much in need of additional attention.

Abnormal Plumages and Plumage Variation

Abnormalities in feather pigmentation include the lack of all pigments and either the lack or overabundance of melanin, the pigment responsible for the black, shades of gray, and full range of browns, red-browns, orange-browns, yellow-browns, and yellows found in the plumages of birds of prey.

Total, complete, or true albinos are extremely rare among diurnal birds of prey. Incomplete, imperfect, and partial albinos are not as rare. The feathers of completely albinistic raptors are white, and their skin and eyes, which lack pigmentation but not circulating blood, are pink. Incomplete albinos have pure white or nearly white feathers and pigmented skin and eyes. White Leghorn Chickens, the principal egg-laying chickens in the United States, are incomplete albinos. Leghorns also lay white-shelled eggs. Imperfect albinos, which often are called leucistic, or diluted-plumage, birds, have light tan plumages and normally or nearly normally pigmented eyes and skin. Partial albinos have both white and normally pigmented feathers and normal or nearly normal skin and eyes. In North America, individuals with diluted plumage have been reported in at least fourteen species of diurnal birds of prey, and partial albinos have been reported in seventeen species. Complete albinos have been reported in only four species of North American raptors, including several ghostlike Turkey Vultures. Harriers and buteos, in particular, appear to be prone to partial albinism and dilute plumages.

Melanism results from excessive deposits of granular, black melanin in the feathers of raptors. In typical amounts, melanin absorbs solar radiation and acts to darken other, more colorful, pigments. When it occurs excessively, melanin produces an all-black appearance. With one known exception, melanism, which is unusual in most raptors, is an autosomal genetic trait (a non-sex-related trait) that can be either dominant or

Box 2.1 Plumage mimicry

Mimicry, the biological phenomenon in which one species resembles another to deceive one or more other species, is most common in invertebrates. Palatable insects, for example, mimic less palatable insects, whereas other insects, including juvenile caterpillar forms, mimic formidable or venomous snakes, to reduce the risk of being eaten. Although vocal mimicry is relatively common in birds—Blue Jays, for example, regularly mimic the call of Red-shouldered Hawks, and I once heard a Northern Cardinal mimic the call of Broad-winged Hawk—visual mimicry tends to be uncommon in birds. There are, however, several instances of visual mimicry in diurnal birds of prey, and I mention them now because they involve behavior and plumage.

Mimicry in raptors falls into two categories. The first is aggressive mimicry, or what some call "wolf-in-sheep's-clothing" mimicry, in which a predator develops a resemblance to a less threatening species so that it can more closely approach its potential prey. The second is Batesian mimicry, or "defensive" mimicry, in which a less formidable raptor develops a resemblance to a more formidable raptor so as to better ward off enemies.

At least two diurnal birds of prey are believed to practice aggressive mimicry. The first is the Zone-tailed Hawk, which is said to mimic vultures in the genus *Cathartes,* including Turkey Vultures. With a distribution that stretches from southern Canada to southernmost South America, Turkey Vultures are by far the most widely distributed scavenging birds of prey in the world. They also are one of the most common vultures in the world, as well as one of the most common raptors. In parts of the species' range, it can be difficult to look up at the sky at midday and not see a flying Turkey Vulture within a minute or so. The almost entirely black plumaged and far less common Zone-tailed Hawk is believed to mimic the far more common vulture. Like Turkey Vultures, Zone-tailed Hawks have relatively slender wings and a longish tail, and both species are roughly the same size. The trailing half of the wings of Zone-tailed Hawks, although barred, are lighter in color than the leading half, just as they are in the Turkey Vulture. And like the vulture, the hawk frequently flies low over vegetation, holding its wings above its back in a V-shaped dihedral, while rocking back and forth from side to side. The hawk rarely associates with raptors other than Turkey Vultures, and at times appears to join flocks of Turkey Vultures deliberately.

Tropical biologist Edwin O. Willis proposed in 1963 that the predatory Zone-tailed Hawk aggressively mimics innocuous Turkey Vultures. Several observers have subsequently argued against this, suggesting that the apparent mimicry in shape and flight style result from aerodynamic considerations, and that the dihedral in the two species was simply an instance of convergent evolution. Nevertheless, raptor biologists who have spent time in the field studying the two species invariably note their extreme resemblance to each other, which goes well beyond the dihedral wing posture, as well as the difficulty of telling the two apart. My own fieldwork on Turkey Vultures in Central and South America, where the two species co-occur, confirms these observations. Although careful studies of the foraging behavior success rates of Zone-tailed Hawks hunting with and without Turkey Vultures nearby are clearly needed, the argument for aggressive mimicry by Zone-tailed Hawks seems reasonable. That Zone-tailed Hawks tend to be far less common than Turkey Vultures also supports the argument for mimicry, in that the model clearly outnumbers the proposed mimic.

A second example of aggressive mimicry in raptors is the somewhat paradoxical "white morph" of Australia's Grey Goshawk that many ornithologists believe has evolved to allow hunting goshawks to hide within flocks of extremely abundant all-white and mainly white species of cockatoos that occur in parts of its range. The idea is simple: white-morph goshawks are more capable of sneaking up on their prey than are darker morphs, appearing to be just another cockatoo in such flocks. Importantly, on islands off Australia where goshawks occur but cockatoos do not, the white-morph goshawk also is missing. Another potential example of aggressive mimicry in raptors is the largely crepuscular Black Baza's passing resemblance to several species of southeastern Asian cuckoos, whose ranges it shares.

Examples of Batesian mimicry in raptors include the widespread and plumage-variable Eurasian Honey Buzzard, a species whose juvenile plumage closely resembles that of the distantly related Common Buzzard. Honey buzzards are preyed on by Northern Goshawks, and the idea is that the plumage of juvenile honey buzzards evolved to resemble that of the far more common and more-formidable Common Buzzard in an effort to reduce the risk of goshawk predation. On the other hand, several ornithologists have suggested that the large amount of plumage variation in European Honey Buzzards results from the fact that different populations of the species mimic the plumages of several species of small eagles that occur in different parts of their wintering ranges in Africa. The latter is also said to be true for both the Crested Honey Buzzard and the Barred Honey Buzzard, two close Asian relatives of the European Honey Buzzard, whose plumages, right down to the lengths of their crests, vary across their respective ranges, again in parallel with those of the several species of local eagles.

Additional instances of Batesian mimicry in raptors involve two species of African serpent eagles, both of which are believed to mimic larger raptors in both plumage patterns and color and their overall body proportions and body sizes. In these cases, the Congo Serpent Eagle is believed to mimic the Cassin's Hawk Eagle in West African tropical forests, and the Madagascar Serpent Eagle is believed to mimic the Henst's Goshawk. In both instances, it is not clear why the less formidable serpent eagles mimic the two bird-eating eagles. Some have suggested that they do so because (1) it enhances the serpent eagle's ability to approach their reptilian prey, who mistake them for their bird-eating models, (2) it reduces the likelihood of the serpent eagle being mobbed by small birds, who mistake them for their bird-eating models, or (3) it makes the serpent eagle less likely to be preyed upon by larger raptors, who mistake them for the more formidable bird-eating models. Whatever the explanation for this mimicry, the two serpent eagles involved clearly do resemble the presumptive models.

recessive. Melanistic Ospreys and harriers have been reported in both the Old World and New World.

Some raptors, including Red-tailed Hawks, Roughlegs, Short-tailed Hawks, and Broad-winged Hawks have stable plumage polymorphisms that result in both light morphs (individuals with low levels of melanin) and dark morphs (individuals with high levels of melanin). Other species, including Swainson's Hawks and Common Buzzards, occur in three color morphs: light, dark, and an intermediate, or rufous, form. At least in some geographic areas, intermediate-morph Common Buzzards achieve greater reproductive success than either light or dark morphs, possibly because of morph-related differences in territorial defense or in blood-parasite levels.

Although both albinistic and melanistic plumages are thought to be genetic conditions that prevail for life, apparently there are exceptions. Normally plumaged juvenile Red-tailed Hawks have been reported to have molted into melanistic adults. And in Africa, melanistic juvenile Augur Buzzards are known to do the opposite, in molting into normally plumaged adults. Partly because it occurs so infrequently, the extent to which abnormal pigmentation affects the behavior, physiology, and survival of individual birds of prey remains largely unstudied. Pigments in the irises of raptors are thought to signal an individual's age, body condition, or both, as well as to facilitate vision, and it seems likely that the lack of such pigmentation in complete albinos affects both mate attraction and longevity. In addition, because plumage color signals age in many raptors and serves to camouflage them, individuals with abnormally pigmented plumages, too, are likely to face hardships. Melanin acts to strengthen feathers, making them less likely to be abraded, which explains why the trailing edges of the wings and wingtips of many raptors are darker than other areas of their wings. Melanistic feathers also are likely to affect thermoregulation. The extent to which these characteristics of melanin affect birds of prey with abnormally high or low levels of this pigment in their feathers remains unstudied.

BARE PARTS

Bare parts, or external surface areas that are not covered with feathers, are largely restricted to the beaks, heads, crops, toes, and feet of diurnal birds of prey. Although the colors of these areas function as social signals, the bareness itself can also have additional functions.

The unfeathered beaks, toes, feet, and legs of raptors, which usually are referred to as a bird's "trophic," or feeding, appendages because they are used to catch and process prey, are featherless largely because it makes them easier to clean up after a meal. The same appears to be so for the featherless heads and necks of vultures, which are often plunged into large carcasses, where they bang up against rotting flesh and associated insect and microscopic fauna. Strictly speaking, only the Lappet-faced Vulture of Africa and the Red-headed Vulture of southern Asia have entirely featherless heads, whereas other vultures have nearly featherless heads with isolated downy tufts of "cotton-wool" feathers. Nevertheless, the cleanliness argument holds, in that it is far easier to rinse debris from bare or all-but-bare skin than it is to remove it from difficult-to-preen head feathers. (Interestingly enough, a thick mat of cotton-wool feathers covers the underside of the patagium on the forewings of the Ruppell's Vulture, where presumably it insulates this lightly vascularized area from the low temperatures that high-flying individuals in this species encounter during their long-distance travels.)

The highly vascularized featherless heads of vultures also function to dissipate excessive body heat, which occurs as needed when blood flow increases to subcutaneous regions in a bird's head, enabling it to cool off more quickly on hot days. Although it seems unlikely that the featherless heads of vultures initially evolved to serve as signals to other individuals, apparently several species now use the bare skin on their heads to do just that. New World Turkey Vultures, Lesser Yellow-headed and Greater Yellow-headed Vultures, King Vultures, and condors all leave the nest with relatively dull, dark heads, and age into more colorful headedness over the course of several years. The same is true of several Old Word species, including Egyptian and Hooded Vultures.

Blushing, the human emotional response that leaves people red-faced, also may occur in a small number of birds of prey, again most prominently in vultures and several facultative scavengers, including caracaras. In most instances, blood flushed to the featherless areas of vultures creates a hemoglobin-mediated reddish color-change in the faces and heads of the birds involved, just as it does in humans. In raptors, richer and brighter colors typically signal excitement or dominance. Lighter colors signal fright or subordination. The Turkey Vultures that my colleagues and I capture, measure, and tag almost always lose their bright-red head coloration immediately upon being handled, as blood drains from the subcutaneous to core areas of their heads. The coloration returns, again almost immediately, after a bird is released. Blood-related color flushes are more ephemeral than are signals produced by pigments deposited in the feathers and skin of birds, making them quite useful in social interactions when individuals can signal dominance or subordination to others quickly, depending on their relative status. In Lappet-faced Vultures, a species in which adults dominate younger birds at carrion, full adults exhibit bright-red heads, whereas subadults sometimes do, and juveniles rarely blush at all, but rather blanch their heads during confrontations with other individuals. A similar pattern of signaling occurs in Hooded Vultures. Northern Crested Caracaras, which feature featherless lores, cheeks, and chins, flush red during interactions with conspecific individuals and drain yellow when frightened. The Palm Nut Vulture of Africa, which, as its name implies, feeds heavily on palm fruit, has a largely featherless orange-to-red face and chin, and it, too, is thought to change its appearance based on the state of its emotions or, possibly, the amount of palm fruit it has eaten.

Blushing is not the only way that raptors change the color of their featherless faces and heads. Many do so by eating foods that are rich in organic carotenoids, colorful yellow, orange, and red pigments that raptors themselves are unable to produce metabolically. The amount of carotenoids in the bare skin feet, toes, ceres, and eye-rings of some species of kestrels has been linked to the number of high-quality small rodents called voles that the birds have been feeding on. That the featherless portions of the faces and feet of kestrels are more brightly colored during the breeding season suggests that carotenoids in these and, presumably, other raptors serve to signal male quality to potential mates. In most instances, raptors acquire carotenoids, which are manufactured in the chloroplasts of plants, by eating herbivores directly or by eating prey that feed upon herbivores. One unusual source of carotenoids occurs in the Egyptian Vulture, which has largely white-and-black adult plumage and a distinctively bright yellow face important in mating displays. Like several avian scavengers, Egyptian Vultures routinely consume the feces of livestock and wild ungulates. Feeding experiments with captive individuals demonstrate that those consuming cow feces have higher blood carotenoid levels and brighter faces than those that do not, suggesting that this dietary component is important in social signaling in the species.

In many large raptors, the boney beak, which is covered with horny keratin, changes in color as an individual ages. Nestling White-tailed Fish Eagles and Bald Eagles, for example, leave the nest with dark gray or black beaks that change to yellow and orange-yellow over the course of several years, whereas those of Steller's Sea Eagle change from yellow to orange-yellow over the same amount of time. The beaks of several vultures also brighten with age, with those of Turkey Vultures changing from a black or dark gray to ivory, those of Black Vultures changing from an all-dark to dark with an ivory tip in subadults, expanding to an ivory "bulb" in adults, and those of King Vultures from black tinged with a bit of red to orange-red with a black base. The bills of caracaras, too, change with age, with those of the Northern Crested Caracara shifting from a grayish white to a pale translucent blue and those of Striated Caracara changing from black to a silvery gray. In many species, the changes begin at the base of the beak and move toward the tip.

Overall, the bare feet and toes of many raptors change slowly from bluish or flesh-colored to yellow, orange, or red with age, presumably as carotenoids build up with time.

WINGS AND TAILS

Wings

Raptors have relatively large wings for their body sizes, in part because their large bodies make flapping flight energetically more costly than in smaller-bodied birds. As a result, raptors rely on gliding and soaring more than do smaller birds, and soaring, in particular, favors the development of oversized wings. Raptors that soar regularly, either when searching for prey or when migrating, tend to have disproportionately large wings, and lighter wing loadings (they carry less body mass per square area of their wing) than do those that soar less often.

Selection for soaring behavior, however, is not the only thing that affects the wings of birds of prey. Wing shapes and sizes, as well as tail shapes, vary considerably among species, depending on both habitat and diet. Recall from Chapter 1 that falcons have relatively long, pointed wings, and long, narrow tails; that accipiters, which include many woodland-dwelling species, have relatively short, rounded wings and long, rudder-like tails; and that buteos and many large eagles, which include many open-habitat species, have broad and relatively oversized wings and short broad tails. The latter also is true for most vultures. Other birds of prey, including many kites, have high-aspect-ratio, long, narrow, and often pointed wings, as well as long, forked tails.

Wings and tails are aerodynamic flight surfaces, and their shapes and sizes determine how quickly or slowly a raptor can fly, as well as its maneuverability and flight efficiency. Overall differences in wing and tail shape are best explained by species differences in behavioral traits, including food choice, habitat use, and whether or not the raptor in question is a long-distance migrant. The long, pointed wings of falcons generate lift more efficiently during flapping flight than do shorter, broader, and more rounded wings. And pointed wings also allow high-speed flight, which

Table 2.1 Wing loadings in raptors compared with those of several large nonraptors

Species	Wing loading (g/cm^2)
Raptors	
Black Vulture	0.77
Black Vulture	0.55
Black Vulture	0.65
Turkey Vulture	0.36
California Condor	1.04
Andean Condor	0.95
Osprey	0.51 (males), 0.65 (females)
Osprey	0.48 (males), 0.61 (females), 0.56 (both)
Osprey	0.51
European Honey Buzzard	0.32
Bat Kite	0.43
Black-shouldered Kite	0.25
Red Kite	0.31
Black Kite	0.28 (males), 0.45 (females)
Bald Eagle	0.63
African Fish Eagle	0.44
White-tailed Eagle	0.68
Hooded Vulture	0.49
Griffon Vulture	0.77
Ruppell's Vulture	0.80
White-backed Vulture	1.08
White-backed Vulture	0.78
Egyptian Vulture	0.60
Bearded Vulture	0.56
Short-toed Snake Eagle	0.40
Bateleur	0.56
Dark Chanting Goshawk	0.46
Pale Chanting Goshawk	0.43
Lizard Buzzard	0.35
Northern Harrier	0.24 (males), 26 (females)
Northern Harrier	0.34
Pallid Harrier	0.21
Montagu's Harrier	0.21
Pied Harrier	0.21 (males)
Marsh Harrier	0.30
Marsh Harrier	0.32
African Marsh Harrier	0.23 (males)
Shikra	0.22
Eurasian Sparrowhawk	0.32 (males), 0.40 (females)
Eurasian Sparrowhawk	0.32
Sharp-shinned Hawk	0.24 (males), 0.31 (females)
Sharp-shinned Hawk	0.26
Sharp-shinned Hawk	0.19 (males), 0.30 (females), 0.26 (both)
Cooper's Hawk	0.37 (males), 0.41 (females)
Cooper's Hawk	0.49
Northern Goshawk	0.53
Northern Goshawk	0.56 (males), 0.58 (females)

(Continued)

Table 2.1 (Continued)

Species	Wing loading (g/cm²)
Black-chested Buzzard-Eagle	0.60
Red-shouldered Hawk	0.78
Broad-winged Hawk	0.38
Broad-winged Hawk	0.40 (males), 0.37 (females), 0.38 (both)
Red-tailed Hawk	0.47 (males), 0.54 (females), 0.51 (both)
Common Buzzard	0.45
Common Buzzard	0.38
Mountain Buzzard	0.41
Long-legged Buzzard	0.45
Roughleg	0.40
Jackal Buzzard	0.45
Lesser Spotted Eagle	0.43
Tawny Eagle	0.53
Steppe Eagle	0.57
Golden Eagle	0.67
Golden Eagle	0.70
Golden Eagle	0.73
Bonelli's Eagle	0.62
Booted Eagle	0.36
Secretarybird	0.66
Northern Crested Caracara	0.56
Long-crested Eagle	0.81
African Pygmy Falcon	0.30
Lesser Kestrel	0.27
American Kestrel	0.32
American Kestrel	0.30 (females)
Eurasian Kestrel	0.31
Eurasian Kestrel	0.24 (males), 0.35 (females)
Eurasian Kestrel	0.30
Greater Kestrel	0.35
Merlin	0.33
Merlin	0.38 (males)
Bat Falcon	0.51
Northern Hobby	0.29
New Zealand Hobby	0.43 (males), 0.55 (females)
Prairie Falcon	0.54
Lanner Falcon	0.43
Peregrine Falcon	0.63
Peregrine Falcon	0.65
Orange-breasted Falcon	0.42 (males), 0.56 (females)
Large-bodied nonraptors	
Common Crane	0.76
Common Pigeon	0.63
Northern Fulmar	0.60

Sources: After Brown and Amadon 1968; Cade 1982; Kerlinger 1989; MacWhirter and Bildstein 1997; Tennekes 1997.
Note: Multiple values for species result from multiple references.

can be important when raptors pursue aerial prey. On the other hand, raptors with broad, rounded, and relatively oversized wings soar more efficiently, particularly at low speeds, than do those with pointed wings, and also soar more and flap less when migrating. Among closely related species, long-distance migrants usually have more pointed wings and lighter wing loading than do short-distance migrants and nonmigrants. A North American example of this phenomenon are the relatively broad and less pointed wings of the Red-tailed Hawk, a short-distance partial migrant, compared with the relatively narrower and more pointed wings of the Broad-winged Hawk and Swainson's Hawk, two long-distance complete migrants.

Raptors with rounded wings, including accipiters and vultures, often have deeply slotted wingtips. Slotting occurs when the outermost primaries on a bird's wing have notched or emarginated tips that create spaces among neighboring flight feathers on fully outstretched wings. Species with slotted wingtips are said to have "fingers" at the tips of their wings. Slotting is thought to reduce turbulence and drag at the bird's wingtips, both of which negatively impact aerodynamic lift during soaring and gliding.

Forest-dwelling raptors, including forest-falcons and many accipiters, have relatively small, short, rounded wings, and long rudder-like tails that act to increase maneuverability, which is important for birds that frequently pursue prey at high speeds while trying to avoid striking trees in woodlands.

Ospreys, harriers, and a few species of vultures have both relatively long and narrow high-aspect-ratio wings and slotted wing tips. Open-habitat-dwelling aerial hunters such as these spend most of their hunting time flapping and gliding at low speeds and their wings appear to be somewhat all-purpose. Harriers also usually hold their wings above their backs, creating an upswept dihedral posture, a posture also used by Turkey Vultures, Lesser Yellow-headed Vultures, Greater Yellow-headed Vultures, and Egyptian Vultures. All of these species routinely search for prey while flying within several meters of the vegetative canopy, airspace with considerable small-scale "boundary layer" turbulence.

In such situations, flying in a dihedral is more efficient than flying on horizontal wings. Consider what happens when one wing in the dihedral receives a small puff of turbulent air and is pushed upward. This repositioning immediately reduces the lift of the upswept wing because it is no longer as nearly parallel to the ground and therefore has lower aerodynamic lift. At exactly the same moment, the laterally opposite downswept wing is more nearly parallel to the ground, increasing its lift overall. The two concurrent shifts in lift act to rock the bird back and forth sideways, allowing it to right itself in small-scale turbulence with little, if any, muscular movement. Viewed aerodynamically, the back-and-forth rocking flight of harriers and other dihedral fliers, far from being uncontrolled or "lazy" flight, actually reflects a form of well-designed, energy-efficient flight. Ospreys and, for that matter, gulls, which also spend a lot of time flying overwater, have noticeable crooks in their wings. The gull-wing profile, which also occurs in aircraft designed to fly low over water, apparently helps stabilize flight, reducing the likelihood of uncontrollable stalling in the turbulence.

Wing and tail shape can vary even within species among different age class and sexes. In Merlins, for example, adult males have greater wing loading than do juvenile males, and both adult and juvenile females have greater aspect ratios *and* wing loadings than both adult and juvenile males. In Sharp-shinned Hawks, adults have shorter tails, longer and wider wings, and greater wing loading than do juveniles. Presumably, differences in the latter result from age-specific shifts in selection pressure, with skilled adults having wings that allow them to fly faster and strike prey harder, and naïve juveniles have wings that increase their maneuverability. A human analogy helps: inexperienced juvenile Sharp-shinned Hawks have wings that function like a child's fat-tired tricycle, whereas experienced adult Sharp-shinned Hawks have wings that function like an adult human's thin-tired racing bicycle. Age and sex differences in the wings of other species of raptors also seem likely, but the phenomenon has yet to be studied in detail.

Tails

As is in other species of birds, the tails of raptors serve both as aerodynamic lift surfaces and as control devices that improve in-flight stability and maneuverability. Long, narrow, rudder-like tails help raptors maneuver in closed forested habitats, whereas broadly fanned tails help soaring raptors generate lift in open habitats. The relatively long and often forked tails of several kites improves flight agility (an ability to turn both rapidly and accurately at high speeds). Agility is particularly important for aerial insectivores such as Swallow-tailed and Scissor-tailed Kites that feed in-flight on flying insects; two species that, like the far smaller Barn Swallow, need to be able to turn "on a dime" in pursuit of prey.

BODY MASS

Diurnal birds of prey, which range in mass almost three-hundredfold, are relatively large when compared with most other birds. That said, they are not as large as most people think. Disproportionately large wings routinely lead people to think that raptors weigh more than they actually do. At Hawk Mountain Sanctuary, both children and adult visitors almost always over-guess the body masses of the sanctuary's nonreleasable raptors used in its education programs. Many people, for example, guess that our Red-tailed Hawk, which actually weighs about 1.5 kg (about 3 lb), weighs between about 2.5 and 12 kg (5 to 25 lb). The species' nearly 1.2-m (4-ft) wingspan, together with its often-ruffled feathers, belies its slight mass. The weights of other raptors are similarly overestimated by most of the public.

In reality, more than two-thirds of the world's 330 raptors weigh less than 1 kg (about 2 lb). And one in five weighs less than 0.25 kg (about 0.5 lb). Only about 4% of all birds of prey weigh more than 7 kg (about 15 lb), and a number of those are nonpredatory New and Old World vultures.

The world's lightest raptors, the White-fronted Falconet of northern Borneo and the Philippine Falconet, each weigh less than 50 g (only about 0.1 lb). North America's smallest falcon, the American Kestrel, weighs less than 130 g (about 0.25 lb). And Europe's smallest kestrel, the Lesser Kestrel, weighs less than 200 g (less than 0.5 lb). Eastern North America's two most common migratory raptors, the Sharp-shinned Hawk and the Broad-winged Hawk, average less than 250 g (little over 0.5 lb) and 450 g (about 1 lb), respectively. And the two most common raptor migrants at the Strait of Gibraltar, the Black Kite and the Western Honey Buzzard, average less than 830 g and 750 g, respectively (less than 2 lb). Even many large raptors are relative lightweights. Golden Eagles, for example, rarely weigh more than 5 to 7 kg (about 12 to 14 lb), and Bald Eagles typically weigh less than 5 kg (about 11 pounds). Given these figures, stories of Red-tailed Hawks and even smaller raptors carrying off 4- to 6-kg (about 9- to 12-lb) farm cats should be viewed for what they are: imaginary.

That said, several diurnal birds of prey are quite large. The California Condor and the Andean Condor both have wingspans of about 3 m (10 ft), and both can weigh as much as 14 kg (30 lb). Although both claim the title of the world's largest flying living birds, the two are small compared with the extinct teratorns mentioned in Chapter 1, one of which is estimated to have been 1.5 m (5 ft) tall and to have had a wingspan of more than 4.5 m (about 15 ft). (Realize, however, that estimating the weight of extinct birds is often a bit more "parlor game" than science.) Additional large living raptors include the Old World Cinereous Vulture, which weighs up to 12.5 kg (about 27 lb), and the Himalayan Vulture, which weighs up to 12 kg (about 26 lb).

Large predatory eagles include Central and South America's Harpy Eagle and the slightly smaller Philippine Eagle, both of which have wingspans of about 2.5 m (about 8 ft) and weigh up to 9 kg (about 20 lb) and 8 kg (about 18 lb), respectively. The world's largest fish eagle, the Steller's Sea Eagle of northeastern Asia and northern Korea, can weigh as much as 9 kg (about 20 lb). Overall, scavenging diurnal birds of prey outweigh the average predatory diurnal bird of prey.

Falcons tend to be relatively small raptors, with the largest, the female Gyrfalcon, rarely weighing in at more than 2 kg (less than 5 lb). Table 2.2 provides the body masses of several birds of prey.

Table 2.2 Body masses and wingspans of selected raptors

Species	Body Mass (kg)	Wingspan (cm)
Turkey Vulture	1.5	160–181
Black Vulture		
Male	2.1	
Female	1.9	
Both		141–160
California Condor		
Both	11.3	250–300
Andean Condor		
Male	12.3	
Female	10.5	
Both		272–310
Osprey		
Male	1.4	
Female	1.6	
Both		149–171
European Honey Buzzard		
Male	0.68	
Female	0.83	
Both		118–144

Species	Body Mass (kg)	Wingspan (cm)
Red Kite		
Male	0.95	
Female	1.2	
Both		143–171
Bald Eagle		
Male	4.1	
Female	5.5	
Both		180–230
African Fish Eagle		
Male	2.2	
Female	3.4	
Both		175–210
Hooded Vulture		
Both	2.1	150–180
Cape Vulture		
Both	8.2	228–250
Northern Harrier		
Male	0.36	103
Female	0.51	116
Western Marsh Harrier		
Male	0.49	
Female	0.76	
Both		115–145
Eurasian Sparrowhawk		
Male	0.15	
Female	0.33	
Both		56–78
Sharp-shinned Hawk		
Male	0.10	54
Female	0.17	62
Cooper's Hawk		
Male	0.35	73
Female	0.53	84
Northern Goshawk		
Male	0.91	101
Female	1.0	108
Red-shouldered Hawk		
Male	0.48	
Female	0.64	
Both		101
Broad-winged Hawk		
Male	0.42	
Female	0.49	
Both		86
Swainson's Hawk		
Male	0.91	
Female	1.1	
Both		128

(*Continued*)

Table 2.2 (Continued)

Species	Body Mass (kg)	Wingspan (cm)
Red-tailed Hawk		
Male	1.0	
Female	1.2	
Both		120
Common Buzzard		
Male	0.78	
Female	0.97	
Both		109–136
Roughleg		
Male	0.85	
Female	1.1	
Both		134
Lesser Spotted Eagle		
Male	1.2	
Female	1.5	
Both		146–168
Golden Eagle		
Male	3.5	
Female	4.9	
Both		180–234
Verreaux's Eagle		
Male	3.6	
Female	4.6	
Both		181–219
Barred Forest Falcon		
Male	0.16	
Female	0.20	
Both		46–60
American Kestrel		
Male	0.11	55
Female	0.12	57
Red-footed Falcon		
Male	0.15	
Female	0.18	
Both		66–77
Merlin		
Male	0.16	57
Female	0.22	64
Bat Falcon		
Male	0.13	
Female	0.20	
Both		51–67
Gyrfalcon		
Male	1.2	115
Female	1.8	127
Peregrine Falcon		
Male	0.61	97
Female	0.95	111

Sources: After Dunning 1993; Clark and Wheeler 2001; Ferguson-Lees and Christie 2001.

Owls, which include no scavenging species, are smaller on average than diurnal birds of prey. The world's 205 or so species vary in body mass one-hundredfold, ranging from North America's tiny 40-g (1.4 oz) Elf Owl to the huge female Eurasian Eagle Owl, which can weigh as much as 4.2 kg (about 9 lb).

By comparison, ostriches weigh up to 135 kg (300 lb), and Emperor Penguins up to 43 kg (about 95 lb). Keep in mind, however, that neither of these birds actually flies.

Reversed Size Dimorphism

One unusual characteristic of both diurnal birds of prey and owls is that females tend to be larger and heavier than their male counterparts. In many raptors, including Broad-winged Hawks, Common Buzzards, and Red-tailed Hawks, the difference in body mass is about 15–20%. In others, like bird-eating Sharp-shinned Hawks, Eurasian Sparrowhawks, and Peregrine Falcons, the difference is about 50–60%. In a few extreme instances, female raptors can be 75%, or more, larger than their mates. Because males are larger than females in most mammals, including humans, as well as in most birds, biologists call this atypical sex-related difference in size "reversed size dimorphism."

Table 2.3 Mean body masses and wing chords of selected female raptors in North America as a percentage of male body mass and wing chord

	Female relative to male	
Species	Body mass (%)	Wing cord (%)
Northern Harrier	152	120
Sharp-shinned Hawk	175	119
Cooper's Hawk	148	114
Northern Goshawk	127	108
Red-shouldered Hawk	127	106
Broad-winged Hawk	117	108
Swainson's Hawk	117	106
Red-tailed Hawk	119	108
Ferruginous Hawk	116	104
Roughleg	124	106
American Kestrel	113	104
Merlin	135	112
Peregrine Falcon	156	114
Gyrfalcon	150	110

Source: After Snyder and Wiley 1976.

Reversed size dimorphism (RSD) tends to be most pronounced in raptors that prey on birds, and is far less pronounced in those that prey on insects or scavenge their food. Raptors that prey on mammals typically fall in the middle. In a handful of species, including the Andean Condor, males are heavier than females. Overall, RSD is less pronounced in owls than in diurnal birds of prey, and the males of several species of Australian owls are larger than their female counterparts.

Raptor biologists have tried to explain reversed size dimorphism in many ways. Most modern hypotheses fall into one of three categories: ecological, behavioral, or physiological-anatomical. Although dozens of explanations have been proposed, none has gained universal acceptance. One reason for the lack of consensus is that although many current hypotheses explain why male and female raptors should differ in size, most fail to explain why it is the female, rather than the male, that is larger, and why it occurs only in raptors and a few other groups of birds, including hummingbirds.

Most of the ecological hypotheses depend on factors that affect adult individuals. One popular explanation for the phenomenon, for example, suggests that RSD acts to reduce competition between members of breeding pairs by allowing larger females to take larger prey and smaller males to take smaller prey. And, indeed, field evidence indicates that females often do take larger prey than males. But, like many other proposed explanations, this one fails to explain why it is the female and not the male that is larger in raptors, whereas the opposite occurs in most other birds. Another explanation, focused on size differences in adults, is that smaller males are more maneuverable than larger females and, thus, are better able to catch smaller and, usually, more numerous prey, an ability that might be important during the breeding season, when males do most of the hunting for their mates and young. Presumably, females are larger than males either to better incubate their eggs and brood their young or to reduce competition with males outside of the breeding season. But, again, the question becomes: why isn't reversed size dimorphism the rule in most other birds?

One existing hypothesis focuses on selection pressure on nestlings, rather than adults. The head-start hypothesis is based on the following lines of reasoning: (1) Male raptors, which typically provide the overwhelming majority of prey during the breeding season, must be particularly proficient hunters if they are to breed successfully. Field evidence

Box 2.2 Several of the more popular hypotheses for reversed size dimorphism in raptors, based on selective pressures acting on adult males and females

Ecological hypothesis

There are many variations, but, in general, this hypothesis suggests that because different-sized raptors feed on different-sized prey, reversed size dimorphism acts to reduce competition within breeding pairs.

Physiological and anatomical hypotheses

Large females lay larger and, possibly, better eggs than do small females.

Large females are better able to protect developing eggs in their ovary during hunting.

Large females are better incubators than are small females.

Large females are better able to withstand periods of food shortage during incubation than are small females.

Small males expend less energy when providing food for their young than do large females.

Behavioral hypotheses

Large females, being more massive, protect their nests better than do small females.

Small males, being more maneuverable, protect their nests better than large females.

Large females are better preparers of food for their nestlings than are small females, who might be overwhelmed by their young.

Large females protect small males from eating their own young.

Large females are better able to form and maintain pair bonds than are small females.

After Bildstein 1992.

suggests that males do hunt more successfully than females. (2) Both females and males are under intense selection pressure to breed as early in life as possible, as this increases their lifetime reproductive success of overall fitness. This means that as the principal food provider during the breeding season, males will need to develop hunting skills as quickly as possible. And, again, field evidence indicates that in most raptors, males do grow up faster than their larger female siblings, and furthermore that they leave the nest earlier than females and begin learning how to hunt earlier than females. (3) If males were larger than nestling females, or even if they were the same size as females, selection for more rapid development in first year of life could place their female siblings at a disproportionate risk of being killed by nestling males, a phenomenon that biologists call siblicide. (4) By being smaller than their female siblings, males reduce this last risk while at the same time enhancing their own rapid development. That several other groups of birds that exhibit reserved size dimorphism, including boobies and skuas, also exhibit siblicide supports this hypothesis. (On the other hand, the fact that hummingbird females are larger than males may be due to selection on females for laying relatively large, and more viable, eggs in this group of super-small birds.) In my mind, at least, this head-start hypothesis, which makes a series of additional predictions that have yet to be tested, remains a leading candidate for explaining reversed size dimorphism in birds of prey. That said, the jury is still out.

GROWTH AND DEVELOPMENT

As with other birds, raptors grow across three distinct periods: in the egg, before hatching; in the nest, after hatching; and during the first several weeks after they leave the nest. Overall, small species grow and develop more quickly than larger species.

Incubation and Hatching

As in all birds, the embryos of diurnal birds of prey develop in eggs encased in calcareous shells laid sequentially over the course of several days to a week or more. In most species, most or all incubation (the warming of eggs) is done by the female. Eggs typically hatch between 1 and 2 months after incubation begins. Larger raptors have longer incubation periods than do smaller raptors. Embryos develop in the classic fetal position in the egg, and, while doing so, develop a temporary, or deciduous, egg tooth, on the tip of their upper beak. Using precociously well developed neck muscles, embryos hatch from their eggs by hammering the egg tooth against the inner surface of shell, repeatedly puncturing it and creating an escape hatch at the egg's blunt end, which hatchlings struggle through to extract themselves. Hatching can take anywhere from several hours to several days, with smaller species hatching more rapidly than larger species. Once out of the shell, the blind and helpless chick dries off while being brooded (warmed) by the female (and, sometimes, male), and protected by both the female and male from intruders. Brooding continues for at least a week in smaller species and considerably longer in large ones, sometimes for more than a month. The egg tooth, which is functionally obsolete after hatching, is sloughed from the beak or reabsorbed several days thereafter.

The Nestling Stage

Initially covered in a coat of insulating down, nestling raptors are fed small pieces of food by the female, who receives intact prey from the male and tears it apart to divide among the young. Like many birds, raptors are altricial at hatching, which means that their eyes are closed and that they are unable to walk and thermoregulate and, therefore, must be brooded by the adults. (The precocial young of ducks and chickens, by comparison, hatch with eyes wide open, a thicker downy coat, and the capacity to walk.) Overall, the nestling period lasts for less than a month to more than several months, with smaller raptors leaving the nest earlier than larger ones (Table 2.4).

Table 2.4 Lengths of incubation and nestling periods of North American raptors

		Length in days	
Species	Body mass (g)	Incubation period	Nestling period
Turkey Vulture	2000	34	60
Black Vulture	2100	38	77
California Condor	8500	57	170
Osprey	1500	37	53
Swallow-tailed Kite	465	28	38
White-tailed Kite	325	31	31
Snail Kite	420	27	38

(Continued)

Table 2.4 (Continued)

Species	Body mass (g)	Length in days	
		Incubation period	Nestling period
Mississippi Kite	275	30	32
Bald Eagle	4700	35	80
Northern Harrier	425	31	31
Sharp-shinned Hawk	135	31	26
Cooper's Hawk	415	34	31
Northern Goshawk	905	32	37
Red-shouldered Hawk	605	33	38
Broad-winged Hawk	400	30	38
Swainson's Hawk	960	34	42
Red-tailed Hawk	1100	31	44
Ferruginous Hawk	1470	32	44
Roughleg	950	31	32
Golden Eagle	4230	42	70
American Kestrel	115	30	30
Merlin	210	30	29
Gyrfalcon	1480	35	48
Prairie Falcon	710	34	38
Peregrine Falcon	850	34	42

Sources: After *The Birds of North America* species accounts, 1992–2002; Sherrod 1983.

Fledging

Raptors are all but fully grown when they make their first flight from the nest, or fledge. Depending upon the species involved, fledging occurs as little as a month to more than 5 months after hatching, with larger species taking considerably longer to do so than smaller species. Smaller raptors develop and grow faster in the nest in part because of their higher rates of metabolism. At fledging, raptors usually weigh as much as adults, with reversed size dimorphism already in place so that fledgling males weigh as much as adult males and fledgling females weigh as much as adult females. Wing feathers and, particularly, tail feathers, are still growing at this time, making for rather clumsy initial flights and landings.

Most raptors are fed by their parents for at least several weeks after fledging, and the young of larger species often receive prey from their parents for several months. Social species like Black Vultures often maintain family groups well into the winter. Throughout this period, parents also protect their young from other raptors, several species of which routinely prey on the recently fledged young of smaller species.

In small raptors, incubation and nestling periods tend to be nearly equal in length, whereas in larger raptors the nestling period tends to be considerably longer than the incubation period. Thus the incubation and nestling periods of American Kestrels each last about 30 days, whereas those of Bald Eagles last about 35 days and 80 days, respectively. Mid-sized raptors tend to have intermediate values (Table 2.4). Why this is so is unclear. Part of the reason may be that incubation is, day by day, the most dangerous period in a young raptor's life, whereas as the nestling period is decidedly less so. As a result, rapid development of the embryo in the egg may be under greater selection pressure than nestling development. This, coupled with suggestions that hunting skills are more difficult to acquire in large compared with small raptors, may mean that the nestlings of large raptors need more protracted care than do those of smaller raptors and thus need to remain in the nest longer to grow and develop than do the nestlings of smaller species. Whatever the reason, the protracted period of parental care in large birds of prey means that several large raptors require more than a year to raise their young to independence and, therefore, are unable to successfully breed each year. That this occurs in both predatory and scavenging birds of prey is clear evidence that size matters during the developmental stages of raptors.

THERMOREGULATION

Like mammals and other birds, diurnal birds of prey are warm-blooded homeotherms that maintain relatively constant internal body temperatures. As is true of other homeotherms, the relatively constant core body temperatures of raptors results from internal heat produced by metabolic processes and muscular movements, as well as from external heat gained from sunning and other behavioral patterns. Feathers help keep raptors warm in cold weather and gular fluttering, or, loosely, "throat panting," helps prevent overheating in hot weather. (Like other birds, raptors lack sweat glands.)

Most raptors maintain internal temperatures of about 40 °C (104–105 °F), compared with 37 °C (98.6 °F) in humans. Small raptors tend to have slightly higher core body temperatures than do large raptors. A delicate thermal balancing act allows for more efficient physiological and muscular processing across a broad range of external temperatures

Box 2.3 Sunning

"Sunning" occurs when a bird spreads one or both of its wings and its tail, positioning itself to maximize the exposure of the wing and tail feathers to direct sunlight. The behavior, which occurs in a large number of unrelated birds, including cormorants, Anhingas, storks, and vultures and other birds of prey, happens almost entirely during periods of strong, direct sunlight.

In waterbirds, the behavior helps dry feathers that are wetted by diving underwater. Ospreys, which make their living plunge diving for fish, typically "rouse," or shake their bodies vigorously, much as dogs shake theirs, after leaving the water. Ospreys also often sun themselves after diving for fish.

Both Turkey Vultures and Northern Harriers frequently sun themselves after leaving their nighttime roosts in winter, sometimes with frost or light snow on their backs and wings, again presumably to both warm and dry their feathers. Griffon Vultures and Turkey Vultures drop their core body temperature by as much as 2–4 °C (4–7 °F) each evening to reduce the metabolic cost associated with keeping warm at night, and other vultures probably do the same. Sunning behavior, together with erecting back feathers, helps vultures raise their core body temperatures back to that needed for daily activities, explaining why most vultures do so before leaving their nighttime roosts each morning. That said, raptors also sun later in the day, even on hot and rainless days, suggesting that there may be more going than just raising core body temperatures and drying one's feathers.

The vulture specialist David Houston has suggested that such midday sunning also plays a role in maintaining the aerodynamic quality of a raptor's feathers. While studying the White-rumped Vulture in India, Houston carefully detailed the sunning behavior of more than a hundred individuals, none one of which had sunned before leaving a large roost each morning. Many of the birds that Houston studied landed next to a carcass and were about to feed on it. As in early morning sunning, the vultures spread and aligned the upper surfaces of their wings at right angles to the direct rays of the sun. In most observations, sunning lasted for 4 to 5 minutes, after which the birds folded their wings, even with the sun still shining, and began feeding. Houston hypothesized that, rather than trying to dry their feathers or warm their bodies at such times, the individuals he watched were warming their flight feathers in an effort to bring the feathers back into their optimal aerodynamic shape, for later flapping and soaring flight. To test his hypothesis, Houston deformed a molted vulture flight feather he had collected by clamping it to a metal stand and forcibly curving the tip upward with a piece of cardboard for 2 hours, an attempt to replicate what occurs when the bird's feathers are forced upward when air passes under them during prolonged soaring flight. When he eventually released the molted feather from the clamp, the shaft of the feather remained bent slightly upward in a smooth curve. After he exposed the feather to direct sunlight, however, it returned to its original shape within 5 minutes, or about the same amount of time the birds were sunning themselves. By comparison, it took 3 to 6 hours for the bent feather to do so when it was placed in the shade. Houston's observations of midday sunning behavior, together with his rather clever feather-bending experiment, led him to conclude that vultures sometimes sun themselves to reshape their flight feathers for more effective flight, an explanation for sunning that had not been considered before.

Many other raptors, including Black Kites, Eurasian Sparrowhawks, and Peregrine Falcons, also sun themselves, often while preening and dust bathing. Taken as a whole, these observations suggest that sunning also may act to help clean feathers and possibly rid them of temperature-sensitive ectoparasites.

than occurs in cold-blooded poikilotherms, including most invertebrates, fishes, amphibians, and reptiles, whose core body temperatures shift depending upon environmental conditions and levels of activity. Warm-bloodedness is one of the reasons that raptors are able to breed in the Arctic and other frigid climates. Life in the metabolic fast lane of warm-bloodedness, however, is relatively expensive, and raptors attempt to "cheat" this high-energy metabolic system whenever possible.

Newly hatched nestling diurnal birds of prey, for example, are not warm blooded, and thus require brooding by their parents until they develop the ability to maintain near-constant core-body temperatures on their own. Because of this, parental raptors need to be particularly careful about when they decide to raise their young. First and foremost, raptors breed when food is most plentiful, which in the Temperate Zone usually means spring into summer. They cannot afford to do so earlier because late-winter or early-spring cold spells would challenge their ability to successfully incubate eggs and brood nestlings. And, in fact, late-winter snows and rainy springs, which can make hunting more difficult and cause parents to be away from their nests and warming duties longer than usual while hunting, do limit reproductive success in many birds of prey. The effect of such events is exacerbated after hatching, when snow and rain can soak and chill helpless nestlings to a point of no return. Later in spring as temperatures rise, parents frequently need to shade their young from the heat of the midday sun.

At least for less-active raptors, another way to "cheat" the system is by having lower resting metabolic rates. Species like accipiters and falcons, for example, which pursue prey in flight, have high resting metabolic rates for their body sizes, but others, like buteos, which hunt for prey from perches or while soaring, are able get by with lower rates of resting metabolism. Laboratory findings indicate that the difference is correlated with the sizes of the hearts and flight muscles of the species involved, both of which are nearly twice as large—relative to their overall body masses—in high-flight-speed accipiters and falcons than they are in relatively low-flight-speed buteos. Living in the fast lane, it seems, does have a cost.

Another way for homeotherms to reduce the metabolic costs of warm-bloodedness is to decrease their core body temperatures nightly by as much as 3–4.5 °C (5–8 °F), a controlled metabolic process called torpor. The downside of this behavior is that the species involved, which include Old World Griffon Vultures and New World Turkey Vultures, need to sun themselves on cold mornings to bring themselves up-to-speed before engaging in their daily activities (Box 2.3). Fortunately, the energy cost of doing so appears to be less than the energy savings involved in the nighttime cooldown. Vultures also seem to have significantly lower overall metabolic rates than do birds of similar sizes and are capable of maintaining lower daytime core body temperatures than do other large diurnal birds of prey. And vultures are not the only raptors known to cool off at night. The Red-tailed Hawk also exhibits a nocturnal decline in core body temperatures of almost 2 °C (3 °F) in the summer. Unfortunately, only a few raptors have been examined in this regard, and the extent to which controlled torpor occurs in other diurnal birds of prey remains unknown.

SLEEP

The subject of sleep in birds has received scant attention from ornithologists. Defined as a period of immobility accompanied by high arousal levels and closed or partially closed eyes, sleep is currently thought to take up about a third of a bird of prey's daily cycle, with owls sleeping mainly during the day and diurnal birds of prey doing so mainly at night. Apparently, sleep functions similarly in birds and people, and sleep deprivation appears to affect raptors just as it does humans. When under veterinary care, captive raptors whose normal circadian rhythm has been upset by a medical condition frequently act "drowsy." In wild-caught raptors, falconers have long used sleep deprivation in the company of a human handler to bond birds to them. The English novelist and grammarian T. H. White, while learning to be a falconer, once endured keeping himself and the Northern Goshawk he was trying to "man," or train to the fist, awake for an extended period. White said it best, when commenting from the hawk's point of view: "I am so sleepy that I will trust you as a perch to sleep on, even though you stroke me, and even though you have no wings and a beak." The term "roosting" is often used synonymously to describe sleeping and resting in birds, and the term "roost site" is used to refer to the place where they regularly go to rest and sleep. Most birds of prey sit or stand when sleeping, sometimes with their head tucked behind one shoulder, and with their eyelids blinking slowly. Young raptors appear to sleep far more than adults. During the first 9 days after hatching, nestling Broad-winged Hawks spend 97% of their time lying down in a "head-droop-sleep" posture and then begin sleeping with their head tucked among developing contour feathers, continuing to sleep for most of the day until they are about 3 weeks old. Nestling Short-eared Owls continue to sleep "stretched out" in a prone position, until they are ready to fledge at 31–36 days of age. A 5- to 6-week-old orphaned Barred Owl observed continually for 14 daylight hours during a single day (the time when most owls sleep), rested with its eyes completely closed 28% of the time, with periods of sleep ranging from none to 82% hourly. The owl was awakened five times over the course of the observations: thrice by vehicles traveling along a nearby road, once by a passing plane, and once by food-begging fledgling Northern Harriers. Seemingly spontaneous awakenings occurred eighteen times.

Adult birds of prey often assume sleeping postures that allow them to either dissipate or conserve heat depending upon environmental circumstances. On sunny summer days, for example, brooding Short-eared Owls often sleep standing up, with their legs and neck slightly extended, and their wings drooped and feathers ruffled both to dissipate heat and to shield their young from the sun. Individuals trying to conserve heat frequently sleep on one leg, with the other retracted, and their neck withdrawn.

Although sleep itself has been little observed in diurnal birds of prey, the impact that it has in increasing the risk of predation, including that by other raptors, has been investigated. Not surprisingly, species facing high risks of predation sleep less than those that do not. In addition, raptors are thought to be capable of unihemispheric sleep, which means that they are able to sleep with one eye open, with its corresponding cerebral hemisphere awake, and one eye closed with its respective hemisphere at rest. Vigilance is thought to be at least one function of monocular sleep. Lateralization, or "handedness" tendencies, of one-eyed sleep in captive domestic chickens suggests that vigilance plays an important role in this behavior. When imprinting as young chicks, chickens preferentially tend to open their right eye during monocular sleep, but switch their preference to the left eye when the imprinting object (another chick) disappears, a phenomenon some believe is associated with antipredator behavior. Furthermore, sleeping ducks, which engage in frequent brief bouts of "peeking" while sleeping, change the rates at which they do so depending on the relative trade-off of the need to save energy and the need to remain vigilant in the face of potential raptor predation. It seems likely, albeit unknown, that raptors also tailor their style of sleeping in response to the predatory landscape they face.

Whether or not raptors sleep in flight, particularly during migration, remains an open question. The common belief is that because birds in general need to sleep regularly, those that fly frequently, such as Common Swifts, indeed may sleep in flight. Like humans, birds exhibit two types of sleep, slow-wave sleep and rapid-eye-movement sleep. Slow-wave sleep can occur unihemispherically, which allows the eye associated with the hemisphere that is awake to remain open, a phenomenon that presumably allows flying birds to orient visually while nodding off. Experiments with dolphins suggest that information received by the hemisphere that is "awake" during slow-wave sleep allows an individual to orient and maintain group cohesion when swimming in a pod. Whether the same is true for raptors soaring long distances in flocks is unclear. At least several species, including seabirds, routinely engage in nonstop flights of several days or more, suggesting that they may be capable of sleeping in flight. Metabolic cheats like Turkey Vultures (see above) would seem likely candidates for in-flight sleeping on migration. And, indeed, a migrating Turkey Vulture fitted with a heart-rate monitor exhibited little, if any, increase in heart rate when flying short and long distances during daytime migratory flights, compared to its heart rate when roosting, suggesting that it may have been nodding off in the air. Unfortunately, the electrophysiological parameters that metabolically define sleep have yet to be measured in raptors, and the truth of the matter is that although it seems likely, we simply do not know if raptors sleep when flying.

SWIMMING

A few species of diurnal birds of prey do swim, but not very well, and not very often. Raptors have several good reasons for not swimming. Their feathers are not particularly well adapted to wetting, and field observations suggest that taking off from water is not easy for them. And unlike most waterbirds that are foot-propelled swimmers, raptors use their wings, and not their feet, to propel themselves on the surface. Indeed, except for Ospreys, which have dense, oily plumage that helps keep them from getting soaked when plunging for fish, diurnal birds of prey lack the water-repellant feathers of seabirds. And even Ospreys, along with other fish-catching raptors, vigorously shake their plumage after diving to remove excess water and, once perched, typically spread their wings to facilitate drying.

Intentional swimming in raptors is rare. In Massachusetts, a group of fledgling Peregrine Falcons was observed "swimming" several meters while ferrying themselves between the shoreline and a semisubmerged log while bathing in a shallow pond, but such events appear uncommon, with most bathing occurring in water shallow enough for wading. Unintentional swimming occurs when (1) exhausted migrants fall into water barriers while crossing them, (2) individuals that catch large prey that is too heavy to transport in flight attempt to paddle to shore, and (3) individuals are forced to swim while being mobbed by smaller birds. In many instances, the individuals involved drown.

The most spectacular documented example of mass drowning in raptors involved at least 1300 migrants representing twelve mainly soaring species whose flight line had been diverted over the Mediterranean, and whose carcasses turned up along the coast of Israel in April of 1980. Soaring migrants appear especially prone to this sort of catastrophic event. Griffon Vultures that ferry across the Strait of Gibraltar between southernmost Spain and northernmost Morocco appear noticeably hesitant to cross the 14- to 25-km (9- to 15-mi) wide body of water. And when individuals do cross, most make the crossing within 2 hours of noon, usually on sunny days with light crosswinds or tailwinds, and even then only after soaring to great heights over land before attempting the passage. Observations suggest the overwhelming majority survive such attempts. Nevertheless, there are many reports of birds failing to do so. In May 2005, for example, birders at Europa Point, Gibraltar, watched as seven of twenty Griffon Vultures traveling together in a single flock were forced into the water by mobbing Yellow-legged Gulls, and drowned. Eleven additional vultures were found washed up in Spain west of Gibraltar the same season. Gulls in the region also aggressively pursue migrating Black Kites, Short-toed Snake Eagles, and Booted Eagles, all of which have been known to put down in the surf, either from exhaustion or while being chased, and thereafter swim or wade up to 100 meters to dry land, where they rest with wings drooped until their feathers dry.

Fish-eating raptors, including Ospreys and Bald Eagles, have been known to swim hundreds of meters after catching fish too heavy to carry in flight. Other species, too, have been known to swim to shore after catching prey over water. The report of a Peregrine Falcon chasing a low-flying Common Pigeon off the coast of Dorset, England, is instructive. The falcon "clipped" the water surface with one of its wings as it tackled its prey, and dropped into the sea 300 meters from shore. Employing what was described as a "butterfly stroke," the bird then swam for 20 minutes, apparently without the pigeon, before reaching land and drying off.

Incidents of raptors drowning in watering troughs for livestock highlight the limited swimming and water take-off abilities of birds of prey. American Kestrels and Prairie Falcons are listed among the birds that have drowned in watering tanks in the American West. The problem appears to be particularly acute in southern Africa, where hundreds of raptors representing twenty-nine species have been reported drowned in 163 separate incidents, including twelve different "mass events" involving a total of more than 150 White-backed Vultures and Cape Vultures. It is thought that the birds involved either were perched on the rim of the tank while trying to drink and accidentally fell in, or had purposefully entered the water, either to cool off, bathe, or, possibly, feed on the carcass of an animal that had previously drowned there. In the Masai Mara National Reserve in Kenya, vultures feeding atop bloated wildebeests that have drowned while attempting to forge rivers are extremely careful in trying not to get wet. And when individuals did slip off of a carcass and fall into the river, they immediately swam to shore to dry out.

ILLNESSES

Like other birds, diurnal birds of prey do get sick, and in the usual variety of ways. Infectious viruses, bacteria, *Chlamydia*, mycoplasma, fungi, protozoa, and worms, among other infectious organisms, all can affect raptors. As do the avian equivalents of many well-known human diseases, including influenza, malaria, cholera, and tuberculosis.

The acute poultry disorder, Newcastle disease, also occurs in at least some raptors, particularly those that consume contaminated pheasants, pigeons, crows, or other birds. Typically, infected birds exhibit neurological tremors within about a week and die several days later. Poxviruses affect many species, including Turkey Vultures, Red-tailed Hawks, Golden Eagles, and Peregrine Falcons. Unlike Newcastle disease, poxvirus is not transmitted from prey, but rather through biting insects or via direct contact with open lesions. The cutaneous form, which is the most common, can leave permanent scars.

Herpesvirus, which is common among birds, strikes rapidly and can kill an otherwise healthy individual. Rabies can affect raptors that feed on rabies-infected mammals. Although some individuals recover spontaneously, others do not. Avian tuberculosis, which is commonly reported among wild birds of prey, can induce muscle atrophy and emaciation, and, eventually, death. The disease is transmitted via contact with fecal material from infected birds. Aspergillosis is an often-fatal fungal disease of captive raptors that severely affects the respiratory system. Avian cholera, a highly infectious disease of waterfowl, has been reported in dead Ospreys in the Chesapeake Bay of eastern North America following an outbreak of cholera in ducks and geese in the 1990s. Apparently, transmission occurred when Ospreys fed on dead waterfowl or lined their nests with scavenged carcasses. Trichomoniasis, or "frounce," is caused by a flagellated single-celled organism widespread in Common Pigeons. Infections, which can induce esophageal lesions and kill raptors, are especially common in bird-eating accipiters and falcons. Avian malaria is caused by several kinds of blood parasites, with mosquitoes acting as vectors. Although many raptors carry these parasites without exhibiting clinical signs, those from Arctic climates, including Gyrfalcons, can die from an infection.

Numerous worms, including flukes, tapeworms, and roundworms, infest birds of prey, in some instances to the point of weight loss, listlessness, and death. Neoplasms, including benign, premalignant, and malignant tumors, have been diagnosed in many captive raptors. The distribution and abundance of tumors in free-ranging birds of prey remains largely unstudied.

The extent to which the conditions mentioned above, either acting separately or together, affect wild populations of birds of prey, remains largely unstudied. That said, field evidence suggests that significant numbers of wild raptors die annually as a result of such infections. A comprehensive study in Great Britain in the 1970s suggested that as many as half of all deaths in raptors were due to poor health, including disease, and that the risk of disease was density dependent (individuals in high-density populations were associated with higher rates of diseases). Blood samples from fifty living individuals of seven species of raptors in Oklahoma indicated the presence of one or more viral infections, but postmortems indicated that only 9% of the birds had died of infectious diseases.

Free-ranging raptors that are clinically compromised by disease are likely to succumb to their illnesses quickly. This, together with the fact that in most cases their carcasses are scavenged or quickly decomposed, means that their remains are not likely to be found by scientists attempting to study their demise. As a result, we know considerably more about the kinds of diseases raptors have than about the ways in which these diseases affect survivorship and population biology. Even so, although individual raptors are regularly infected, with the notable exception of newly emerging diseases, population-level effects appear to be uncommon.

The West Nile virus is a case in point. Evidence suggests that this Old World disease, which was first diagnosed in North America in 1999 and has been documented in many species of North American birds of prey, including Bald Eagles, Red-tailed Hawks, and Great Horned Owls, is known to have affected local populations of at least one immunologically naïve North American raptor, the American Kestrel. The numbers of kestrels breeding in nest boxes surrounding Hawk Mountain Sanctuary declined in the area by 40% within 2 years of the appearance of the disease. Experimental infections of American Kestrels by the highly pathogenic avian influenza H5N1 virus suggest that it, too, has the potential for decimating at least some populations of previously unexposed North American birds of prey.

Many raptors routinely migrate long distances and, therefore, potentially could act as vectors of both livestock and human diseases. Not surprisingly, there is growing interest in both the global agricultural and public health communities for expanding studies of avian infectious disease. Assuming that this interest continues, there is reason to believe that we will soon learn more about raptor illnesses and the roles that they play in the general ecology and population biology of diurnal birds of prey.

MORTALITY FACTORS

Given the number of illnesses, ecological hazards, and human threats raptors face, it is not surprising that they die in many different ways. In a handful of species, particularly large eagles, close to half of all raptor deaths can be said to be biologically calculated and "preprogrammed." In many others, the principal causes of death have changed over time considerably in light of changes in human behavior. Here I detail several of the more well known "natural" causes of raptor death and touch briefly upon those that result from human threats, a subject I expand on in Chapter 8.

Many parental raptors begin incubating their clutches before completing them. Consequently, clutches hatch asynchronously in nests, and older chicks have both a size and developmental advantage over younger sibs, which can be a problem for the latter when food becomes scarce. Hatching asynchrony tends to be greater in larger than in smaller diurnal birds of prey, with the eggs of several large species hatching five to eight days apart. Physical contests in which older sibs bully younger sibs can result in younger nestlings receiving less food from their parents than older sibs, sometimes to the point of starvation. Size differences among brood-mates also can result in older chicks killing and sometimes eating younger chicks. Parents do not interfere at such times, even when cannibalism is involved. Siblicide, or "Cain-and-Able" behavior, is said to be "facultative" in species when it occurs only during periods of food shortages, and "obligate" in species where it occurs regardless of food availability. Facultative siblicide occurs in dozens of raptors, including kites, harriers, accipiters, buteos, sea eagles, and booted eagles. It also occurs in several owls. In such situations, as many as two, three, or four young may be killed, depending on the extent of food shortages and the "personalities" of the larger chicks. Obligate siblicide, which is far less common, occurs in large species that almost invariably lay two eggs but raise only one young, with the second egg, and the chick inside, serving as "reproductive insurance" in situations when the first egg does not hatch or produce a healthy offspring. Reported in Bearded Vultures, Lesser-spotted Eagles, Verreaux's Eagles, and Crowned Eagles, calculated, preprogrammed, obligate siblicide is the leading form of mortality, both in the nest and overall.

Additional causes of nestling mortality include congenital defects, predation (which in most cases results in the loss of the entire clutch), and starvation. Nestling mortality can exceed 20% of all chicks born in a single year.

The next major hurdle in a young raptor's life is fledging. Fledglings, which by definition are inexperienced in the ways of the world outside of the nest, immediately confront a number of new mortality factors. Not surprisingly, this naïve cohort faces high rates of death, particularly during the first few weeks postfledging. Significant factors influencing mortality at this time include predation by larger raptors that find the fledglings easy targets, human persecution whenever and wherever it occurs, accidents that include high-speed flights into both moving and stationary objects, and, sometimes, misguided predation attempts directed at prey that are far too big and dangerous to handle.

Because actual deaths are seldom seen, much of what we know about the factors influencing raptor mortality comes from postmortem examinations. People are more likely to find dead raptors where they, themselves, are active and, therefore, information concerning how raptors die tends to be biased toward events occurring where people live and work. Because of this, raptors dying from human persecution and from striking manmade objects are overrepresented in recovery data, whereas those dying from more "natural" events in areas far from people are likely to be underrepresented.

An example involving the fates of nestling eagles serves as an example. Charles Broley, a retired bank manager from Winnipeg, Canada, started banding nestling Bald Eagles in central Florida in the late 1930s. By April 1946, forty-eight of the 814 birds Broley had banded were recovered in thirteen states and four Canadian provinces. Of those recoveries, 71% had been found either shot or wounded or had been captured. An additional 19% were reported simply as having been found dead, leaving little doubt that human persecution remained an important mortality factor for eagles at the time. The same appears to be true for twelve of thirteen other species of birds of prey, based on a report from the US Bird Banding Laboratory published in January 1936. The report and others indicate that even American Kestrels were vulnerable to shooting at the time. The ferocity with which birds of prey were at risk from shooting is evidenced by a report of a banded Osprey that had been "recovered" while attempting to alight in the mast of a ship 73 miles east of Cape Hatteras, North Carolina, in autumn 1933. A detailed accounting of the recoveries of 535 Cooper's Hawk nestlings that had been banded in the United States and Canada between the late 1920s and the late 1950s estimated the first-year mortality from shooting at approximately 38% in the 1930s, 20% in the early 1940s, and 16% in the late 1940s through late 1950s, the decrease over time likely reflecting increasing protection over the course of the study. A study of 231 Bald Eagles recovered between 1966 and 1974 suggested that 43% of the eagles had been killed as a result of direct persecution, 21% from accidents, 12% from "natural" causes, and 10% from pesticide poisoning. More recent reports, both from the United States and Europe, suggest significant reductions in numbers of raptors being recovered after they had been shot. In North America nowadays, collisions with windows near bird feeders probably remove as many, if not more, Sharp-shinned Hawks and Cooper's Hawks from their increasing populations as do misguided shooters. Even so, deaths from "natural" causes are still likely to be underrepresented overall.

A second significant source of bias in mortality information from band-recovery data occurs when migrants spend some of the year in areas with high reporting rates and other times of the year in areas with lower reporting rates. Deaths of long-distance migrants such as Broad-winged Hawks, which breed in North America but overwinter in Central and South America, for example, are less likely to be reported on the wintering grounds, biasing information regarding both when and how individuals die.

Notwithstanding such considerations, starvation, disease, predation, electrocution, shooting and trapping, poisoning, and collisions are all believed to kill many birds of prey. Starvation, in particular, almost certainly is underreported, as starving raptors often seek shelter before death and thus their carcasses are less likely to be found. Numerous avian diseases, including those mentioned earlier, can be common and widespread causes of death, especially when they affect "immunologically naïve" populations of birds of prey.

Endo- and ectoparasites also can kill raptors. Nematode worms, for example, which are common but usually nonproblematic in the air sacs of several species of raptors, are known to have killed Prairie Falcons, whose respiratory systems were infested with hundreds of adult worms. One problem faced by veterinarians who examine such cases is that many diseases and parasites become problematic only after the immune system of an individual is compromised by a less obvious underlying condition such as malnutrition, making the actual cause of death difficult to determine.

Not surprisingly, species often differ considerably in their abilities to survive certain threats. Botulism, which is caused by the bacterial toxin botulinum, is a type of food poisoning that episodically kills thousands of waterfowl feeding in contaminated wetlands. Remarkably, Turkey Vultures that feed on botulism-killed ducks and geese suffer little if any ill effect, and there are suggestions that their blood actually detoxifies botulinum. Similarly, lead toxicity, a major factor in the decline of the California Condor—and one that is significantly slowing recovery in portions of its former range—is far more problematic in some species than in others. Unlike California Condors, Bald Eagles, and American Kestrels, for example, Turkey Vultures again demonstrate considerable tolerance to lead ingestion, to the extent that toxicologists consider them a poor model for risk assessment of lead exposure in most raptors.

Another example of significant interspecies effects of environmental toxicants involves the veterinary drug, diclofenac, which, arguably, has killed as many diurnal birds of prey in the past 20 years as any single environmental toxicant. This nonsteroidal anti-inflammatory veterinary drug, which came into widespread use in southern Asia in the 1990s, quickly initiated catastrophic declines in populations of three common vultures in India, Pakistan, and Nepal. Approved doses of this livestock drug killed White-rumped Vultures, Indian Vultures, and Slender-billed Vultures within two days of their feeding on tainted carcasses; Turkey Vultures, experimentally exposed to doses at one hundred times the level that killed 50% of these Old World vultures, survived. Although all of the examples above involve Turkey Vultures, toxicologists report considerable variability among species of birds of prey in their responses to environmental contaminants.

Satellite tracking, which originally was developed to track the geographic locations of migratory raptors (see Chapter 7 for details), is now contributing information about how diurnal birds of prey die. Tracking units typically provide temperature as well as location data, and together these data allow scientists to determine both where and when a tracked individual has died. Examples of this new source of mortality information include the deaths of juvenile Peregrine Falcons on their first migrations resulting from collisions with vehicles, predation by other raptors, and electrocution; and the deaths of migrating Turkey Vultures as a result of collisions with vehicles and high-tension power lines, and of being blown up by ordnance on a US Army bombing range (presumably because the vulture had been attracted to wildlife previously killed by ordnance at the site). Although the death of satellite-tracked raptors is rarely welcomed by the researchers who are following the bird, the growing number of deaths, together with the ability to get to the carcasses quickly and diagnose the situation, provides new insight into this critical aspect of raptor population biology. This is important. Because the death of a raptor is seldom witnessed, and because their relatively small carcasses decay rapidly, we still know less about how raptors die than about almost any other aspect of their life history.

LONGEVITY

Given the veritable gauntlet of mortality factors faced by diurnal birds of prey, it might seem that life would be all too short for many, if not most, of them. And, in fact, first-year mortality for raptors can range as high as 50–60%. Mortality, however, drops off significantly after the first year of life, and many individuals live longer than might be expected.

In spite of their high metabolic rates, core body temperatures, and blood glucose levels, all of which can contribute to an animal's early demise, birds, including raptors, typically live to be quite a bit older than similarly sized mammals. Although the reason for this remains open to question, birds, like humans, produce a quarter to half as many chemically reactive free radicals per unit of oxygen consumed than do most mammals, which some believe is why birds and humans beat the mammalian average. Whether birds purposefully seek out and consume antioxidants to further extend their life spans has yet to be investigated.

The best longevity data for captive diurnal birds of prey come from the records of zoos and aviaries. Wedge-tailed Eagles and Tawny Eagles have lived for 40 years in zoos, Bald Eagles for 47 years, Eastern Imperial Eagles for 56 years, California Condors for 36 years, and Bearded Vultures for 40 years. By comparison, African Grey Parrots can live for more than 60 years in captivity, and macaws can do so for 90 years.

Based on the recoveries of free-ranging birds banded in North America, Ospreys live for at least as long as 22 years in the wild, Bald Eagles for at least 21 years, and Golden Eagles for at least 17 years. Smaller raptors, including Mississippi Kites, at 8 years, and American Kestrels, at 13 years, tend to have shorter lifespans in the wild. By comparison, banded Least Terns have lived for at least 24 years in the wild, Cliff Swallows at least 10 years, and Canada Warblers at least 8 years. Banded owls appear to live about as long as similarly sized diurnal birds of prey.

Importantly, because only about 3–8% of all raptors banded in North America and the British Isles are ever recovered, longevity records based on band recoveries are likely to be underestimates overall. That said, recoveries of banded birds of prey can tell intriguing stories. Consider, for example, the adult Red-tailed Hawk banded by Hawk Mountain Sanctuary volunteer banders Bob and Sue Robertson. When the two placed US Fish & Wildlife Service aluminum band #0877–17127 on the leg of this southbound migrant on the afternoon of 21 October 1972, they had little hope of ever seeing the bird again. Raptor banding, particularly when migrants are involved, tends to be long-shot, lottery science. Only about 5% of the more than 135,000 Red-tailed Hawks that have ever been banded in North America have been recovered, and almost all of them within a few years. Persistent and extensive banding, however, does pay off. And having banded more than 10,000 thousand raptors over five decades, Bob and Sue Robertson were certainly persistent banders. So you can imagine their surprise in 1994 when Len Soucy called the sanctuary with news of #0877–17127. Soucy, the head of the Raptor Trust in Millington, New Jersey, had trapped the bird at his Kittatinny Raptor Banding Station, 60 miles northeast of the sanctuary on 12 November. Bob and Sue's 1972 Red-tail turned out to be one lucky bird. The time between its initial banding and Soucy's recapture was more than 22 years. Because the bird was in adult plumage when the Robertsons had banded it, it must have hatched no later than the spring of 1971. Thus, the bird was at least 23 years and 5 months old at the time of its recapture. At the time, #0877–17127 was the oldest Red-tailed Hawk ever banded. Since then, an adult-plumaged Red-tailed Hawk banded on 21 November 1976 in Maryland was found starved in New York State on 13 March 2005, making it the world's oldest Red-tailed Hawk, at something more than 29 years of age.

Recoveries such as these beg the question of just how long birds of prey live, on average. For Red-tailed Hawks some of the best data come from an avian actuarial table extrapolated from the survivorship patterns of birds that were banded as nestlings in Alberta, Canada, in the late 1960s. The data indicate that only about 52% of all nestlings manage to reach one year of age, and that although 40% reached 10 years of age, only 4% reached 20 years of age. Assuming these data are representative of the species, only one Red-tailed Hawk in a hundred reaches the quarter-century mark.

Information gleaned from more than half a century of the US Biological Service's extensive banding records suggests an important rule of thumb regarding raptor longevity: within species, older and more experienced birds are more likely to be alive a year later than are recently hatched, naïve birds. Finally, there appears to be no clear rule regarding female compared with male survivorship.

SYNTHESIS AND CONCLUSIONS

1. The feathers of diurnal birds of prey provide a rigid airfoil for flight, protect and insulate the raptors' bodies from solar radiation and low temperature, and shield them from dust, debris and, to a limited extent, water.
2. Featherless areas of raptors, which are usually on the head, neck and throat, and feet and toes, serve both thermoregulatory and signal functions.
3. Raptors have fourteen to twenty-five secondary flight feathers on each wing, and ten primary flight feathers. Body feathers usually number in the hundreds to low thousands.
4. Most raptors initiate molt in late summer. Migratory species typically suspend their molt during migration.

5. The normal plumages of raptors are darker and less patterned above than below, presumably because countershading makes the birds more difficult to see.
6. The plumages of juveniles and adult females are often more cryptic than those of adult males.
7. The plumages of raptors sometimes mimic those of other birds, including other raptors.
8. Abnormal plumages, including albinistic and melanistic morphs, are rare. That said, several species, including Grey Goshawks and Gyrfalcons, have relatively common white morphs that are believed to enhance hunting success.
9. Raptors are relatively massive birds, but not as massive as most people believe. Over-guessing the body masses of raptors is due, in part, to their disproportionately large wings, which make them appear to be heavier than they are.
10. At hatching, nestling raptors are blind and unable to walk or thermoregulate, a condition ornithologists refer to as altricial.
11. Raptors are warm-blooded homeotherms, but several species engage in nightly torpor.
12. Raptors rest or sleep on a daily basis, and several soaring species may sleep during lengthy migratory flights.
13. Primary causes of death in raptors include starvation, disease, predation, electrocution, shooting and trapping, poisoning, and collisions with both moving and stationary objects.
14. Raptor mortality is high during the first year of life and much lower thereafter.
15. In the wild, raptors live well into their twenties, and a few live 40 years or more.

3 SENSES AND INTELLIGENCE

The study of senses—and bird senses in particular—has had a chequered history. Despite the abundance of descriptive information accumulated over the past few centuries, the sensory biology of birds has never been a hot topic.

Tim Birkhead, 2012

THE SENSORY SYSTEMS of birds of prey are their windows to the outside world. A raptor that is blinded in one eye may be able get along at least for a while, and in fact there are examples of individual birds of prey doing exactly that. But a fully blinded raptor is unlikely to survive for more than several days. Unfortunately, as Tim Birkhead points out above, information about the senses of birds, including raptors, remains in its academic infancy. Much of what we "know" about the senses of diurnal birds of prey depends on extrapolations from more fully studied species, including mammals. And indeed a true field of avian sensory biology has yet to evolve. Part of the problem is that ornithologists as a group have never been as intrigued with the sensory abilities and physiology of their subjects as they have with their behavior, ecology, and evolution. Consequently, progress in what I call the "skin-side-in" part of raptor biology lags pitifully behind its "skin-side-out" counterpart. Except for the many excellent studies of the biomedicine of birds of prey that have been driven largely by the falconry community, and which largely focuses on keeping captive birds healthy, little is known about how free-ranging raptors feel. How it actually feels to soar like a Rüppell's Vulture or to stoop like a Peregrine Falcon, as opposed to what it feels like to be a person strapped into an ultralight or hanging from a parachute with a raptor in tow, remains unknown. But while we may never really know what it feels like to be a bird of prey, understanding more about the sensory capacities of raptors helps build an appreciation of what they see, hear, smell, and taste on a daily basis.

Below, I detail some of what we know, and think we know, about the visual, olfactory, and auditory systems of birds of prey, as well as their sense of taste. I leave the details of their navigational capacities to Chapter 7.

SIGHT

Raptors are incredibly visual organisms. Their eyes often take up at least 50% of the volume of their skull, as compared with less than 5% in humans. Consequently, the eyes of a Red-tailed Hawk, for example, are almost as large as those of a person. Large eyes and the large retinas they contain allow for improved visual acuity (how well the eye resolves two finite points in a visual landscape). As a result, bird-hunting Cooper's Hawks and Eurasian Sparrowhawks are capable of chasing darting songbirds at high speeds, adjusting their movements to the diversions of their potential prey, while simultaneously avoiding woody vegetation. Similarly, perched Red-tailed Hawks are capable of detecting and catching rabbits that are feeding in the tall grass hundreds of meters away. Unlike most mammals—but like humans—raptors can see in color, a capacity that considerably increases their ability to spot and follow potential prey. In addition, some raptors can see into the ultraviolet spectrum, which has important implications for detecting the presence of small mammals.

Because eye charts are not especially useful in testing the visual acuity of diurnal birds of prey, biologists have relied heavily on anatomical comparisons and two-choice discrimination tests to assess the relative visual abilities of hawks, eagles, and falcons.

Raptors' oversized eyes are relatively fixed in their sockets (torsional movements in most species are limited to 2° to 5°), and so to see well in any direction, a raptor needs to turn its head, which is one of the reasons that they appear to be so alert all the time. Because their eyes do not face as far forward as those of humans, raptors' binocular vision ranges from about 35° to 50°, compared with about 140° in people. On the other hand, each eye has about a 150° field of view; the blind spot behind them, where raptors cannot see without turning their head, is only about 20°, far smaller than the almost 180° blind spot in people. Recent studies of the visual fields of White-backed Vultures and Griffon Vultures suggest that in foraging flight, when their heads are looking down, both species have excellent visual coverage of the ground below as well as laterally. But both species also have a blind spot in front of them, a situation exacerbated by the prominent brows above their eyes. Unfortunately, this blind spot increases the likelihood of their striking unseen objects, including the blades and stanchions of wind turbines that intrude into the birds' airspace.

Unlike those of mammals, the muscles controlling the pupils and lenses of raptors are mainly striated, or "voluntary," which allows them to adjust pupil size and shape more rapidly than do the smooth, involuntary, muscles of mammals. The pronounced supercilliary "eyebrows" of many raptors, together with specialized feathers, help shield the eyes from glare. Some raptors, including many falcons, have a malar, or "moustachial," stripe of dark feathering that runs down from the corners of their eyes that also helps reduce glare. Osprey have a similarly functioning dark, horizontal eye stripe that reduces glare from the water's surface as they search for fish.

In addition to having both upper and lower eyelids to protect their eyes, raptors have a third, thin, translucent eyelid, or nictitating membrane, that helps keep the surface of their eyes clear of debris while the birds are flying at high speeds or subduing prey. A specialized gland associated with this membrane produces an antibody that protects the lens surface against infections.

Like humans, raptors have a light-sensitive retinal surface at the back of their eyes that is covered with low-light-sensitive rods and color-sensitive cones. But compared with those of humans, their cone density is much greater. The retinas of Common Buzzards, for example, have a cone density that is five times that of humans, as well as more than twice that of House Sparrows. Raptors also differ from humans in being bifoveate, or having two foveae, or cone-rich depressions, on their retinas, which enhances both acuity and binocular depth perception, two aspects that help diurnal birds of prey lock on to and successfully pursue fast-moving prey. The more densely packed and deeper central foveae function in monocular vision, whereas the forward-facing temporal foveae function in binocular vision. In some species, the two foveae are connected to each other by a by a thin, densely cone-packed, ribbon-like trench, which functions to help keep moving objects on the horizon in focus. When a raptor cocks its head sideways and looks skyward with one eye, it is using monocular vision and its central fovea. When it looks forward intently and bobs its head, it is using binocular vision and its temporal foveae. The eyes of owls, which are more forward facing, are tubular shaped and are nearly immobile (torsional movements generally are limited to less than 2°). Owls do have temporal foveae and great binocular vision, but they are believed to lack a central fovea.

The low densities of retinal rods in diurnal birds of prey suggests poor visual acuity in low light. And it appears that raptors take longer to adjust to low-light environments than do humans. (Not surprisingly, rod density is much greater in owls than in humans and in raptors.) That said, some Peregrine Falcons routinely hunt at night around artificially lit areas, and so do Lesser Kestrels. In fact, an assortment of diurnal birds of prey, including Bald Eagles, Black-shouldered Kites, Common Kestrels, Brown Falcons, and Australian Hobbies, often hunt crepuscularly, nocturnally, or both. The largely nocturnal Letter-winged Kite of Australia is said to have eyes that in some ways resemble those of owls, and many raptors, including Ospreys, Northern Harriers, Levant Sparrowhawks, Amur Falcons, and Peregrine Falcons, routinely migrate into or through the night. But, of course, overall, owls are more active at night than are diurnal birds of prey, and their visual abilities in low light reflect that fact.

In addition to having the three classes of cones found in the human color spectrum, many diurnal birds of prey also possess a class of cones that allows them to see in the ultraviolet spectrum, thereby enhancing both their signaling and prey-detection capabilities. Many species of small mammals, including mice and voles, scent-mark their trails and territories with urine, feces, and hormones that function to attract potential mates and space-out intraspecies competitors. Many of these markings are visible in the ultraviolet spectrum. Recent studies involving Common Kestrels, Roughlegs, and Tawny Owls indicate that these species, and presumably other raptors that feed heavily on small mammals, use these visible markings to detect areas with abundant populations of potential prey. An ability to "screen" large areas for prey abundance quickly is likely to be of particular importance to migratory raptors returning to their breeding grounds in spring, when they are scrambling to find the best feeding sites for their breeding territories.

The ability to see ultraviolet light also may be important to raptors seeking high-quality mates. Studies of the ultraviolet reflection of feathers reveal that many groups of birds use this plumage "color" in mate choice. Although the phenomenon has yet to be examined in detail in diurnal birds of prey, it appears to be important in at least one owl. Male and female Eurasian Eagle-Owls have a bright white throat badge that is visible only during territorial and mating displays. Reflectiveness of the badge in the ultraviolet range, a characteristic that peaks during courtship and mating, apparently signals individual status. Whether this also occurs in closely related species, such as the North America's Great Horned Owl, is not known.

HEARING

Although extremely well sighted relative to other birds, raptors are similar to most birds in terms of hearing. The acoustic abilities of birds are thought to be better than ours. For aerodynamic and, possibly, mass reasons, the external ears of birds, including those of raptors, are relatively simple, with specialized feathers known as ear coverts making up for the lack of large skin-covered cartilaginous external funneling devices. Like most birds, raptors communicate to conspecifics (members of their own species) acoustically as well as by sight, and, in some instances, excessively so. The African Fish Eagle is a case in point. The species' scientific name, *Haliaeetus vocifer*, or literally "noisy sea-eagle," says it all. Abundant in wetlands throughout much of sub-Saharan Africa, the species' loud yelping "wheeah-hyo-hyo-hyo" creates an instantly recognizable dawn chorus throughout its large range and was said by the raptor biologist Leslie Brown to be "the best known and best loved" of all African bird calls. Most pairs begin their day in duet, both inside and outside of the breeding season, and subsequently respond in kind to their neighbors. In densely settled neighborhoods of eagles, vocal conflicts can continue for most of the day, and, in one study, pairs that spent most of their days "yelling" at each other produced fewer young than those that did not. Although African Fish Eagles take courtship and territorial calling to the extreme, other species of diurnal birds of prey, including Bald Eagles, Red-tailed Hawks, Common Buzzards, American Kestrels, and Common Kestrels, also call frequently in similar circumstances. Most calls sound emphatic to humans, presumably because they serve notice of an individual's presence to mates and neighbors, as well as to conspecific intruders and potential predators.

Diurnal birds of prey also call to one another to improve flock cohesion during migration. Migrating Ospreys passing along the mountainsides of southeastern Cuba in late summer and early autumn en route to crossing the 600-km (370-mi) Caribbean Sea on their way to wintering areas in South America, regularly call to one another when their soaring flight carries them into thick clouds. Presumably, the birds are doing so not only to avoid collisions within the clouds, but also to keep individuals together in anticipation of the lengthy overwater passage, during which traveling in groups while searching for sea-thermal updrafts is especially important.

Unlike owls, raptors are ill equipped to catch prey in complete darkness, in part because they lack the asymmetrical ear openings (i.e., one ear opening being higher than the other) of their nocturnal counterparts that allow the latter to detect the sounds of moving prey in three-dimensional acoustic space. Even so, several raptors have converged evolutionally with owls in other aspects of their facial structure in ways that significantly enhance their ability to hunt by sound as well as by sight. Harriers, a near-cosmopolitan group of thirteen species of long-winged and long-legged diurnal birds of prey belonging to the genus *Circus*, differ from most other raptors in having owl-like feathered facial disks. The feathered disks of both harriers and owls are made up of several rows of tightly packed, specialized, solid-vaned feathers that sit atop a ridge of semicircular skin behind their ears. The structures serve to funnel high-frequency sounds from in front of the bird into its ear canals, in much the same way as our external ears focus and amplify the sounds we hear.

Harriers typically hunt for small mammals, such as mice and voles, while flying over open habitats, usually within several feet of the vegetative canopy. Although it has long been suggested that harriers do so to increase their chances of hearing as well as seeing potential prey, it was not until the 1980s that this was confirmed, when the raptor biologist Bill Rice conducted a series of lab and field experiments that tested the harrier's ability to locate prey acoustically in the absence of olfactory or visual cues. Rice conducted his experiments on the hunting abilities of the widespread Northern Harrier in the American West. His work, which focused on "choice experiments" with captive birds, led him to conclude that harriers were capable of detecting voles by sound with a strike accuracy of 2°. More recently, the Australian Swamp Harrier has been shown to have a strike accuracy of 4°. By contrast, Red-tailed Hawks, American Kestrels, and Brown Falcons have relatively meager strike accuracies of about 10°, whereas Barn Owls, with their asymmetric ear openings and large facial disks, have a strike accuracy of 1°.

To test whether harriers actually used their enhanced acoustic abilities in the wild, Rice created an ingenious underground network of miniature loudspeakers hidden below an old field that free-ranging harriers routinely hunted over. He then sat back and waited until a harrier flew over the field before broadcasting a synthetic "vole vocalization," consisting of a series of mouse-like squeaks, from one of the loudspeakers, and watched to see how the bird reacted. Both perched and flying harriers routinely pounced on the activated speakers. Interestingly, American Kestrels hunting the same field failed to do so, even when the speakers were turned up well beyond the natural volume of voles. The extent to which other species of harriers locate prey acoustically has yet to be tested. That said, all harriers have disproportionately large, oval ear openings—those of the Montagu's Harrier, for example, are the same size as those of the eight-times-larger Golden Eagle—suggesting that others in the genus *Circus* also hunt by enhanced hearing as well as by sight. Hooded Vultures and Turkey Vultures also have large ear openings, suggesting they, too, may search for carcasses by sound as well as by sight, with the former scavenger possibly tuning into to the bone-crushing behavior of feeding hyenas.

At least some prey species are thought to play on the limited hearing abilities of diurnal birds of prey when warning conspecifics of a raptor's presence. Many species of songbirds, for example, employ two distinctively different types of distress calls to warn conspecifics of the presence of a hunting raptor. The first is an easily locatable, "broadband" mobbing call that attracts conspecifics to a perched raptor that is then mobbed by a group of diving and scolding birds until it leaves the area. The second is a less-locatable "narrowband" distress call that is used to alert others to the presence of an in-flight and, potentially, far more dangerous raptor. Narrowband calls are far more difficult for raptors to localize than are broadband calls, and are thought to take advantage of the hearing limitations of the birds of prey while at the same time getting the message out to conspecifics regarding their presence.

SMELL

For some time, it has been fashionable among ornithologists to say that birds have a poor sense of smell, and that they make up for it with keen eyesight and hearing. But, in fact, many birds, including several raptors, have remarkable senses of smell. And, on closer consideration, it makes a lot of sense.

Although many birds appear have a limited capacity for olfaction, others, including Kiwis, petrels, shearwaters, albatrosses, Whip-poor-wills, and oilbirds, have exceptionally large olfactory bulbs in their forebrains that, in some instances, make up more than a third of the weight of their brains. And all appear to use them extensively. The same is true for several species of New World vultures in the genus *Cathartes*.

Until recently, the olfactory abilities of raptors have been questioned. A case in point is those of the Turkey Vulture. The early American illustrator and ornithologist John James Audubon had this to say in a letter to a Scottish friend in December 1826:

> As soon as, like me, you shall have seen the Turkey Buzzard [*sic*] follow, with arduous closeness of investigation, the skirts of the forests, the meanders of creeks and rivers, sweeping over the whole of extensive plains, glancing his quick eye in all directions, with as much intentness as ever did the noblest of falcons, to discover where below him lies suitable prey;—when, like me, you have repeatedly seen that bird pass over objects calculated to glut his voracious appetite unnoticed, because unseen; and when you have also observed the greedy vulture propelled by hunger, if not famine, moving like the wind suddenly round his course as carrion attracts his eye,—then you will abandon the deeply-rooted notion that this bird possesses the faculty of discovering, by his sense of smell, his prey at an immense distance.

It turns out that Audubon was mistaken. Although several contemporaries quickly published arguments against his lengthy dismissal of a sense of smell in the species, it was not until the early 1960s that the question of olfaction was finally tested professionally, when Kenneth Stager, senior curator of ornithology at the Los Angeles County Museum, performed a series of experiments on the subject.

After careful studies of the "flight, food-locating habits and predatory behavior" of five species of New World vultures involving both field observations and experiments involving concealed carcasses, Stager concluded that Turkey Vultures were, in fact, capable of finding decomposing carcasses by smell alone. Using a blower-fan that dispensed the odors of fresh and decomposing animal tissues that were hidden from sight, Stager unequivocally demonstrated that Turkey Vultures had the capacity to locate rotting carcasses by smell alone. He also showed that ethyl mercaptan (CH_3CH_2SH), a highly volatile and disagreeably odorous organosulfide produced by microorganisms as a by-product of decomposition, alone attracted scavenging Turkey Vultures. Intriguingly, Stager learned that engineers at the Union Oil Company of California had been putting ethyl mercaptan into their gas lines, so that Turkey Vultures could help workers locate leaking lines, since the late 1930s. In fact, small amounts of the organic compound are intentionally added to liquid natural gas, as well as to butane and propane, to produce a noticeable smell in these odorless fuels to alert humans to the presence of these explosive materials. It seems that the engineers in the petroleum industry were on top of this ability.

Subsequent lab work indicated that Turkey Vultures have the largest nostrils of all New World vultures, including those of the far more massive California Condor and Andean Condor, as well as the largest olfactory bulbs—the next largest being those of the King Vulture, a species that does not appear to use olfaction in its search of carcasses. Two other Latin American species of vultures, both in the same genus as the Turkey Vulture, the Lesser Yellow-headed Vulture and Greater Yellow-headed Vulture, apparently also have a keen sense of smell, and both presumably hunt by smell as well as by sight. By comparison, Old World vultures show no evidence of locating carcasses via olfaction. Although Stager found no evidence that Black Vultures were able to locate carcasses by smell directly, this species actually is able to locate carcasses by smell indirectly. A socially aggressive species, Black Vultures routinely seek out and follow Turkey Vultures to carcasses, capitalizing on the latter's olfactory capacity.

The Turkey Vulture's ability to find carcasses by smell appears to function as a key innovation in the species, enabling individuals to search successfully over both forested and more-open habitats for carcasses, thereby expanding their feeding habitats considerably. This species, along with the Black Vulture, is currently the most abundant vulture in the world, as well as the most wide-ranging, and it is fair to say to say that at least in part this is due to its keen sense of smell.

Intriguingly, recent preliminary experiments in Taiwan suggest that the Crested Honey Buzzard uses olfaction to locate supplemental food called "pollen dough" that beekeepers place in their apiaries for their honeybees. Although the apparent ability to do so works only within distances of several meters, the studies suggest that this species, too, searches for and locates food by smell as well as by sight.

Although other species of raptors also may have as-yet undiscovered keen senses of smell, several appear to be quite unfazed by nocuous odors. In North America, for example, both the Red-tailed Hawk and, especially, the Great Horned Owl routinely feed on striped skunks, and museum workers report that the prepared skins of horned owls often smell of skunk, this in spite of the skunk's widely respected olfactory defense. Remarkably enough, one nineteenth-century observer mentioned that Turkey Vultures, too, ate skunks, although they did not consume the mammal's scent gland.

TASTE

Perhaps more than any other sense, taste has been little studied in birds. And this is certainly true for raptors. That said, the sense of taste functions to encourage feeding nutritionally and to avoid feeding on potentially toxic items. Taste should therefore be as important in birds as it is in other animals, including humans. Birds do have taste buds, but they usually number in the low hundreds, whereas humans have 10,000 of them. Unlike those of people, the taste buds of birds are located throughout the oral cavity, and not only on the tongue and around its sides and tip. And indeed some birds, including Mallards, do not have any taste buds on their tongues. This begins to make sense when one realizes that the relatively small, arrow-shaped, stiff, horny-sheathed tongues of birds are little more than shovels that push the unchewed food farther on down the digestive track, where, in diurnal birds of prey, a sacklike holding area called the crop begins enzymatic digestion. (By comparison, crops are lacking in owls, where food passes directly through the esophagus and into the muscular gizzard for mechanical digestion.)

Unfortunately, studies of the sense of taste have been an ornithological backwater until recently, and I have been unable to find much in the way of useful information on taste in raptors. Nevertheless, a growing body of knowledge regarding taste in other species of birds provides guidance for understanding its likely parameters in diurnal birds of prey.

Although a contentious issue in the first half of the nineteenth century, evidence for a sense of taste in birds has been mounting for some time.

Naturalists since the time of Darwin and Wallace have been intrigued by the fact that insectivorous birds typically avoid feeding on brightly colored caterpillars, while devouring those that are camouflaged. The aposematic, or warning, colors of the boldly marked caterpillars presumably were conveying their distastefulness to the birds. But how do birds make the connection? Recent tests with naïve individuals of four species of birds, including the ubiquitous Common Starling, show that inexperienced birds pick up but release the distasteful species from their bills so quickly that they rarely damage them, and that once "tricked," individuals subsequently avoid aposematic prey like the plague. The fact that many of the small number of taste buds that birds have are found near the tips of their horny bills and beaks begins to make sense.

Raptors, too, appear to have taste buds near the tips of their beaks. Many falconers observe that captive raptors tend to pick up and immediately reject food even when the birds are holding it only by the tip of their beak. Several times, I have watched American Kestrels capture and carry a short-tailed shrew to a feeding post, only to toss it aside quickly after placing it in their beak while attempting to pull it apart with their feet. The short-tailed shrew's legendary neurotoxic saliva, together with its odiferous and likely distasteful skin-gland secretions, apparently draw an immediate negative response from the small falcons.

Both Old and New World vultures appear to prefer recently dead carcasses over those that have decomposed substantially, most likely because decomposers create noxious digestive by-products to ward off large avian and mammalian scavengers. In fact, the distasteful nature of these decomposers appears to be a natural example of "chemical warfare."

Perhaps the best evidence for a sense of taste in raptors is that the Hooded Pitohui, a New Guinea medium-sized passerine and member of the avian family Pachycephalidae, has feathers laced with several neurotoxic alkaloids called batrachotoxins, the same toxins Central and South American dart-poison frogs use to deter predators. The discovery of these substances, more deadly than strychnine, in the feathers of pitohuis and at least four other New Guinea birds has led ecologists to suggest that toxicity in birds is related to predator deterrence, as it is in frogs. Although the birds in question are known to be eaten by snakes, especially as nestlings, and although snakes find batrachotoxins distasteful, no one has yet seen a raptor catch one of these birds. That the plumage of the Hooded Pitohui, which is boldly patterned in rufous-chestnut and black feathers, is aposematic and functions to deter raptors from feeding on them remains a distinct possibility.

In sum, it is fair to say that we have much to learn about a sense of taste in raptors, and that studies of their abilities in this area are long overdue.

LEARNING

Although they are believed by many to be an enormously successful instinctive group of animals, birds historically have been thought to have rather limited learning abilities. Recent work with parrots, crows, ravens, and their relatives, along with several other groups, however, suggests that birds are not as hardwired as once believed, and that, at their best, at least some species can act every bit as intelligent as many large primates.

So what about diurnal birds of prey? Only a few species of raptors interact in complex social groups, a situation that is believed to select for flexibility and intelligence. And on the surface, most recently fledged young appear to receive relatively little training from their parents. Such characteristics suggest that raptors are largely incapable of experiential learning. A quick review of the literature (albeit mostly anecdotal), however, suggests that learning—the ability to modify behavior in light of personal or, in some instances, a nearby cohort's experience—is both common and widespread in at least some diurnal birds of prey, particularly with regard to their hunting and social behavior.

Tool use, a "gold standard" of animal intelligence initially described in chimpanzees by the primatologist Jane Goodall in the 1960s, occurs in at least several diurnal birds of prey. And, in fact, Egyptian Vultures are one of only a relatively small number of animals reported to use tools to secure food. This relatively small Old World vulture regularly uses stones to break the shells of Common Ostrich eggs in regions where the two species co-occur. And two other Old World vultures, the White-backed Vulture and the Lappet-faced Vulture, also have been seen trying to do so, and both of these species routinely feed on Common Ostrich eggs opened by Egyptian Vultures. Typically, Egyptian Vultures break eggs by picking up and holding a single stone in their beak, carrying it to the egg, and then raising their bill straight up and slamming the stone forcefully downward onto the egg with their head and neck. Although some eggs are opened with a single strike, others require up to a dozen tosses.

The Old World Bearded Vulture, a close relative of the Egyptian Vulture, also use stones to prepare its food, but in a decidedly different way. Bearded Vultures, some of which feed largely on the long bones of ungulates, grasp bones in their talons and carry them to traditional, large, slab-like stone ossuaries. Once there, the vultures repeatedly thrust the bones downward from overhead onto the stones below, eventually breaking the former into smaller, bite-sized pieces, which they consume on-site or carry back to awaiting young. An anecdotal report of a flying Red-tailed Hawk repeatedly smashing a live snake it was carrying in its talons into a cliff until the reptile was limp and apparently lifeless, suggests that this species, too, uses stones to prepare prey for consumption. Observations of Northern Harriers and Cooper's Hawks forcibly submerging and drowning birds in shallow water suggests that water, too, can be used to disable prey.

Other, albeit less spectacular, cases of learned behavior in raptors include individual diurnal birds of prey following plows in agricultural fields, presumably in search of exposed and vulnerable prey. Two of the world's most common small falcons, the Common Kestrel and the American Kestrel, regularly engage in this behavior, as do Common Buzzards and both Red Kites and Black Kites. One of the more impressive examples of this behavior occurs among the relatively common Chimango Caracara of South America. This small, omnivorous raptor, which normally feeds on insects, small vertebrates, livestock dung, and fruits, congregates by the hundreds in recently plowed fields, where individuals gorge on recently exposed food items. The species' larger cousin, the Striated Caracara, an inhabitant Tierra del Fuego and the Falkland Islands, goes one step farther—literally. Populations of this species on at least several small islands routinely scrape peaty soil with their toes and talons to excavate earthworms, which they then consume. And they do the same when raking kelp wracks for buried invertebrates. How these traits came about and spread among individuals remains a mystery.

Observations suggest that early experience plays a key role in establishing individual differences in prey selection within species of diurnal birds of prey. As a result, the diets of widespread, feeding generalists can vary both geographically and seasonally (Chapter 7). Bird-eating Peregrine Falcons breeding at water holes in the outback of Western Australia feed all-but-exclusively on regional endemics, including parrots and cockatoos, whereas individuals breeding on bridges in the North American cities of New York and Philadelphia feed largely on Common Pigeons and

Common Starlings. Populations elsewhere in the species range likewise specialize on common and available prey. The widespread and migratory Steppe Eagle, which breeds in central Asia and overwinters in southeastern Asia, India, and East and southern Africa, feeds almost exclusively on small ground squirrels called susliks in central Asia during the breeding season, feeds heavily on carrion when migrating, and, when overwintering in Africa, feeds on termites, nestling songbirds, flamingos, and mole rats.

Geographic and seasonal dietary shifts such as those mentioned above require extensive modifications in searching, pursuit, and capture behavior, suggesting enormous flexibility and leaning in raptors, at least as far as hunting is concerned.

Some of the best evidence for learning in raptors comes from falconers, who spend large amounts of time in the field with their birds and get to know them well as individuals. Almost unanimously, falconers who hunt with raptors speak of enormous variability among different individuals within the same species in terms of both prey choice and hunting techniques, and, for that matter, overall hunting success. The fact that falconry birds can be trained to seek out and capture species preferred by their handlers, and not necessarily themselves, also supports the idea of considerable plasticity in hunting. One example of raptors' ability to change their hunting behavior in light of a single event is that they often become "trap-shy" if caught in a trap, making them sometimes all-but-impossible to recapture using the same trap. Fortunately for the scientists, trap-shy individuals often can be captured a second time using a different type of trap.

Another striking example of the degree to which being trapped can affect a raptor's behavior involves a study of a breeding population of Northern Harriers in central Wisconsin. Working as part of team of scientists interested in learning about the behavior of individual adults during the breeding season, I spent a lot of time catching and individually color-marking numerous individuals. One unintended effect of trapping and marking the birds with small, colored leg flags was that when we returned in the same vehicle several days later to check on them, we could expect the harriers in question to fly out and "greet" us as we parked the vehicle, while at the same time simultaneously distress-calling and exposing their new leg flags. Clearly, these birds had remembered their experiences and were behaving differently toward us as a result.

Although the social behavior of diurnal birds of prey has received relatively little attention—in part because of its apparent rarity—social behavior enhances the likelihood of learning. In highly social species like the Black Vulture, for example, fledglings continue to follow and associate with their parents in family groups for 6 months or more, far longer than fledgling Turkey Vultures spend with their parents, apparently because the decidedly more social Black Vulture needs more time to learn how to enter vulture "society" than does the less-social Turkey Vulture.

A second highly social raptor, the Striated Caracara of the Falkland Islands, provides another example of protracted learning in raptors. The Striated Caracara's idiosyncratic behavior closely resembles that of another unusual island endemic, an enigmatic parrot, New Zealand's Kea, suggesting behavioral convergence stemming from similar ecological circumstances. Like the caracara, this crow-sized parrot lives on relatively remote islands at the edges of human-dominated landscapes in that country, and is considered to be both a vicious killer and an incorrigibly curious bird. Both birds, which have evolved in some of the least human-populated parts of the world, appear to be what the evolutionary biologist Ernst Mayr called open-program species: animals that live together socially and that are constantly learning, both from their own experiences and from those of conspecifics. In 2010, I began a long-term study of Striated Caracaras as part of a team of ornithologists studying them in the Falkland Islands. During our studies, we have color-banded more than a thousand individuals, which means that nearly all of the birds we observe are now recognizable as individuals, whose personal histories we can follow. Although the investigation remains in its infancy, it is now apparent that individual wild caracaras behave in predictably different ways, with some being more vocal, others more aggressive, others bolder, and still others decidedly more curious. Although the individual "personalities" emerging from our initial observations may simply reflect genetic traits, similar observations have been made in Keas and other highly social birds, as well as in many social mammals, including primates. And, in all of these instances, researchers have suggested that experience helps create these individual differences.

SYNTHESIS AND CONCLUSIONS

1. Raptors possess disproportionately large eyes and retinal surfaces that in some species make up at least 50% of the volume of

their crania. In some species of raptors, the eyes are so large that they touch each other inside the skull.

2. The eyes of raptors are somewhat fixed in their sockets. Because of this, raptors turn their heads frequently when scanning the visual landscape, making them appear to be especially alert.
3. The eyes of raptors have two cone-packed depressions called foveae on their retinal surfaces that help keep the horizon in sharp focus, and that help individuals lock on to fast-moving prey while pursuing them.
4. At least some raptors can see in the ultraviolet spectrum. This allows them to see the scent trails of their small-mammal prey.
5. Raptors possess a third eyelid called the nictitating membrane that helps protect the cornea and keeps it free of debris during flight.
6. The hearing of raptors is thought to be better than that of humans. Some raptors, most notably harriers, possess facial disks made up of tightly packed feathers that amplify sound, allowing the birds to hunt by sound as well as by sight.
7. The olfactory capabilities of most raptors are limited, but several species of New World vultures, including the Turkey Vulture, have a keen sense of smell that allows them to detect decaying carcasses kilometers away.
8. The taste buds of raptors are found in the oral cavity, but not mainly on the tongue, and allow raptors to reject distasteful food as soon as it touches the tip of their beak.
9. Raptors show ample evidence of trial-and-error learning especially in their hunting behavior. That raptors typically become trap-shy after having been trapped is but one example of trial-and-error learning in the group.
10. At least one diurnal bird of prey, the Egyptian Vulture, engages in tool use, when it picks up and throws stones at ostrich eggs to break their shells and feeds on their contents.
11. Social raptors, including the Striated Caracara, exhibit learning skills similar to those of other highly social birds, including parrots, ravens, and crows.

4 DISTRIBUTION AND ABUNDANCE

I have lately been especially attending to geography [and] distribution, & most splendid sport it is,—a grand game of chess with the world for a board.
Charles Darwin, 1856

NUMEROUS FACTORS AFFECT the distribution and abundance of plants and animals, so much so that Darwin spent two complete chapters of *On the Origin of Species* reviewing many of these influences, which he considered foundational to understanding the process of speciation. That many other biologists believe so as well confirms the idea that the distribution and abundance of individual species shape biodiversity globally. Where species are found and how many exist globally are keys to understanding the biology of diurnal birds of prey. As with other birds, both intrinsic and extrinsic factors shape the species distribution and abundance of raptors. Intrinsic factors include species-specific aspects of breeding, feeding, and movement ecology. Extrinsic factors include the distribution and abundance of the other organisms with which they interact, together with the distribution and juxtaposition of the world's landmasses, topography, global and regional climates, and other physical features.

In this chapter, I summarize the overall species distribution and abundance of diurnal birds of prey and review the principles, processes, and complexities that influence them.

GEOGRAPHIC DISTRIBUTIONS OF RAPTORS

Global Geography

Diurnal birds of prey are a diverse assemblage of migratory and sedentary species that occur on all continents except Antarctica, as well as on many continental and oceanic islands. The ancient origins of raptors—diurnal birds of prey appear in the fossil record of northern Europe more than 50 MYA—together with the fact that many are able to fly long distances at relatively low costs, have given the group both the time and the mechanism to spread out across a majority of the world's landscapes and, thereafter, adapt to them. Perhaps not surprisingly, several species are well distributed globally. Two of the world's six naturally occurring cosmopolitan land birds—the Osprey and Peregrine Falcon—are diurnal birds of prey, and a third, the Barn Owl, is a nocturnal bird of prey. (Other naturally cosmopolitan land birds include three long-legged wading birds, the Great Egret, Cattle Egret, and Glossy Ibis, all of which have large coastal populations and a propensity to fly over water. Other widely distributed land birds, such as the Common Pigeon, House Sparrow, and Common Starling, owe their near-cosmopolitan distributions largely to human introduction and are therefore not necessarily naturally occurring cosmopolitan birds.)

Compared with their sedentary counterparts, migratory raptors are particularly widespread. The breeding distributions of all but three of the world's twenty-two species of complete migrants, for example, extend across at least two continents.

Table 4.1 Breeding distributions of the world's twenty-two species of complete migrant raptors

Species	Geographic regions in which the species breeds
Osprey	Eurasia, North America, Middle America, Africa, Oriental region, Australasia
European Honey Buzzard	Eurasia
Mississippi Kite	North America
Short-toed Snake Eagle	Eurasia, Africa, Oriental region
Pallid Harrier	Eurasia
Pied Harrier	Eurasia
Montagu's Harrier	Eurasia
Levant Sparrowhawk	Eurasia
Chinese Sparrowhawk	Eurasia, Oriental region
Grey-faced Buzzard	Eurasia
Broad-winged Hawk	North America
Swainson's Hawk	North America
Roughleg	Eurasia, North America
Lesser Spotted Eagle	Eurasia, Oriental region
Greater Spotted Eagle	Eurasia
Steppe Eagle	Eurasia
Lesser Kestrel	Eurasia, Africa
Red-footed Falcon	Eurasia
Amur Falcon	Eurasia
Eleonora's Falcon	Eurasia, Africa
Sooty Falcon	Eurasia, Africa
Northern Hobby	Eurasia, Oriental region

Although a species' ability to cross large bodies of water affects the likelihood that a bird will occur in both the Old and New World, all nine species of diurnal birds of prey and six species owls that do occur in both have distributions that extend north of the Arctic Circle (66.5° N), where the two worlds (the Eastern and Western Hemispheres) come closest together and where water-crossings are shortest and often ice-covered.

That all of these species are migratory, nomadic, or both, also enhances the likelihood of their traveling between the Old and New Worlds. An additional ten species of migratory birds of prey—Black Vulture, Turkey Vulture, Osprey, Swallow-tailed Kite, Snail Kite, Black Kite, Black-shouldered Kite, American Kestrel, Common Kestrel, and Peregrine Falcon—breed well into the temperate and tropical regions of both the Northern and Southern Hemisphere. Overall, the global ranges of complete and partial migratory raptors average 13 million km^2, whereas those of sedentary raptors average about 2 million km^2. Table 4.3 provides some of the largest and smallest ranges of diurnal birds of prey. Notice how no sedentary species are among the species with the largest ranges and that only two migratory species, both of which are high Arctic breeders, are among species with the smallest ranges.

As is true of all terrestrial organisms, the extent and locations of Earth's landmasses determine the maximum possible distributions of the world's diurnal birds of prey. About 71% of our planet is covered by water, and

Table 4.2 Northern limits of the distributions of diurnal birds of prey and owls that occur in both the New World and the Old World

Species	Northern limit of species range (in degrees)
Diurnal birds of prey	
Osprey	71
White-tailed Eagle	74
Northern Harrier	68
Northern Goshawk	71
Roughleg	76
Golden Eagle	71
Merlin	71
Gyrfalcon	84
Peregrine Falcon	78
Owls	
Snowy Owl	81
Great Grey Owl	68
Northern Hawk Owl	70
Boreal Owl	68
Long-eared Owl	67
Short-eared Owl	71

Table 4.3 Largest and smallest raptor species ranges and world population estimates

Species	Migration status	Total range (km²)[a]	Estimated world population[b]
Largest ranges			
Old and New World species			
Peregrine Falcon	Partial	81,900,000	500,000–1,000,000
Osprey	Complete	68,900,000	500,000
Golden Eagle	Partial	43,100,000	100,000–500,000
Merlin	Partial	42,500,000	500,000–1,000,000
Northern Harrier[c]	Partial	41,700,000	>1,000,000
Old World species			
Black Kite	Partial	59,200,000	>5,000,000
Common Kestrel	Partial	57,600,000	>5,000,000
Common Buzzard	Partial	34,800,000	>4,000,000
Western Marsh Harrier	Partial	34,300,000	500,000–1,000,000
Eurasian Sparrowhawk	Partial	33,200,000	>1,000,000
New World species			
Turkey Vulture	Partial	27,700,000	>5,000,000
American Kestrel	Partial	26,100,000	>1,000,000
Sharp-shinned Hawk	Partial	21,000,000	500,000–1,000,000
Black Vulture	Partial	19,700,000	>5,000,000
Red-tailed Hawk	Partial	15,700,000	>1,000,000
Smallest ranges			
Old and New World species			
Roughleg	Complete	24,400,000	100,000–1,000,000
Gyrfalcon	Partial	21,900,000	10,000–30,000
Old World species			
Seychelles Kestrel	Sedentary	250	500
Nicobar Serpent Eagle	Sedentary	1400	Unknown
Nicobar Sparrowhawk	Sedentary	1700	100–1000
Mauritius Kestrel	Sedentary	2040	250–300
Andaman Serpent Eagle	Sedentary	5500	1000–5000
New World species			
White-collared Kite	Sedentary	4800	50–250
Galapagos Hawk	Sedentary	7700	300
Hawaiian Hawk	Sedentary	7000	1100
Gundlach's Hawk	Sedentary	12,700	270
Striated Caracara	Sedentary	17,200	1000–3000

[a] Total range includes breeding ranges for sedentary birds and both breeding ranges and wintering areas for migratory species.
[b] Estimated world population is based on Ferguson-Lees and Christie, as updated by BirdLife International and migration counts at well-established migration watch sites.
[c] Includes Eurasia population that is sometimes referred to as Hen Harrier.

the 29% that is land is unevenly distributed across the planetary surface. Fully 72% of all land lies north of the equator, and 75% of that is in northern Africa, Europe, and Asia. Excepting Antarctica, a continent without raptors, terrestrial habitats north of the equator extend much farther poleward than do their terrestrial habitats south of it. As a result, 94% of all land more than 40° north and south of the equator occurs in the

Northern Hemisphere. Raptors, like other birds, tend to be more migratory at higher latitudes (see Chapter 7 for details), and therefore it should come as no surprise that all but four of the world's twenty-two species of complete migrants breed partly or entirely in northern Eurasia, and that all twenty-two migrants breed mainly or entirely north of the tropics (Table 4.1).

Despite the widespread nature of several species of migratory raptors, a recent analysis of their global distributions confirms that a majority are distributed in areas less than half the size of continental Australia, and that only eleven species of diurnal birds of prey have ranges that exceed the size of continental Africa. The analysis also shows an increase in range size with increasing latitude north of the equator and an opposite trend with increasing latitude south of the equator, with the two trends most likely reflecting latitudinal differences in overall land mass availability north and south of the equator.

In the New World, thirty-six species of raptors breed in North America, including the Caribbean Islands, and 103 breed in Latin America, including Mexico and Central and South America and their associated islands. In the Old World, eighty-four species breed in Africa; about fifty breed in Europe and Asia north of the Himalayas, about fifty-three breed in the Oriental Region, which includes Southeast Asia and its associated islands; and about fifty-four breed in Australasia, including Australia and New Zealand and their associated islands. In addition to these, one raptor, the Hawaiian Hawk, whose scientific name, *Buteo solitarius*, translates as the "solitary" or "lonely" hawk, breeds on the Hawaiian Islands, the world's most remote oceanic archipelago. (See Appendix for the geographic breeding ranges of individual species of diurnal birds of prey and owls.)

Although most raptors breed north of the equator, many migrants overwinter in tropical and temperate areas in the Southern Hemisphere. As a result, the numbers of raptors living south of the equator increases substantially during northern winter. The extent to which this seasonal latitudinal population-compression phenomenon affects the northern and southern ecosystems these birds inhabit has yet to be studied in detail. Certainly the "big squeeze" must affect countless numbers of diurnal birds of prey, including first-year migrants, which during northern winter find themselves competing for food and other resources with more experienced adults from their breeding grounds, as well as with other migrants and local sedentary birds of all ages. It also must affect the prey on which these birds feed during northern winter.

Island Raptors

About fifty species, or approximately one in six of all raptors, are island-only, or insular, species. At least some of these appear to be closely related to continental migrants, some of which most likely lost their way on their migrations, having either misnavigated or been blown out to sea by storms, and thereafter sought refuge on islands and evolved into a new species. Although the extent to which this has happened has not been studied closely, mounting evidence suggests that it may be substantial.

Box 4.1 Migration-dosing speciation

Although most raptors breed and overwinter on six of the seven continents, about fifty species, or approximately one in six of all diurnal birds of prey, occur only on islands. At least some of these "insular" forms appear to be closely related to continental migratory species that may have either lost their way during their migrations or been blown out to sea during storms and then settled on "raptor-uninhabited" islands and evolved into new species. The extent to which this has happened remains unknown. That said, both the Galapagos Hawk and Hawaiian Hawk appear to be closely related to the completely migratory Swainson's Hawk, a species that migrates through Central America at the height of the Caribbean hurricane season. Almost certainly, members of this species are driven out over the Pacific Ocean at least occasionally during such storms. Although the overwhelming majority of these misbegotten migrants are likely to die at sea, in exceptional circumstances a few could reach the Galapagos and even the Hawaiian Islands via trade-wind thermals and thereafter become ancestral stock for new species on them.

The extent to which such infrequent "migration dosing" of islands tangential to major raptor migration flyways plays a role in the biogeography of island raptors remains unstudied, but the occurrence of nineteen species of island-inhabiting accipiters in the Wallacean South Pacific, just beyond the end of the greatest accipiter migration route in the world, suggests that it may be substantial.

As I see it, migration dosing is orchestrated vagrancy that occurs when "doses" of potential colonists in the form of flocks of misbegotten migrants

(1) simultaneously arrive on islands "off the beaten track," (2) subsequently fail to return to their breeding grounds the following year, (3) manage to survive and breed while marooned, and (4) over time evolve into new non-migratory species.

One of the best examples of speciation in raptors that follows this scenario involves the island-dwelling South Pacific accipiters mentioned above. With only two complete migrants, accipiters are one of the least migratory groups of all raptors. Most species in the genus *Accipiter* that do migrate are short- or medium-distance continental partial migrants that travel alone or in small groups. There are two notable exceptions, however: the Chinese Sparrowhawk and the Levant Sparrowhawk. The first migrates along two transequatorial, East Asian flyways that spill into the islands of the South Pacific. The second migrates along the Eurasian–East African Flyway. Both species migrate synchronously in flocks that often number in the multiple thousands. Each autumn, close to half a million Chinese Sparrowhawks travel south along these flyways into peninsular Malaysia, the Philippines, the Indonesian Archipelago, and their associated islands with the help of seasonal northwesterly monsoonal winds. Migration geography and climatic conditions such as these set the stage for infrequent but inevitable occurrences of misdirected movements, during which flocks of Chinese Sparrowhawks are blown off course and onto islands from which they typically fail to return the next spring.

Almost certainly, such "dosed" propagules provided the seed stock for the region's large number of island-endemic accipiters, which today account for 38% of all accipiter species globally.

Another example of migration-dosing speciation in the genus *Accipiter* involves three South and Central American endemics, the White-breasted Hawk, Plain-breasted Hawk, and Rufous-thighed Hawk, that almost certainly evolved from misguided and nonreturning Sharp-shinned Hawks that episodically migrate in small numbers into northern portions of the region. Other examples involving members of the genus *Falco* include both the Oriental Hobby and the Australian Hobby, which are likely derived from the highly migratory Northern Hobby, and several insular endemic kestrels in the Indian Ocean that likely owe their origins to flights of tens if not hundreds of thousands of Amur Falcons that routinely cross that body of water in monsoonal tailwinds in autumn each year.

Box 4.2 The tropics and their raptors

Geographically, the tropics include the regions between 23.5° N and S latitude, i.e., the tropics of Cancer and Capricorn, respectively. These are latitudes within which the sun is directly overhead during at least one day each year. Land-based ecosystems within the tropics are characterized by nearly constant day lengths, relatively high rates of sunlight, or solar radiation, and relatively high daytime and nighttime temperatures. Although the tropics have a somewhat more complex ecological definition that takes regional climate, along with geography, into account, the "geographical" and "ecological" tropics largely overlap.

Ecologists recognize four distinct tropical regions: the Neotropics, the Afrotropics, the Indomalayan tropics, and the Australotropics.

The **Neotropics** include parts of central and southern Mexico, most of the Caribbean Basin, Central America, and South America south to northern Chile and Argentina, central Paraguay, and southern Brazil. The largest rainforests in the Neotropics are the Amazonian forests of Brazil, Peru, Colombia, and six other South American countries. Overall, approximately 40% of the Neotropics is tropical rainforest. Twenty-six genera of raptors, representing more than fifty species, occur in the Neotropics and in no other tropical region.

The **Afrotropics** include the southern half of the Arabian Peninsula and most of Africa, including portions of Mauritania, Algeria, Libya, and Egypt, south to central Namibia, Botswana, northern South Africa, southern Mozambique, and most of Madagascar. Africa is a particularly dry continent with many deserts, including the Sahara in the north and Kalahari in the south. Only 10% of the Afrotropics is tropical rainforest. Twelve genera of raptors, representing nineteen species, occur in the Afrotropics and in no other tropical region.

The **Indomalayan tropics** extend from central India, Bangladesh, central Myanmar (formerly called Burma), and southernmost China, south through Indochina, most of Indonesia, the Philippines, and southern Taiwan, including many South Pacific Islands. Approximately 50% of the Indomalayan tropics is tropical rainforest. Five genera of raptors, representing more than ten species, occur in the Indomalayan tropics and no other tropical region.

The **Australotropics** extend from the Moluccas of eastern Indonesia, south to central Australia, and east through all of New Guinea. Most of mainland Australia is desert or desert-like steppe. As a result, only 30% of the Australotropics is tropical rainforest. Four genera of raptors, representing five species, occur in the Australotropics and in no other tropical region.

Overall, the tropics, which make up approximately 35% of the earth's land surface, are home to close to 80% of the world's approximately 330 species of raptors. Most raptors that occur wholly or mainly within the tropics are either sedentary species or short-distance migrants. In fact, only 15% of all entirely tropical raptors migrate on a regular basis, compared with more than 90% of all European species, and almost all North American species.

As is true for many birds, raptor diversity increases substantially at lower latitudes and, therefore, the tropical regions of the world host a disproportionate number of the world's raptors, including most of its island-dwelling endemics. The section below describes this diversity, along with the general conservation status of the birds involved.

Tropical Raptors

The Old and New World tropics, particularly their rainforests, are some of the most biologically diverse habitats on earth. Although they make up little more than one-third of the world's terrestrial landscapes, the tropics are home to close to four out of five of the world's approximately 330 species of diurnal birds of prey. In fact, the distributions of all but ten of the world's raptors occur, at least in part, within the tropics. Raptor diversity in the tropics is such that three countries in tropical South America—Colombia, Ecuador, and Venezuela—and four in tropical Africa—Ethiopia, Kenya, Sudan, and Uganda—each host more than seventy species of diurnal birds of prey, and, overall, half of all tropical countries host more than fifty species of diurnal birds of prey. By comparison, the United States and Canada together, and all of continental Europe, each host fewer than forty species.

Most ecologists recognize four distinct tropical regions, the Neotropics of Central and South America, the Afrotropics of Africa, the Indomalayan tropics of southern Asia, and the Australotropics (Box 4.2). Tropical diversity is largely region-specific, in that more than 90% of all mainly or wholly tropical raptor species occur within a single region, and that, excepting the contiguous Indomalayan tropics and Australotropics, there is little, if any, species overlap among tropical regions.

Because of their sedentary and island nature, tropical raptors are more likely to have small or limited ranges than are nontropical raptors. In fact, more than half of the world's approximately eighty species of entirely tropical raptors occur in only one or, at most, two countries, compared with about 10% of all mainly nontropical raptors. The vast majority of tropical endemics are found on islands, mainly in the Indian and South Pacific Oceans. Three island nations alone—Indonesia, with sixteen; Papua New Guinea, with fourteen; and Madagascar, with eleven—are home to thirty-four limited-range species, or 44% of all endemic tropical raptors. Because of their relatively small global ranges, wholly and mainly tropical raptors are at disproportionate risk.

Compared with their nontropical relatives, little is known about most tropical birds of prey. Many are forest-dependent species, most of which are secretive and extremely difficult to locate and study. Even so, several significant patterns emerge regarding the conservation status of these birds. As is true of raptors outside of the tropics, land-use change, which affects 46% of all species, is, by far, the major human threat. Environmental contaminants and shooting, respectively, threaten 11% and 19% of all tropical raptors. Fully 17% of all tropical raptors are threatened by two of these factors, and 2% are threatened by all three. (See Chapter 8 for details about how threats from two or more factors significantly compound the problems that face diurnal birds of prey.) Regionally, 42% of all Neotropical, 60% of all Afrotropical, 60% of all Indomalayan, and 77% of all Australotropical raptors are currently threatened by one or more factors.

As a result of such threats, in mid-2013, fifty-one species of wholly or mainly tropical raptors were considered to be Critically Endangered, Endangered, or Vulnerable by world experts, and an additional forty-five species were considered to be Near Threatened. Approximately forty-five of these threatened species are forest-dependent, thirty-eight are found only on islands, and thirty-two are both forest-dependent and island-restricted species. Overall, endemic, or limited-range, species (species with relatively small populations to begin with) are more likely to be threatened than are more widespread tropical raptors, and Old World tropical raptors appear to be more threatened than New World species.

One of the greatest challenges in protecting tropical raptors is dealing with the lack of knowledge regarding their conservation biology. In many cases, conservationists lack reasonable population estimates for the species in question, as well as little information on factors that currently limit their distribution and abundance. As a result, aside from protecting forested habitats, particularly those on islands, and educating local populations about the need to protect tropical raptors, one of the greatest needs in tropical raptor conservation today is the need for additional fieldwork.

RARITY IN RAPTORS

Understanding why some diurnal birds of prey are rare whereas others are common remains a topic of considerable debate among ecologists and conservationists. Although some raptors are rarer than they otherwise would be because of the human threats, some raptors are inherently, or "naturally," rare species, with limited populations and distributions, whereas others are naturally abundant, common, and widespread species. Why this so is not always obvious. And, in many situations, the reason for rarity remains open to discussion.

But before I get into the sticking points about natural rarity, I offer a refresher on ecological terminology. Scientists use two terms that are of significance in understanding species rarity, *habitat* and *niche*. Ecologists define habitat as an organism's ecological "address," its location, or, simply put, where one would go to find it. By contrast, niche is defined as an organism's ecological role, or "profession," in nature, or, as the ecologist Paul Colinvaux puts it, "its place in the grand scheme of things." Like other organisms, raptors are categorized, and, to a certain extent, defined by their habitats and niches, and these two defining characteristics are thought to play major roles in determining a species' distribution and abundance, which, in turn, should shape the relative commonness or rarity of a species. "Should," however, is the operative word here, and things are not as simple as they seem. Part of the complexity lies in how most people think of an organism's habitat and niche. Although both terms are defined in the strict sense as characteristics of an individual species, we often think of the two as interchangeable among species.

An analogy from professional baseball is helpful. Consider two players ("species," if you will), one that pitches for the New York Mets, and a second that pitches for the New York Yankees. Most people would define the two as baseball pitchers whose professional "habitats" include Major League baseball parks. But since one is in the National League (the pitcher for the Mets) and the other is in the American League (the pitcher for the Yankees), the actual ballparks you need to visit to find them, although overlapping, will differ overall. Now consider the two players' professional "niches": although both are pitchers, let's make the Yankee a relief pitcher, and the Met, a starting pitcher. The professional niches of the two, although similar and somewhat overlapping, will be temporally different within each game. Making one a lefty and the other a righty would also make them different. And this is only during the regular season; think about possible differences in the two players during the playoffs and World Series. This complexity in the two hypothetical baseball players' professional habitats and niches is exactly the kind of detail regarding ecological habitats and niches of species that makes it so difficult to explain why some are relatively rare and others are relatively common. But enough said about baseball. Let's get back to raptor distribution and abundance.

In general, a raptor that does well in relatively common habitats, and shows substantial flexibility and breadth in its niche, should be more widespread and common than one that is limited to rare habitats, or that has a narrow suite of behavioral traits and exhibits minimal flexibility regarding its niche. This, for example, is why the Red-tailed Hawk, which inhabits both uplands and lowlands, and which regularly hunts both while perched and while soaring, is more widespread and common across North America than is the closely related Red-shouldered Hawk, whose habitats are more limited and whose hunting occurs almost entirely from a perch. But this is not always as simple as it sounds.

For a long time, it was believed that, everything else being equal, raptors that were capable of exploiting a broad range of environmental conditions would be widespread, and that because widespread species tended to be more abundant that those inhabiting limited ranges, they would be more common as well. Although that is generally true, everything else is rarely equal in ecology, and some widespread species are not nearly as common as others. Nor are species with broad ecological niches necessarily more common than those with narrow niches. It is one thing, for example, if a species' broad niche is the result of flexibility among individuals within it, and quite another if a species' broad niche means that all individuals belonging to it can exist only in areas with high numbers of different resources. Opportunistic flexibility at the individual level is one thing, whereas required variability at the species level is something else entirely. So where does all of this complexity leave us in answering the question of why some species are rare and others are common?

Over the years, ecologists have come to realize that part of the answer as to why some species are rare whereas others are common lies in understanding that not all species are rare for the same reason or in the same way, and that there are, in fact, different types of rarity. I describe three of these types of rarity, "geographic," "habitat-specific," and "low-population-density," below.

"Geographic" rarity occurs when an organism occupies a relatively small or limited-range area, and is said to be endemic to that area. Although the organism may be quite common where it does occur, its small geographic range limits its overall global population. Some of the best examples of geographically rare raptors are insular, or island-restricted, species. As mentioned above, one in six species of raptors occur only on islands, and many of these inhabit a single island whose size determines the range of the species. Such species are called "island endemics." A good example of this type of rarity involves three closely related species of serpent eagles, the Andaman Serpent Eagle, the Sulawesi Serpent Eagle, and the Crested Serpent Eagle. All three belong to the oriental genus *Spilornis* and all three differ little in their feeding and breeding ecology. All are small- to medium-size sedentary tropical eagles that live in or near forests in the Indomalayan tropics and that feed mainly on reptiles, rodents, and small birds.

As its name suggests, the Andaman Serpent Eagle is endemic to the Andaman Islands, a tiny archipelago of approximately 200 small islands that runs north-to-south in the Bay of Bengal, off the western coast of the Malay Peninsula. Although common throughout the archipelago, the Andaman Serpent Eagle has a geographic range of only 15,500 km^2 (about 6000 mi^2), the total land area of the archipelago, and, consequently, a global population estimated to be in the low thousands, making it one of the world's rarest diurnal birds of prey. By comparison, the Sulawesi Serpent Eagle, which inhabits the South Pacific island of Sulawesi (the eleventh-largest island in the world), together with several smaller islands nearby, has a geographic range of 190,000 km^2 (about 73,000 mi^2) and a world population of between 10,000 and 100,000 individuals, making it a relatively common, albeit endemic raptor. The continental counterpart of these two island-restricted species, the Crested Serpent Eagle, inhabits an enormous range of more than 7 million km^2 (nearly 3 million mi^2) that includes parts of India, most of Southeast Asia, and the islands of Borneo and western Indonesia, and an estimated world population of between 100,000 and 1,000,000 individuals. Although many aspects of the ecology of these three species have yet to be studied in detail, all three occur at somewhat similar densities in appropriate habitats within their ranges, and differences in their global populations—and the degree to which they are rare or common—reflect differences in sizes of their distributional ranges more than anything else. Other examples of geographically rare raptors include the Indian Ocean endemics, the Mauritius Kestrel and the Seychelles Kestrel, which have island-limited distributions of 2000 and 245 km^2 (770 and 105 mi^2), and estimated world populations of about 1000 and 800 individuals, respectively.

The second kind of rarity, "habitat-specialist" rarity, occurs when a species distribution is limited to a rare habitat type. One example of a habitat-specialist rarity is the White-collared Kite of northeastern Brazil, a species that has a tiny continental range of about 12,500 km^2 (about 4800 mi^2), and an estimated world population of not more than 250 mature adults. The Critically Endangered White-collared Kite's small continental range, and hence its rarity, is due to the fact that it is a habitat specialist limited to remnants of the Atlantic Forest of northeastern Brazil, a habitat that is being lost to deforestation. As in other habitat-specific rarities, the kite is rare because the habitat it depends on is exceptionally unusual and limited in extent. Another raptor that is rare because of habitat specificity is the Black Harrier of southern Africa, a species with a relatively small continental range of 1.6 million km^2 (620,000 mi^2) that relies heavily on Mediterranean fynbos and adjacent karooid scrub-and-grasslands, two limited open-habitat types in southern Africa. With a world population estimated to be fewer than 1000 breeding adults, the Black Harrier is currently listed as globally Vulnerable. By comparison, the Northern Harrier, an ecological equivalent of the Black Harrier that lives in many kinds of open habitats across North America and Eurasia, has a global range of about 40 million km^2 (16 million mi^2) and a world population of more than one million individuals.

The third type of rarity, "low-population-density" rarity, occurs when a species has a large geographic range, but is nowhere common within it. Examples of raptors that are rare because of low population densities include the somewhat aptly named Crowned Solitary Eagle of South America. The species has a huge and largely Neotropical range of more than 4.6 million km^2 (1.8 million mi^2), but appears to be nowhere common within it and, consequently, has an estimated world population just over 1000 individuals, and is listed as globally Endangered. Why some species are rare because of this factor is unclear. Although population density tends to decrease in larger predatory raptors, several species of equally large or larger eagles, including Africa's Verreaux's Eagle, occur at far higher densities within their ranges than do Crowned Solitary Eagles and, as such, are far more numerous.

Of course, combinations of two or three of these three types of rarity also can create rarity in a species and, as a result, some of our most widely distributed raptors, including the Osprey and Peregrine Falcon, two species with breeding ranges of about 30 million and 45 million km^2 (12 million and 17 million mi^2), respectively, have relatively small global populations, estimated at below 500,000 birds each. Although neither is currently globally threatened, with notable exceptions, both are considered by many to be naturally rare, or at least uncommon in many parts of their geographic ranges. The Osprey is somewhat peculiar in this regard in that it is loosely colonial in many of its principal breeding areas, including, for example, the 11,600-km^2 (4,500-mi^2) Chesapeake Bay of eastern North America, where the species had a growing population estimated at 3500 pairs in the mid-1990s. For Ospreys and several other species of raptors, the presence of conspecifics seems to be one of the most significant "habitat features," meaning that at least some otherwise appropriate areas lack nesting Ospreys simply because an initial "pioneering pair" of breeders has yet to settle in. The proclivity for Ospreys to nest near each other is so great that successful reintroductions in areas from which it was extirpated during the DDT era of the twentieth century have been restricted mainly to sites where large numbers of fledglings have been released simultaneously in an attempt to rapidly re-create breeding colonies.

Another factor that increases the likelihood of rarity in raptors is that the birds are predatory. And again, the explanation involves basic ecology. In most ecosystems, including all that are available to raptors, all energy, including food energy, initially derives from the sun. Plants get their food energy, and are able to grow and reproduce, by converting solar energy into plant material via the chemical process known as photosynthesis. Herbivores, or plant-eating animals, derive their food energy from plants via digestion and metabolism. Carnivores, or meat-eating animals, including predatory birds such as raptors, derive their food energy from herbivores in the same manner. This is how food chains work in ecosystems. Unfortunately, the movement of food through the different links of an ecological food chain is not terribly efficient. At each step, transfer inefficiencies overwhelm the movement of energy, resulting in an ecological pyramid of decidedly less available energy along the way, which, in turn, sets limits on the biomass and the numbers of individual organisms that can be supported at the top.

Conceptually introduced by the English ecologist Charles Elton in the 1927, the inefficient transfer of energy in food chains because of individual growth, death, and other biological and physical inefficiencies was quantified by the American limnologist Raymond Lindeman in 1942. Lindeman's inefficiency calculations suggest a relatively constant loss, or "leakage," of available energy of around 90% across each link in a food chain. This means that only 10% of the energy inherent in sunlight entering the system makes it into plant-available energy for herbivores, and only 10% of that makes it into herbivore-available energy for primary carnivores.

Because of this, plant biomass is overwhelmingly greater than that of herbivore biomass, and, in turn, herbivore biomass is overwhelmingly greater than that of carnivore biomass. Holding species size constant for comparisons, because most raptors act as primary, secondary, or tertiary carnivores, their biomasses, and their respective numbers, are lower than those of many herbivorous birds, making them relatively uncommon and rare to begin with. That regional populations of many large scavenging birds of prey, including many vultures, sometimes exceed those of most large predatory raptors, reflects the fact that the former feed on dead animals, which represent part of the "leakage" in Lindeman's efficiency calculations, whereas the latter are limited to the 10% available as "throughput." And, in fact, the highest breeding density of large raptors ever reported across a relatively large area was an estimated 2.7 pairs per square kilometer of mainly White-rumped Vultures in then-150-km^2 New Delhi, India, in the late 1960s. Implausibly, this species is now Critically Endangered globally, its populations having collapsed by more than 99% because of the introduction of the veterinary drug diclofenac in livestock in the species range. (See Chapter 8 for details.)

All of these explanations for rarity aside, the question of why some diurnal birds of prey are more common than others remains one of the most enigmatic of all aspects of raptor biology. Two final examples add to the puzzle. The first involves a pair of widespread, common, and closely related New World vultures; the second, a generally rare but widespread Eurasian eagle.

Consider first the distribution and abundance of the Black Vulture and Turkey Vulture. Black Vultures have an enormous breeding range of 19,700,000 km^2 (7.6 million mi^2), Turkey Vultures an even larger breeding range of 27,700,000 km^2 (11 million mi^2), or about 45% and 65% of the entire Western Hemisphere, respectively. Overall, Turkey Vultures occur everywhere that Black Vultures occur and then some. Much of the difference in range size occurs because Turkey Vultures breed farther north in North America and farther south in South America than do Black Vultures. Within areas where the two species do co-occur, breeding populations of Black Vultures tend to be higher than those of Turkey Vultures in the tropics, but the reverse is largely true in areas where the two co-occur in the mid–Temperate Zone. Why this is so remains an open question. Some argue that Black Vultures are more tropical and less migratory than are Turkey Vultures, but this is something of a "chicken-and-egg" answer. Others argue that Black Vultures, which are human commensals in many parts of their range—and are decidedly more social than Turkey Vultures—are both absolutely and relatively more common in the tropics because the species is better able to take advantage of the more lax conditions of human sanitation near the equator, where nonsanitary slaughterhouses and garbage dumps abound. Whatever the reason for the distributional and abundance differences in the two species, they appear to be cryptic.

Compare this to one of the more perplexing distributions of any raptor, the Bonelli's Eagle. This species is a smallish to medium-sized eagle with an enormous, but largely disarticulated, geographic range that stretches from Portugal in the west to southern China and the Indonesian archipelago of the lesser Sundas in the east. The 5.4-million-km^2 (12-million-mi^2) range, which includes parts of Eurasia, the Afrotropics, and the Indomalayan tropics, is represented as occurring on four islands and across twenty-nine separate continental areas in one recent publication and on four islands and across ten separate areas in a second, less recent, publication. Conservationists believe that the species is declining in many areas due, at least in part, to a variety of human-related threats, including the misuse of pesticides, habitat loss, declines in its prey base, collisions with power lines, and direct human persecution. Despite its large range, Bonelli's Eagle is uncommon almost everywhere in its range and currently ranks as one of the rarest diurnal birds of prey in Europe.

Why the eagle's range is disjunct is not immediately obvious. The species appears to be a diet generalist that preys on a variety of birds, mammals, and reptiles. Furthermore, its habitat, although difficult to define precisely, is not rare, and includes low- to mid-elevation mountainous and, in the Far East at least, generally forested areas, with an abundance of sunny days, crags, and cliffs. Some have suggested that its distribution is restricted by Golden Eagle competition both for nests sites and for prey, but there are few data supporting this view. The Bonelli's Eagle has been in significant decline in both the Mediterranean Basin and Middle East for many years, where it is said to be threatened by a number of human actions (see above). Whether these threats have created a series of fortress-like strongholds for the species remains to be studied rangewide.

The epigraph to this chapter is a quote by an excited Charles Darwin in a letter written to the English naturalist Charles Burnbury in the spring of 1856. I finish with a second quote from Darwin. This one, written merely three months later in a letter to longtime friend, the English botanist Joseph Dalton Hooker:

> I *really* hope no other chapter in my book will be so bad; how *atrociously* bad it is I know not; but plainly I see it is too long, & dull, & hypothetical. (Emphasis is Darwin's.)

Not much has changed in the last 160 years. Today, the distribution and abundance of raptors remain two of the most appealing but frustrating aspects of their biology. I have little doubt that when we finally understand what determines both, we will also finally understand the ecology of the birds themselves.

SYNTHESIS AND CONCLUSIONS

1. Raptors are a diverse assemblage of migratory and sedentary species that occur on all continents except Antarctica.
2. Migratory raptors are more widespread geographically than are sedentary raptors.

3. Two species of diurnal birds of prey, the Osprey and the Peregrine Falcon, have near-cosmopolitan ranges. Nine species occur in both the Old and New World, and ten species breed well into the Temperate Zone both north and south of the equator.
4. Raptors are more migratory at higher latitudes. Because 72% of all land lies north of the equator, and because land extends farther north in the Northern Hemisphere than in the Southern Hemisphere, migratory raptors are far more numerous north versus south of the equator.
5. Because many Northern Hemisphere species migrate south to overwinter, the numbers of raptors living south of the equator increases substantially during northern winter.
6. Fifty species of raptors, or approximately one in six of all diurnal birds of prey, live only on islands, most of which are tropical.
7. Raptor diversity increases substantially at lower latitudes, which is where most endemic and limited-rage species occur.
8. Understanding why some species of raptors are rare and others are common remains a topic of debate among raptor ecologists and conservationists.
9. There are three types of rarity in raptors: geographic, habitat-specific, and low-population density.
10. Excepting scavenging raptors, most species of diurnal birds of prey are rarer than they would otherwise be because they are predatory and exist at or near the tops of "ecological pyramids."
11. Why some raptors are rare remains enigmatic for many species.

5 BREEDING ECOLOGY

> Reproduction is the most significant event in the life
> of any animal, because only through this process
> can the animal perpetuate its genes in future generations.
> **Ian Newton, 1986**

THE RAPTOR BIOLOGIST Ian Newton had it right: reproduction is an essential part of raptor biology. Without reproduction, there will be no next generation. And without a next generation, extinction will follow.

As is true for most birds, raptors are largely monogamous, at least within a single breeding season. And, like most birds, raptors breed during the time of year that food is most available. In fact, the breeding biology of raptors largely mirrors that of most other birds. Most raptors court, pair, build a nest, produce and incubate young, brood and feed their young, and thereafter protect them at least for several weeks after they leave the nest. The breeding ecology of raptors is as nuanced and as variable as any aspect of their biology, with numerous intrinsic and extrinsic factors determining how, when, where, and whether individuals will breed.

By their very nature, breeding diurnal birds of prey are difficult to study. Many live at low densities and are secretive, especially when nesting, and many nest in remote and relatively inaccessible locations. Nevertheless, knowledge of their breeding ecology is more advanced and robust than is our understanding of their migratory movements and their wintering ecology. There are several reasons for this. First and foremost, much of their breeding ecology centers on the nest. Once found, this ecological magnet helps researchers concentrate their fieldwork on a relatively small ecological neighborhood for the duration of the breeding season. A second reason is the academic calendar. Although there are many notable exceptions, most research on breeding raptors has been conducted by college professors and their students during the academic break in the summer, which is when most Temperate Zone raptors breed and are readily available for study.

Box 5.1 The value of long-term and multispecies studies

The rate at which knowledge accumulates in raptor biology depends on many factors. Emerging technologies and new techniques allow us to study the raptors in new and different ways and help grow information in the field, as does asking new kinds of questions about the birds, the answers to which provide fresh insights into their biology. Often underappreciated is the critical role that comprehensive, long-term, multispecies studies play in helping develop a better understanding of the biology of birds of prey.

Much of what we know about the breeding ecology of birds of prey comes from relatively short, 2-to-4-year, single-species, single-site studies conducted by graduate students working under the guidance of senior scientists. Overall, this work is admirable, and the results of such studies have contributed significantly to the field. But short-term, single-species, single-site studies are snapshots of ecological reality that are inherently limited in their ability to detect and appreciate slow ecological processes, and rare and episodic events, and often miss capturing long-term variability in ecological systems, as well as the implications of subtle and complex factors that influence the ecology of birds of prey. Fortunately for the discipline, a number of first-class, comprehensive long-term studies, many of them self-supported by the researchers involved, have been conducted, and the results of many have been published, either as series of papers or as book-length manuscripts. Below, I briefly highlight several of these studies that have contributed immeasurably to knowledge in the field.

Sparrowhawks (1986). Ian Newton's 14-year comprehensive investigation of Eurasian Sparrowhawks in two areas of southwestern Scotland increased substantially our understanding of multiyear trends in breeding populations of the species, as well as our understanding of significant human impacts. The study is particularly noteworthy for its focus on factors responsible for population regulation and on establishing the degree to which individuals differ considerably in lifetime breeding performance. Newton's meticulous observations of differences associated with the marked size differences in male and female sparrowhawks provide significant insights into the question of why female raptors are larger than their male counterparts. His careful analyses of interannual variability in food availability clearly demonstrates that it, more than any other factor, determined breeding numbers in the populations he studied. As the author concludes, "the contribution of long-term studies to evolutionary theory comes largely from the detailed knowledge of individuals, and the factors that govern their lifespan and breeding rates." This long-term study clearly confirms that fact.

African Fish Eagles (1980). Leslie Brown's more than two decades of somewhat less structured but decidedly no less detailed investigations of the African Fish Eagle across much of Africa, including Nigeria to Ethiopia and south into Kenya and Botswana, provides a comprehensive look into both the breeding and nonbreeding biology of this, arguably, most common of all fish eagles. Painstaking observations of a dense population of eagles in and around Lake Naivasha, in southwestern Kenya, revealed remarkable consistency in both the spacing and sizes of breeding territories, with most individuals apparently remaining paired and on territory year-round, presumably to defend their relatively small several-hectare feeding areas from would-be intruders. Unlike Newton' sparrowhawks, Brown's fish-eagle populations appeared to be more limited by competition for nesting territories than by prey availability, with the latter seemingly ample enough so that the birds spent most of the day loafing and chasing intruders rather than fishing. Unfortunately, Brown's studies, unlike Newton's, lacked individually marked birds, and much remains to be learned about individual differences in long-term reproductive success of the species.

Verreaux's Eagles (1990). One of the most comprehensive long-term studies of any raptor, Valerie Gargett's 21-year investigation of a dense population of Verreaux's Eagles (a.k.a. Black Eagles) in the Matobo Hills of southwestern Zimbabwe ranks as what Leslie Brown called the first "statistical" long-term study of eagles. Gargett's findings, which were discussed in several dozen technical papers, as well as an encyclopedic book-length monograph, focus on many aspects of the species biology, including breeding ecology, nestling dispersal, and both adult and subadult mortality. Breeding performance in the eagle was clearly related to the relative availability of the species' principal prey, rabbit-like hyraxes (a.k.a. dassies), small, primitive ungulates, whose vulnerability was inversely linked to magnitude of the previous season's rainfall. The study also detailed the ferocious nestling rivalry and siblicide that restricts fledgling success to a maximum single individual per nest. Unfortunately, Gargett's study, like many long-term efforts, focused on the biology of a single dense population and therefore may not likely be representative of other, less-dense, populations. Nevertheless, the information that Gargett accumulated and published across more than two decades of fieldwork make Verreaux's Eagle species one the world's best-studied and best-known eagles.

Golden Eagles (1997). Based on a 15-year study of Golden Eagles in Scotland, together with up-to-date information on the species from North America and continental Europe, Jeff Watson's monograph on the species provides an enviable global perspective on this, the world's most widely distributed large eagle. Included in the monograph is detailed information on the species' ranging behavior, nest spacing, and nest density throughout much of Scotland. Especially intriguing are data on the dispersal habits of recently fledged young and the influence of spring weather, including late-winter and early-spring storms on breeding success in this circumboreal breeder. The role of food availability, especially higher populations of grouse and hares in eastern Scotland, in affecting regional differences in reproductive productivity in lowland compared with highland habitats is especially enlightening, as are the larger-scale geographical differences in the species' ecology, which may explain the birds' enormous range.

Northern Harriers. Two coincidental long-term studies of Northern Harriers (a.k.a. the Hen Harrier in the United Kingdom), one by Donald Watson (1977) in southwestern Scotland, and a second by Frances Hamerstrom (1986) in central Wisconsin, together and alone have provided considerable significant information of this open-habitat polygynous raptor. Watson's studies, which began with a single nest in 1959 and concluded with many more in the 1970s, saw the ground-nesting populations he studied initially increase and then "move" increasingly from moorland habitat into recently forested habitat, with peaks in occupancy occurring when the evergreen trees in question were 6 to 12 years old. Hamerstrom's studies, which began in 1957 and continued for 27 years, initially focused on a single question: do harriers mate for life?—a question that Hamerstrom at first thought would be easy to answer, simply, "catch the male and catch the female and band them," then follow their breeding activities in subsequent years. The question, however, took years to answer, and included the discovery that in some but not all years some males simultaneously mated with several females and the fact that interannual divorce turned out to be more common that interannual mate fidelity. Hamerstrom's work, which bridged the DDT era in twentieth-century north-central North America, also successfully captured the organochlorine's impact on the species' breeding population numbers and breeding success. In addition, both Watson and Hamerstrom concluded that prey availability was critical in determining annual differences in reproductive success, much as Newton had found with his sparrowhawks, and both concluded that much also still needed to be learned about the species they had spent so much time with.

Neotropical Birds of Prey (2012). As rare as long-term, single-species studies are in raptor ecology, coincidental multispecies studies at a single location are even more so. One of the most ambitious and best of such studies was an 8-year program conducted by the Peregrine Fund of Boise, Idaho, in and around Tikal National Park, Guatemala, in the late 1980s through the mid-1990s. This groundbreaking effort developed a number of new methods for detecting, counting, and studying the breeding ecology of eighteen species of tropical diurnal birds of prey and two species of owls, including focused efforts on nesting behavior and habitat use. The impressive monograph, *Neotropoical Birds of Prey*, resulting from this effort represents a fundamental benchmark resource for understanding the ecology of tropical forest raptors, and also serves as a model for future multispecies studies both within and beyond the tropics. Edited by David Whitacre, the report's more significant findings included confirming the long-held belief that the leading cause of nest failure in tropical raptors is predation both on eggs and nestlings, and that predation on adults also sometimes causes nest failure. Intriguingly, the study also documented that tree-nesting tropical raptors, like their counterparts in the Temperate Zone, also prefer nesting in isolated and emergent trees rather those within the canopy proper, and that tropical breeders, like those in the Temperate Zone, tended to breed seasonally, with most of the tropical species studied initiating breeding late in the dry season, and most nestlings being reared in the rainy season. Although the representative nature of this multispecies study's findings awaits additional investigations elsewhere, the work represents an essential first step in developing a better understanding of the breeding ecology of tropical nesting raptors.

In sum, studies such as those described above have played important roles in expanding our knowledge base not only regarding the breeding ecology of birds of prey, but also regarding raptor ecology in general.

Below, I try to summarize the essential aspects of the breeding ecology of diurnal birds of prey. Most of my examples describe the breeding behavior and ecology of northern Temperate Zone species, whose biology is far better known than that of tropical and Southern Hemisphere species. Whenever possible, however, I do include breeding aspects of the latter, particularly when those aspects differ from those of their more northern counterparts. As a result, my coverage is indicative and illustrative more than encyclopedic, and this chapter should be viewed with this in mind.

MATING SYSTEMS

Monogamy

Unlike some birds in which females (e.g., Gallinaceous birds, including chickens and ducks) or males (e.g., shorebirds, including Spotted Sandpipers and some phalaropes) provide all post-laying parental care, raising raptors usually requires full-time work for both parents. The vast majority of diurnal birds of prey do so as monogamous pairs, with females serving as the primary, if not only, incubators and brooders, and males serving as prey providers, both for their mate and their offspring. In many raptors, chores merge as nestlings grow and fledge, with both males and females serving as food providers and protectors of their naïve and vulnerable young. In some cases, parental females leave the nesting area and cease protecting and providing for their young before parental males do.

Monogamy is the norm, especially in nonmigratory populations, but migratory species often practice monogamy "light," a phenomenon in which breeding pairs are monogamous, but only during individual breeding seasons. In eastern Pennsylvania, for example, migratory American Kestrels breeding in Hawk Mountain Sanctuary nest boxes, engage in scramble competition for mates when they return to breed in spring. Males vie for the best breeding sites, which, presumably, are determined by both prey density and nest-box availability, because a "good kitchen" and "safe bedroom" are essential features of successful breeding. Shortly thereafter, females, which return to the breeding area several days later than their potential mates, vie for males with the best territories. Competition is keen, and returning males typically breed in different nestboxes across years, and returning females almost always pair with different males from the previous year. The system is so fluid that one or more members of presumptive kestrel "pairs" trapped and banded at specific nest boxes during courtship and mating are often replaced by competitors who arrive after them, and who subsequently mate and raise young in the same box. Although the extent of scramble competition for nest sites and mates can be difficult to study, careful observations that follow banded birds through the nesting season suggest that surplus, nonbreeding "floaters" quickly replace females killed by humans or taken by natural predators. And even among presumably monogamous populations, a handful of nests can be occupied by two unrelated females in a single season, with both individuals laying in the same nest box.

Monogamy is substantially more pronounced in larger, longer-lived raptors. Migratory Ospreys breeding in eastern North America typically retain mates across years, except when one member of the pair fails to return to the nesting area. In this species, like many diurnal birds of prey, breeding success increases as the birds age, until senility sets in. In addition, individuals that retain mates have higher reproductive rates than those with new mates. This, together with the fact that mates probably stay together because both are attracted to areas they are familiar with and have been successful in, leads to high rates of mate-retention in Ospreys. The same appears to be true of Bald Eagles, although there are reports of "divorce" in the species, and, in a few instances, of three birds attending a single active nest. In cases of divorce, courtship and mating with new partners most likely does not happen until the start of the next breeding season.

The literature suggests that nonmigratory individuals are more likely to remain paired across years than are migratory individuals. In eastern Pennsylvania, for example, pairs of both year-round resident Red-tailed Hawks and the few sedentary American Kestrels that remain on territory and close to one another during autumn and early winter tend to be monogamous across years. But keep in mind that raptors reabsorb their gonads after mating, and the recrudescence of these organs typically does not occur until mid-to-late winter. At that point, the avian equivalent of "happy hour" follows, as pairs rebond sexually while undergoing what the nature writer Pete Dunne refers to an annual avian "puberty" of sorts.

Polygamy

Polygamy, which also is called cooperative breeding, occurs when more than two parental birds care for and raise their broods, typically in a single nest. In most species, polygamy involves trios of birds. There are two types of polygamy, both of which are relatively rare in raptors. Polygyny, which, literally, means many females, occurs when a male is mated to two or more females simultaneously. Polyandry, which, literally, means many males, occurs when a female is mated to two or more males simultaneously. Polygamy has been reported in at least forty-two species of raptors, including the Osprey, several eagles, kites, vultures, harriers, accipiters, and falcons and their relatives. In most species, it is uncommon or rare. Exceptions include several harriers, and Madagascar Fish Eagle, Galapagos Hawk, and, possibly, Mississippi Kite, and Javan Hawk-Eagle. In species in which the genders of the birds have been identified, polyandry appears to be slightly more common than polygyny. Because polygamy is easy to overlook in little-studied species, it probably takes place at least infrequently in other species as well.

Polygynous behavior in raptors has been best studied in harriers. The Northern Harrier, for example, is polygynous in many parts of its range. In northern Scotland, for example, males averaged two females each, with some males attending three or more females. Second-year males, which often do not breed, were monogamous when they did breed. In central Wisconsin, sex ratios of the adult breeding population were closer to 1:1, with polygyny occurring only in some years. The details, however, are more interesting than the long-term average indicates. Polygyny in Wisconsin is loosely related to highs in the harriers' local principal prey, the meadow vole. Polygyny was relatively rare during the 1960s, when few harriers nested in the region, because populations had declined due to the misuse of DDT and other organochlorine pesticides. (See Chapter 8 for details on pesticide effects on raptors.) The harrier population increased substantially, however, in the late 1970s, after North American bans on the widespread use of DDT, and by 1979, when the long-term study high of thirty-four nests was identified, twenty-one of the nests were occupied by females mated to polygynous males. And even a few second-year males bred. Thus, even at this relatively small 40,000-acre study site, polygyny appears to have been a flexible and, apparently, adaptive mating phenomenon, as the population continued to increase well into the 1980s.

Island-inhabiting raptors are more likely to exhibit polygamy than are continental populations. The record number of mates for a male Northern Harrier, for example, was a harem of seven for an island-inhabiting supermale. But the best-studied example of an island-dwelling polygamous species is the polyandrous Galapagos Hawk.

A resident of several islands on the Galapagos Archipelago, the Galapagos Hawk was once common across most of the larger islands. (Unfortunately, today, this hawk ranks as Globally Vulnerable, with an estimated population of fewer than 1000 individuals.) As is true of many island endemics, the Galapagos Hawk is a diet generalist that feeds on a variety of vertebrates, invertebrates, and carrion. Genetic analyses suggest that individual hawks do not move regularly among islands. Most breeding groups consist of a single female and two to four unrelated males. On a per-nest basis, cooperative breeders are reproductively more successful than are monogamous pairs. Polyandry is thought to occur because females in polyandrous groups are able to save needed energy by having more helpers, and because males in such groups get to breed at an earlier age than if they otherwise waited to acquire a territory on their own. Although many questions remain regarding why polygamy is more common in island raptors than in their continental counterparts, the physically isolated nature of these populations severely limits the dispersal abilities of young birds, making it challenging to disperse into less densely populated areas where they might otherwise be able to breed monogamously at an earlier age.

Another intriguing case of raptor polyandry involves the Harris's Hawk, a largely open-country bird whose range stretches south from Arizona and New Mexico in North America, into Central America and central Chile and Argentina, all the way into the southern cone of South America. Social breeding groups include as many as seven individuals, including both adults and immatures. Depending on geographic location, populations can be monogamous, polyandrous, and, in some cases, even polygynous. Many researchers believe that the species' social breeding habits are tied to its cooperative hunting behavior. The species is both a single- and multi-individual perch and flight hunter, in which duos, trios, and larger teams of individuals engage in surprise pounces, converging on vulnerable prey in the open from various directions, thus increasing the likelihood of success. A second hunting strategy involves one hawk entering cover to flush prey while others wait for and capture the escaping prey when it makes a break for it. Although helpers at the nest do not appear to affect clutch size, social breeders raise heavier nestlings than monogamous pairs, and they may be able to acquire and defend better nesting territories.

COURTING AND MATING

Although normally secretive during the breeding season, especially near the nest, breeding raptors are atypically conspicuous when "courting." This is the time when females are busy assessing and selecting their mates, and when males are busy demonstrating their ability to secure sufficient prey for their mates and potential young. In sedentary populations, courting begins in mid- to late winter and builds to a crescendo just before mating. In the Mid-Atlantic region of the United States, for example, sedentary Red-tailed Hawks begin spending more time with their mate of the previous season in December and January, when, on warm sunny afternoons, pairs can be seen soaring above last year's nesting territory and subsequently landing close to one another, often on the same branch of a large tree. Their gonads are recrudescing, or regrowing, at this time, and the "friendship" of the previous autumn is blossoming into the raptor equivalent of true love, at least for the upcoming breeding season. In migratory populations, courtship is decidedly more rushed, amounting to little more than scramble competition for the "best of the best" as waves of returning males and females perform the raptor equivalent of speed dating. In many, but not all, species, males arrive anywhere from several days to several weeks before females, although, in some populations of accipiters, the reverse has been reported. In general, experienced males attempt to reestablish previously held territories upon arrival and wait there for females to arrive, at which time competition for mates begins. In the ensuing rush, females often test the abilities of several potential suitors simultaneously or in rapid succession, and what appears to be a mated pair one day can become old news the next, as last year's mate, or a more aggressive female, ousts a male's initial choice. My experiences with American Kestrels in eastern Pennsylvania offer an example of the confusion that can ensue.

In the early 1990s, Jurgen Wiehn, a graduate student from the University of Turku, in Finland, studied mate selection in American Kestrels near Hawk Mountain Sanctuary. The young biologist was surprised to learn that many of the males and females he trapped and banded at nest boxes soon after the first wave of migrants arrived were not the same birds that eventually nested and raised young in those boxes. Indeed, during the first few weeks of the breeding season, female kestrels were often seen in the company of two, three, or even four males simultaneously, and mating can be fluid, to say the least. In both sedentary and migratory populations, male competition for the best territories is keen, because males will need to secure prey not only for themselves but also for their mates and developing young. Similarly, so is competition for the best females, because finding the right mate is likely to have significant consequences as well.

Unlike in songbirds, where singing "rules" the competition for territories and mates, in raptors, the competition involves aerial displays that demonstrate both a willingness to fight for a territory and sufficient hunting prowess. That said, calling—if not actual singing—can play an important supporting role. In many species, aerial displays are used both to attract mates and to repel same-sex interlopers, and determining which is occurring can be confusing. The cartwheeling flights of medium-sized and large raptors, in which adults lock toes and talons high in the air before swirling downward, are a case in point. Cartwheeling is well documented as aggressive behavior in a number of species in situations involving territorial intruders, and sufficient field evidence exists to confirm its use for this purpose. It is less well documented as courtship behavior. Perhaps the best example of its use in the latter comes from careful observations over three years of studies involving eleven pairs of color-marked Bearded Vultures conducted by Christopher Brown in southern Africa. During Brown's observations, all cartwheeling flights were initiated by the male in late afternoon when the pair was nest building, and all typically preceded chasing and diving by both pair members. Cartwheeling occurred when the lower of the two birds, while flying in tandem, flipped over and presented its talons to the bird above. Pairs then locked toes and cartwheeled for about 200 meters, parting when they were 100 to 200 meters over the ground. The two then flew to the nest site, roosted together and sometimes allopreened. These observations, along with similar reports in other species, indicate that cartwheeling, at least in some instances, does function as courtship behavior.

Other types of courtship display include undulating flight, which can be accompanied by deliberate and exaggerated wing flapping, and "sky-dancing," in which one or both members of the pair engage in roller coaster–like, and sometimes even corkscrew-like, up-and-down displays unlike anything the birds do normally. In Northern Goshawks, undulating courtship flights usually occur on sunny, windless days. Individuals performing such flights sometimes appear to engage in harrier "rowing-like flights" with tails unflared but with their white undertail-coverts and femoral-tract feathers spread wide and fluttering conspicuously. The flights, which typically take place above the forest canopy, are usually performed as a duet, with both members of the pair participating. Similar courtship flights occur in both Cooper's Hawks and Sharp-shinned Hawks and, probably, in other accipiters as well. Booted Eagles engage in even more spectacular display flights, or sky dances, in which they dive repeatedly with half-closed wings, with their heart-shaped, folded-back tips touching the tops of their tails, before returning to great heights with less tightly tucked wings. Such aerial dances, which often are performed by both members of the pair, can continue for half an hour or more in a tight series of exaggerated roller coaster–like moves spanning 50 meters or more.

Equally impressive—albeit somewhat slower—sky dances are performed by harriers over prospective breeding territories. Sky dances have been described best by the harrier biologist Frances Hamerstrom regarding such events on the species breeding grounds in her Buena Vista Marsh study site in central Wisconsin: "Two [male] harriers are sky-dancing over the marsh. One is almost pure white and the other is slate gray. The white one swings in giant loops like a stunting aviator. The gray one gives a twisting back flip at the top of the loop-the-loop, but seems to have less style—perhaps because his black-tipped wings are merely gray against the early morning, April sky, and his loops are smaller—or perhaps because he is younger. These dancing males are presumably establishing nesting territories."

Female harriers also engage in aerial display, but less frequently and less dramatically. The events that occur as soon as the birds arrive in spring continue as a wave northward as individuals initiate migrants even before settling in on their eventual nesting territories.

Serial displays are sometimes accompanied by loud and seemingly belligerent calls. Such calls most likely serve to announce to non-mates their presence on an occupied territory, as well as by softer coos, chips, chirps, and whistles, and pre-nuptial feeding, when mates are involved. In the end, however, it is the aerial acrobatics that provide the most memorable part of the birds' arrival on the breeding grounds each spring.

The mating that follows courtship can be a touchy affair in raptors—both literally and figuratively—as females are larger than their male counterparts and, as such, are superiorly "well armed." Not surprisingly, males approach prospective mates tentatively, often gifting them with prey before copulating. In most vertebrates, copulations occur several times during the breeding season, most typically shortly before the fertilization period, which occurs across a period of a few days to several weeks. There are exceptions, however, and birds, including raptors, are one of them. A careful study of American Kestrels in southern Quebec documented the phenomenon during two breeding seasons in the early 1990s. Graduate student Mark Villarroel, working with his supervisor, David Bird, conducted the systematic observations of a population of kestrels in southern Quebec. The two found that copulations involved few preliminaries except for solicitation. Once a male approached and perched near a female, the latter typically moved toward him and lifted her tail to expose her cloaca. Copulations were considered successful when the male jumped on the female and gently crouched on her back while dipping his tail to join the female's cloaca. To say that the kestrels did this frequently is an understatement. Shortly after pairs formed, they copulated at a rate approaching twice an hour for more than a month, with the rate declining to zero two-and-a-half-weeks after egg laying. More than 80% of the copulation attempts were considered successful. Thirteen percent were considered unsuccessful. The remainder were not seen well enough to determine outcome. Most of the copulations occurred during the fertile period of about 20–25 days shortly before and after the first egg was laid. Males and females were equally likely to solicit copulations. Approximately two-thirds of failed attempts were initiated by males, whereas only one-third were initiated by females. Pairs copulated an average of 454 times each, at a rate of approximately 14 times daily for about a month-and-a-half.

Extra-pair Copulations

In addition to intra-pair copulations between members of monogamous pairs, many species of birds engage in what ornithologists objectively call extra-pair copulations, behavioral events during which non-paired males and females copulate with each other, often at times when both are mated to other individuals. Although American Kestrels are known to engage in extra-pair copulations, I watched a mated female do so at least twice when being courted simultaneously by three males, none of which were mated to her. Somewhat surprisingly, Villarroel and Bird observed extra-pair copulations only once during their study, leading them to conclude that paternity assurance (i.e., making certain that the nestlings a male was caring for were indeed his own) was not the only reason for the frequent intra-pair copulations they observed, and that additional functions including both "pair bonding or mate assessment" also were involved, something that humans might identify with.

Nevertheless, paternity assurance is important, at least in some species, and experimental studies support this function. One such investigation involved Red Kites breeding on Corsica. In this study, males with high numbers of close neighbors spent more time guarding their mates during pre-laying than did males with fewer close neighbors. More importantly, when the researchers simulated male territorial intrusions with a plastic decoy painted to resemble an adult kite, only males attacked the decoy. Males spent significantly more time in their territories near their mates when decoys were present than before the decoys had been placed in their territories, but this occurred only during pre-laying period at times when the female was fertile and shortly after the female had laid her eggs. A similar experiment involving Montagu's Harriers in western France yielded similar results.

Obviously, non-mated males can benefit genetically from extra-pair copulations, but why females willingly engage in extra-pair copulations (forced copulations are rare in raptors) is not so clear. Several explanations have been advanced. The first is that females are lured into such behavior by males that have "better genes" and, as such, are potentially "fitter" than their mate. After all, the argument goes, such males have both the time and energy to engage in such activity, which may indicate superior fitness over their stay-at-home male counterparts. On the other hand, raising nestlings fathered by two rather than one male may offer a bit of "bet hedging" for the female if she is not in the position to directly evaluate the fitness of her mate versus extra-pair suitors. Whatever the reason, genetic evidence indicates that extra-pair copulations occur in at least some species of diurnal birds of prey.

NESTING SITES

Diurnal birds of prey raise their young in a variety of nest settings. Some build no nest at all, but merely scrape out a bare spot on a relatively flat surface; some reuse the nests of other birds, including other raptors; some nest in cavities; whereas still others build some of the largest stick nests of any bird. Falcons and owls typically build no nest, although many use the abandoned stick nests or cavity nests of other birds, sometimes usurping still-active nests. On the other hand, many large species return to their old nests and refurbish them year after year. Because nest-selection occurs quickly at the beginning of the breeding season, choosing a nest site has been little studied overall. In many raptors, including Ospreys, kites, hawks, and eagles, the female typically selects the site and does most of the nest building, often with the help of the male.

In raptors that reuse the nests of other species, those species are usually common. In North America, Great Horned Owls often usurp the previously used stick nests of other similarly sized birds, including the Red-tailed Hawk. Owls nest earlier in the season than do the hawks, and the latter typically build and subsequently maintain multiple stick nests in their territories, which they use alternately across years depending on where the owls nest. In the Old World, a similar arrangement occurs between Common Kestrels and Eurasian Magpies, with the kestrals nesting in previously used magpie nests. In most cases, the breeding distribution of the usurpers often mirrors that of the nest builders, and may in fact be limited by it. In parts of North America, for example, declines in populations of the primary cavity-nesting Northern Flickers have been suggested as a possible reason for declines in populations of American Kestrels, as the latter often nests in old flicker nests.

Intriguingly, nonraptors sometimes benefit from having raptors nesting nearby. In Japan, Azure-winged Magpies are more successful when they build their nests near rather than far from those of Japanese Sparrowhawks, a species whose own nesting defense provides a protective shield for magpies nesting nearby. The same kind of relationship apparently occurs among nesting Red-breasted Geese and both Snowy Owls and Peregrine Falcons in northern Siberia. The geese, which begin nesting later in the region than do Snowy Owls and Peregrine Falcons, are far more likely to build their nests near the two predatory species. Experiments involving a "surrogate" Arctic fox (actually a pet German Schnauzer) indicated that fox predation, rather than other habitat features, best explained the nesting preference of the geese. Other examples of this predatory-bird "halo-effect" include improved nesting success of Fieldfares, a colonial songbird nesting in northern Sweden near breeding Merlins, and Red-billed Choughs, nesting within breeding colonies of Lesser Kestrels in northern Spain. In the latter, researchers reported that the choughs chose to nest in buildings where colonies of the kestrels already had set up shop. In Italy, a study of nesting Lesser Kestrels and Western Jackdaws indicated that both species appeared to benefit when nesting close to each other, in terms of both reduced vigilance and nest defense.

Another commensal relationship involving the stick nests of raptors occurs when other species of birds build their nests in, or attach them to, those of large raptors. Examples include House Finches and other songbirds nesting in the nests of Swainson's Hawks, and House Sparrows and Monk Parakeets nesting in the nests of Ospreys. Not all raptors are compliant, however. African Fish Eagles have been known to pull off and discard weaver-bird nests attached to their own nests, sometimes using the former as a lining for their own.

The sizes of the stick nests raptors build can depend on the local availability of building materials. Golden Eagles, for example, build smaller nests in treeless areas than in areas where sticks are plentiful. Birds that nest relatively close to the ground, including the Lappet-faced and Cinereous Vultures, build enormous nests, possibly to deter tree-climbing vertebrate predators. Species that lay single-egg clutches, including Hooded Vultures and snake eagles, tend to build relatively small stick nests for their body size. On the other hand, some species including eagles build oversized nests that are refurbished and added to annually. A Bald Eagle nest in Vermillion, Ohio, that was occupied for 34 years measured almost 2.7 m (9 ft) across and 3.3 m (11 ft) deep, and was estimated to have weighed close to two tons. An even larger nest in St. Petersburg, Florida, measured more than 2.7 m (9 ft) across and almost 5.6 m (20 ft) deep. Typical Bald Eagle nests average 1.5 m (5 ft) across and 1 m (3 ft) deep. These oversized nests, which can be visible for several miles, may function to advertise the presence of the territorial pair to potential interlopers. And as the American eagle biologist Francis Herrick put it, such huge nests also serve as both "gymnasium and flying practice-field" for the developing eaglets.

Many raptors line their nests with leafy green sprigs, which they bring to the nest after basic construction is complete, and continue such additions for some time thereafter. In Broad-winged Hawks, where this behavior was studied in detail by the behavioral ecologist Bernd Heinrich, the parental birds added an average of two fern fronds or white cedar branches daily until the young were almost 3 weeks old. The sprigs appeared to offer no structural support or cushioning for the young, and Heinrich concluded that they served an anti-pathogenic function. That green boughs brought to the nests of Eurasian Starlings have been shown to suppress bacterial growth, as well as increase nestling immune function, support this potential sanitizing function.

One thing that certainly does not sanitize a raptor's nest is the "junk," including fragmented glass, pottery, plastic, and metal, along with metal bottle caps, small rocks, and sticks that appear in the nests of many species of both New and Old World vultures. A bit of natural history helps explain the situation. The slow growth rate of nestling vultures is often attributed to their typically boneless, low-calcium, parentally regurgitated diets, and at least some researchers have suggested that the junk found in vulture nests represents mistaken attempts by parental birds to feed their young fragmented, calcium-rich bones, which the junk superficially resembles. Although nestlings presumably regurgitate at least some of the undigestible junk without issue, its ingestion recently has been linked to the deaths of at least two California Condor nestlings. The problem is such that nest junk is now considered the primary cause of nesting failure in the species. With this in mind, bone fragments are now included in the food offered at most supplemental food "vulture restaurants" in the condor's range, with the hope of reducing the rate of junk consumption in developing young.

Many raptors build more than one nest in their breeding territories, with some pairs building five or more in a single year. Because nest building takes considerable time and effort, researchers themselves have spent considerable time trying to understand why the birds do so. Shifting nests immediately prior to egg laying makes sense if the abandoned nest has been discovered by a predator or has been usurped by another pair of raptors. Other possible functions include broadcasting the presence of an active pair to potential competitors or reducing the risk of ectoparasites, which can kill a nestling via exsanguination in instances where ectoparasites routinely overwinter in old nests. Another possible explanation for nest switching is the building of what have been called "frustration nests" by raptors, including Ospreys, following nesting failure in earlier-built primary nests, the idea being that getting a head start on a new nest for next year makes sense if parental behavior is no longer needed in the current breeding season.

Many tree-nesting raptors also nest on cliffs, particularly when trees are not readily available. The opposite is also true. The common and widespread Golden Eagle, for example, which usually nests on cliffs, also nests in trees, as well as on the ground and on human-made structures, such as power-line towers, windmills, and many other elevated structures, including cliff-nesting platforms built for them. Like tree nesters, cliff nesters typically demonstrate selectivity in the cliffs that they use. In Arctic Alaska, for example, Peregrine Falcons nest more often on exposed ledges with little protection from weather, whereas Gyrfalcons typically choose sites with protective overhangs, the latter being common in other cliff nesters. Many cliff nesters also favor nesting on cliffs with specific compass orientations, either to take advantage of increased or decreased exposure to solar radiation, depending on the local climate. Not surprisingly, cliff accessibility often factors into nest-site choice, with relatively "predator-inaccessible" locations being favored, particularly in human-dominated landscapes and at cliff faces close to roadways. (As a result, many researchers that study the nesting biology of cliff-nesting raptors use technical equipment to reach the nest sites they study.)

In addition to reducing the risk factors associated with potential human and other mammalian predators, cliff nesters also appear to site their nests to reduce competition and predation from other raptors. A recent study of a cliff-nesting raptor community in southeastern Spain, for example, suggests that both intra- and interspecies interactions affected the nest-site distribution in several species. Four species, Golden Eagles, Bonelli's Eagles, Peregrine Falcons, and Eurasian Eagle-Owls, made up the community studied. Only Golden Eagle nests were spaced more regularly than expected, presumably because of territoriality in this, the largest cliff nester in the region. The other three species seemed to cluster their nests, with smaller Bonelli's Eagles tending to avoid areas with Golden Eagles, which is a superior competitor and a relatively intolerant species when it comes to other cliff-nesting raptors, and Peregrine Falcons tending to avoid areas with high densities of Eurasian Eagle-Owls, which are known to prey on recently fledged falcons. A similar predator-prey situation between Great Horned Owls and cliff-nesting Peregrine Falcons led conservationists to repopulate nesting populations of peregrines in treeless and relatively owl-free coastal areas in the Mid-Atlantic States, including on buildings and bridges in cities where the owls were relatively few and far between.

Competition among cliff nesters for nest sites sometimes mirrors that seen among tree nesters, with cliff-nesting Griffon Vultures often usurping the nest sites of Egyptian Vultures, Bearded Vultures, Bonelli's Eagles, and, less commonly, Golden Eagles, in areas where this widely distributed, larger-bodied, earlier nester is common. Griffon Vultures typically refurbish used nests each breeding season because their young from the previous season largely destroy their nest structures and, therefore, usurping the reasonably intact nests of other species most likely saves them considerable time and energy in building their own nest.

Although some raptors nest exclusively on cliffs or on man-made structures throughout most of their ranges, several are known to nest in trees, either in nests previously constructed by other species, in hollow tree trunks, or in nests that they themselves build. In the forested lowlands of central and eastern Europe, typically cliff-nesting Peregrine Falcons nest in abandoned raven nests and those of other large stick-nesting birds, as well as in abandoned Bald Eagle nests in coastal British Columbia, Canada, where doing so brings them into close proximity to breeding auklets, a favorite prey. One of the more remarkable aspects of cliff-nesting raptors is their long-term use of particular sites. In East Africa, for example, colonies of cliff-nesting Rüppell's Vultures have used the same nest sites for decades, and in England, where populations of Peregrine Falcons were decimated during the pesticide era of the mid-twentieth century, many of the recolonizing individuals used the very same nesting ledges as had their predecessors decades earlier. Perhaps the most extreme example of nest-site tenacity comes in the form of carbon-14-dated guano found at Gyrfalcon nesting ledges in central-west and northwest Greenland that dates nest occupancy to approximately 2500 years before the present, or about the same time that the Greenland Ice Sheet retreated.

Australia offers something of an anomaly regarding cliff nesting. Although the continent is home to several dozen species of diurnal birds of prey, only four—Osprey, White-bellied Sea Eagle, Nankeen Kestrel, and Peregrine Falcon—routinely nest on cliffs, rather than in trees. The Australian biologist Jerry Olsen speculates that the continent's abundant eucalyptus trees may play a role. Where eucalypts such as the Blue Gum have been introduced elsewhere in the world, many raptors have chosen them as preferred nesting sites, and Olsen suggests that such "superior" nest sites may render cliff nesting unnecessary. Another possibility involves geology. Australian cliffs are older than those in the Northern Hemisphere, and Olsen speculates that they are too eroded and no longer suitable for raptor nest building. That said, there are lots of cliffs in Australia that are similarly structured to those used by raptors elsewhere, and they too are not used as nest sites. Olsen's third explanation posits that Australian raptors have been isolated from their ancestral stock for so long that they have lost the ability to construct nests on cliffs. Although none of these explanations strikes me as particularly likely, the fact remains that cliff nesting is less common in Australian raptors than it is in other geographic locations. Something appears to be going on in Australia, but what it is remains elusive.

Ground Nesting

Presumably, most raptors choose to nest in trees, cliffs, or elevated human-made structures to reduce mammalian predation. Nevertheless, a number of raptors do nest on the ground, particularly on predator-free islands or High Arctic tundra, where alternatives are not necessarily available. Arctic Peregrine Falcons, for example, nest on the ground in bogs, often on low mounds of land near shrubs or stunted trees. Ground-nesting falcons also have been reported in Britain, but the trait has not persisted, most likely because of a high rate of nest failure. Other widespread species, including Bald Eagles, Common Buzzards, Common Kestrels, and Merlins, also nest on the ground, again particularly at high latitudes, and on islands free of mammalian predators. The Egyptian Vulture, a relatively small scavenging bird of prey, is a ground nester on the Canary Islands, where ground predators are rare. Some species, including Roughlegs and Snowy Owls, nest on the ground in the High Arctic, where cliffs are lacking, particularly in areas with abundant prey. The widespread and near-cosmopolitan Osprey, which typically nests in trees and human-made structures to avoid ground predators, nests on the ground on predator-free islands in both North America and the Red Sea. In Northwestern Mexico, Osprey nest on the ground by the hundreds on predator-free desert islands surrounded by the fish-rich waters of the Sea of Cortez, creating what amounts to some of the largest concentrations of this species on the planet.

The true masters of ground-nesting raptors, however, are thirteen raptors of the genus *Circus*, the harriers of the world. Occurring on all six of the world's raptor-inhabited continents, harriers have taken to ground nesting like fish to water. Although several species occasionally nest in bushes and shrubs, most nest directly on the ground. Whereas facultative ground nesters like Ospreys and Bald Eagles do so largely when the threat from mammalian predators is reduced, this clearly is not so with harriers, which frequently nest in areas where mammalian predators pose a significant threat. The harrier specialist Rob Simmons has thought long and hard about this, and his ideas, and mine, are summarized below.

As a group, harriers are long-winged, long-tailed, and lightly wing-loaded aerial hunters that search for small-mammal and small-bird prey while quartering low over open habitats with few if any trees. This relatively treeless predatory lifestyle, together with the relatively small sizes of prey taken, may have shaped the harrier's tendency for ground nesting. The harriers' open-habitat "kitchens," it seems, are typically a bit too far from potential elevated "bedrooms." Even so, ground nesting can be risky. Of the seven species of harriers for which we have good information, 9 to 52% of all nesting attempts are interrupted by predators. As a result, antipredator behavior is a key feature of this group's breeding strategy. Harriers often nest on islands, many of which lack ground predators. Secondly, harriers often nest in what are called "loose colonies," where nests may be less than 100 meters (330 feet) from each other. Although such spacing tends to be great enough to reduce the likelihood of serial predation by ground predators, the relative proximity of nests does allow cooperative mobbing by adults from neighboring nests, which helps deter mammalian predators. And indeed, I have seen effective mobbing behavior many times while watching Northern Harriers in central Wisconsin. Researchers working with Montagu's Harriers in Spain have found that parental alarm calls not only bring in mates but also neighboring harriers, which together mob potential predators, thereby strengthening the defensive attack.

A third antipredatory tactic involves stealth and secrecy near the nest. As in most diurnal birds of prey, male harriers do almost all of the hunting during the breeding season, especially when eggs and nestlings are in the nest. Unlike many species of diurnal birds of prey, however, parental harriers pass prey to their mates via aerial talon-to-talon transfers or aerial prey drops, up to several hundred meters from the nest, with males initially calling to females to signal their approach. Presumably, reduced commotion at the nest during prey transfers reduces the likelihood of a nest being detected by potential predators. Finally, the nest sites are themselves quite cryptic. I found nests in central Wisconsin difficult to locate even after I had watched females return to them with prey after aerial transfers. Most of the nests were situated far from potentially useful landmarks such as small trees or shrubs, and incubating and brooding females almost always remained immobile and silent unless approached to within a meter or less, a behavior that has been reported by harrier biologists elsewhere.

With the exception of the behavior in harriers, ground nesting appears to be a relatively uncommon phenomenon in birds of prey. But that does not mean that it is of little consequence. All five of the world's most widely distributed raptors, the Osprey, Northern Harrier, Golden Eagle, Merlin, and Peregrine Falcon, nest on the ground at least occasionally, and the tendency to do so almost certainly helps increase their distribution and abundance.

Nests on Human-Made Structures

Although many people tend to think of raptors as wilderness denizens that only rarely inhabit human-dominated landscapes, and although raptors evolved long before human-built structures became available to them as nest sites, many species now nest on and in human-made structures. It seems as long as a landscape offers a safe nesting place and available prey, raptors will breed there, including in rural, suburban, and even urban habitats, where they often nest on, as well as in and under, human-built structures. Some of the best examples of this comes from two of the most cosmopolitan of all birds of prey, the Peregrine Falcon and the Osprey, with the former having been recorded as nesting on buildings in towns and cities since the Middle Ages. One of the more famous examples of Peregrine Falcons nesting on urban buildings in North America is the Sun Life Building in Montreal, where from 1936 until 1952, a pair nested on a ledge that had been "enhanced" with a sandbox that provided a nesting scrape. In the Old World, Salisbury Cathedral, in Wiltshire, has hosted nesting Peregrine Falcons on and off since the mid-1860s.

In the wake of reintroduction efforts for Peregrine Falcons following the DDT era, many populations have now been "urbanized" to a greater extent than ever before. By 1993, a third of all pairs in the midwestern United States were nesting in and around cities in human-made structures, including buildings, smokestacks, and grain elevators. By 2000, the number of such nests had increased to 71% of 129 known nest sites. In New York City, twenty-three pairs were nesting in westernmost Long Island in 2010. Similar situations currently exist in many other cities, both in Canada and the United States. Much the same situation exists in Europe. In Germany, for example, 118 pairs were nesting on buildings in 1997, and their numbers have increased since.

The near-cosmopolitan Osprey, too, tends to favor human-made structures as nesting sites. In the Chesapeake Bay region of the eastern United States, where an estimated 1450 pairs of Osprey nested in the mid-1970s, less than a third nested in trees, with the remainder nesting on duck hides (29%), navigational aids, including channel markers (22%), and assorted other human-made structures, including nest platforms purposely built for them (18%). Their dependence on human-made structures in the area is such that geographical locations of nesting concentrations have shifted with time as new structures have appeared. Elsewhere in the United States and Canada as well as in Europe, Ospreys have taken to nesting on structures associated with power lines, which in Germany host more than 75% of all nesting pairs.

Numerous other raptors also have taken to nesting on power-line poles and stanchions. In western North America, one study reported that at least ten species, including Bald Eagles, Harris's Hawks, Red-tailed Hawks, Swainson's Hawks, Ferruginous Hawks, Zone-tailed Hawks, Golden Eagles, American Kestrels, and Prairie Falcons, were known to nest on structures associated with power lines. It is almost as if raptors follow the adage "if you build it, we will nest on it." Not surprisingly, bird lovers have taken that notion to heart and have built structures specifically for nesting birds of prey. Nesting platforms for Ospreys are popular in many parts of North America, with homeowners placing them in wetlands within eyeshot of their homes for much the same reason people place bird feeders for songbirds in their backyards.

Conservationists, too, have used nesting platforms and nest boxes to bolster local and regional populations of raptors, both for conservation benefits and to create accessible breeding populations for scientific study. The latter is particularly true for hole-nesting species, including kestrels and small falcons. In North America, the American Kestrel has been occupying nest boxes for more than 100 years, with an early report from Iowa in the American Midwest dating from 1912. Hawk Mountain's local kestrel nest-box program that began in the early 1950s with six boxes on a single

Pennsylvania farm near the sanctuary has grown into a hundred-plus box, 1300-km^2 (500-mi^2), long-term study of numerous aspects of kestrel breeding, feeding, population, and roosting biology. Over the years, this study has resulted in a series of more than a dozen publications in the ornithological literature, along with four master's theses and a PhD dissertation on topics ranging from parasites and diseases to habitat selection, nesting phenology, nesting success, mate choice, and winter roosting behavior. One study in southern Idaho involving sixty nest boxes erected in a largely treeless shrub-steppe habitat with few natural cavities averaged over 99% occupancy during four years in the early 1990s.

By far, the most ambitious conservation-oriented nest-box program for birds of prey is one that involves erecting 5000 artificial nests for the globally Endangered Saker Falcon in Mongolia. This ongoing project aims to create a large breeding population of falcons from which individuals can be taken for use in falconry, mainly in the Middle East. Other species whose populations have been bolstered by nest boxes, either for conservation, scientific study, or both, include the Common Kestrel, Lesser Kestrel, the globally Near-Threatened Red-footed Falcon, and the globally Vulnerable Mauritius Kestrel.

RAISING YOUNG RAPTORS

Most raptors are monogamous for good reason. It typically takes two parents to raise a brood of young birds of prey successfully, and that happens only when everything goes right. As mentioned in Chapter 2, larger raptors tend to produce fewer young annually than do smaller raptors, with some large eagles fledging, on average, less than a single offspring annually. Food plays a major role in breeding success, with annual shifts in the numbers of young fledged paralleling food availability. The latter is particularly evident in species that feed on small-mammal prey, which in high latitudes often follow 4-year population cycles. In Arctic-breeding Roughlegs, for example, high percentages of all pairs breed, and most do so successfully, when vole populations reach plague proportions, whereas few individuals breed, and many of those are unsuccessful, in years when vole populations crash to low levels.

The same is true for Northern Harriers. In central Wisconsin, where a breeding population was studied for more than 20 years, numbers of nests fluctuated annually from a low of four and five nests to highs of twenty-five, twenty-seven, and thirty-four, in direct correlation with the number of meadow voles available. Polygyny was more common in high-vole years as well, and so was nesting success. In northern Britain, the numbers of Common Kestrels produced annually fluctuated in parallel to vole densities across a 35-year period stretching from the mid-1920s through the late 1960s. In Western Australia, average clutch sizes of Brown Goshawks, Little Eagles, and Wedge-tailed Eagles all increased after introduced European rabbits became common in the region. Habitat differences in prey populations also affect breeding success. In southern Scotland, for example, Eurasian Sparrowhawks laid earlier, produced larger clutches, and fledged more young in lowlands, where populations of avian prey were abundant, than in higher elevations, where avian prey was relatively scarce.

Weather, too, often plays an important role in breeding success, particularly in the Arctic and northern Temperate Zone, where cold spring and late-winter snowstorms can significantly affect hatchability and nestling survival. A 13-year study of Peregrine Falcons breeding at Rankin Inlet on the northwest coast of Hudson Bay in the Northwest Territories of Canada documented significant relationships between clutch size and snowfall during the pre-laying period, as well as increased chick mortality during years with increased precipitation associated with storms. In Zimbabwe, a 13-year study involving Verreaux's Eagles, where the percentages of breeding pairs fluctuated between 46% and 89% annually, breeding effort was higher in years following below-average rainfall. High winds, too, can pose a threat, especially for stick nesters, including Ospreys that build nests in exposed supercanopy snags, as do landslides following heavy rains for species such as Roughlegs, which often nest on talus slopes and screes.

Predation, both natural and human-related, also can play an important role in nesting success, with stick-nesting birds of prey at greater risk of nestling predation than those nesting in cavities. Owls, in particular, rank high among natural predators, even for relatively large raptors. In Florida, Great Horned Owls were responsible for more than thirty nest failures at 619 Bald Eagle nesting attempts, and the same species appears to affect the nesting success of both Red-tailed Hawks and Swainson's Hawks whenever hawks nest close to owls. In southern Scotland, Tawny Owls were a main predator on nestling Eurasian Sparrowhawks, and in the Arctic, Snowy Owls affect both nest-site selection and nesting success in Roughlegs. Northern Goshawks, Cooper's Hawks, and other similarly large-size accipiters also can pose a serious threat to nesting birds of prey. In southern Scotland and northern England, for example, Northern Goshawks are important predators on breeding Eurasian Sparrowhawks, and in Pennsylvania, Cooper's Hawks routinely take American Kestrels. Such bitrophic interactions benefit the larger species in two ways: first, by providing nutrition and, second, by reducing competition with a predator whose diet often overlaps that of the larger species.

Humans also can reduce raptor breeding success and affect nest-site selection. Nests of Bald Eagles, for example, were routinely destroyed in eastern North America during the first half of twentieth century. The American ornithologist Witmer Stone, writing in 1937 in *The Birds of Old Cape May,* had this to say on the subject regarding eagles of southern New Jersey:

> Most of the Cape May County [New Jersey] eagle nests have been located in more or less inaccessible spots in the heart of deep swamps or dense stretches of woodland, due perhaps to the antipathy of the farmers who regard the eagles as enemies of the poultry and cut down the eagle trees when they come upon them. The persistence of egg collectors, too, has much to do with the difficulties that the birds have experienced in raising their broods. . . . In March, 1935, three of the occupied nests in Cape May County were robbed by "oölogists."

In the past, egg collectors, too, have threatened raptors, particularly rare species and those that produce especially attractive eggs. A rather famous case involves Ospreys in Scotland. In 1954, a nesting Osprey was spotted at the southern end of Loch Garten, but the nest was kept secret to reduce the risk from egg collectors. Unfortunately, the secret got out a year later and the eggs were stolen. The same thing happened to a pair in Rothiemurchus in 1956. Determined protection efforts began in 1957. Although a collector was spotted and chased off near the Loch Garten nest in 1958, the eggs were found smashed at the base of the nest later in the season. Protection efforts continued and several young hatched in 1959. Unfortunately, egg collectors remain active in the United Kingdom as of this writing, and eggs of raptors remain at risk there even now.

Although raptors have become increasingly tolerant of human disturbance, many individuals remain prone to nest abandonment in the face of human approach. As recently as the mid-1960s, a study of factors responsible for nest abandonment in largely cliff-nesting Golden Eagles in the American West estimated that at least 8% of all nest losses resulted from human disturbance. Disturbance at nests when incubation is under way is especially problematic, as even a brief absence from the nest can chill eggs and significantly affect the behavior of incubating adults, leading to subsequent abandonment. The presence of nestlings (as opposed to eggs) typically emboldens parents, increasing their willingness to stand fast and protect their young. In North America, many raptors exhibit both active and aggressive nest defense particularly against potential avian predators, with many species exhibiting less aggressive defense against potential human predators, probably with good reason.

A recent report on the subject by the raptor biologists Joan Morrison, Madeline Terry, and Pat Kennedy suggests that small species with fast and more maneuverable flight are more likely to engage in nest defense than are less maneuverable species, although the latter may be able to deter potential predators before an attack happens. Ferruginous Hawks, which often nest in vulnerable sites on or near the ground, are more likely to defend their nests aggressively than are non–ground nesting species of the same size and agility. Low levels of nest defense in areas where people pose a real threat may simply result from the fact that active nest defense is likely to be both futile and dangerous. One the other hand, in areas where the threat from humans is minimal, raptors are more likely to actively defend their nests. Indeed, suburban nesters in eastern North America, species including Mississippi Kites, Broad-winged Hawks, and Red-shouldered Hawks, often swoop on and sometimes strike people that approach too closely. And in Florida, the previously much maligned Bald Eagle now nests and raises young in suburban backyards.

Even so, raptors seem particularly sensitive to potential predators, especially humans, which they are ill equipped to deal with. Because of this, it is best to observe them at a distance during the breeding season, as nest abandonment remains common in many species. In North America, many raptors flush from their nests when humans approach closer than 50 to 100 m (160 to 330 ft). Appropriate nondisturbance buffer zones differ among species and even within populations: recently recommended buffer zones for Osprey nests ranged from 400 to 1500 m (1300 to 4900 ft), whereas those for Bald Eagles ranged from 250 to 800 m (820 to 2600 ft), and those for Golden Eagles ranged from 200 to 1000 m (660 to 3300 ft). Finally, it should be kept in mind that human disturbance can be particularly damaging when it occurs near the time when nestlings ordinarily fledge from their nests, as premature fledging induced by human approach increases substantially the likelihood of predation at this especially vulnerable time.

Indeed, as dangerous as being a nestling is, the post-fledging period and the first few weeks after a young raptor leaves the nest may be the most dangerous period in a raptor's life. Although this remains a relatively little known aspect of raptor breeding biology, several studies offer insights as to what occurs during this period. A careful investigation of several dozen Red Kites observed in the Doñana National Park, in southwestern Spain, in the 1980s found that young kites fledged at about 55 days post-hatching and remained near the nest and depended on their parents for food for an additional 25 days. During the latter period, fledglings spent little time in the air, and only gradually increased their flight times over the course of more than three weeks, beginning initially with short flapping flights, followed by gliding flights, and then moving on to soaring flight. Hunting and play behavior were "very rare" during the post-fledging period, a time during which six of thirty-seven (14%) of the young died, with one electrocuted on a power line 20 m (65 ft) from the nest, one killed most likely by a Spanish Imperial Eagle, another killed by a mammalian carnivore, two others by starving, and one disappearing without a trace. A study involving eighty-nine radio-marked Northern Goshawks fledged at forty-eight different nests in northern Arizona, in the late 1990s and early 2000s yielded similar results. Fledgling goshawks remained near their nests for 33 to 66 days post-fledging, during which eight (9%) of the fledglings died. Three were eaten by Great Horned Owls, three starved, and two were lost to unknown causes. All three starvation deaths occurred in a single year, when prey were particularly scarce. A detailed study of sixty-one radio-tagged American Kestrels fledging from forty-seven nest boxes along an Interstate Highway in central Iowa in the late 1980s amplifies the situation for fledglings. During their first week out of the box, the young kestrels spent less than 1% of the time in flight, with the overwhelming amount time being spent perched on trees, fences and fence posts, or on the ground in short vegetation. They began hunting in earnest during the second week after fledging, at a time when they began feeding almost exclusively on grasshoppers and other insects. All but one of sixteen fledglings that died did so the first week after fledging. Six of the birds fell victim to mammalian predators, including red foxes, three were killed by unidentified avian predators, two were hit by vehicles, one starved, one died of dehydration, one most likely died after hitting a tree, and two died of unknown causes. Most of the deaths occurred during a year in which heavy spring rains delayed crop planting, which may have made the fledglings less able to conceal themselves from predators when ground perching.

My own studies of the post-fledging behavior of Northern Harriers in Central Wisconsin in the 1970s confirm the relatively fragile nature and naïveté of recently fledged birds of prey. I followed the behavior of fourteen unmarked fledglings at five nests in 1974, two individually marked fledglings at one nest in 1977, and twenty-nine individually marked birds at seven nests in 1979. In the latter two years of fieldwork, 64% of marked males and 88% of marked females survived the post-fledging period. Once they had fledged, both male and female fledglings remained within 400 m (1300 ft) of their nest for two to three weeks, when they were fed by both their male and female parents before leaving the area. In all instances, their first flights consisted of brief (less than a few seconds), 1 to 3 m (3 to 10 ft) "helicopter-like" vertical springs into the air above the nest as a parent, usually the female, returned with food. Within several days of their initial flights, the fledglings were flying up to 30 m (100 ft) toward their returning parent, which transferred the prey (usually small birds, mice, or voles) that they were carrying talon-to-talon in the air to the first fledgling to reach them. Many of these flights appeared to be exercising events as well, with fledglings flying circuitously back to their nest. Fledglings spent little time hunting and, although a few pounced on and "played" with prey-sized inanimate objects, I never saw a fledgling catch live prey other than insects. Male fledglings progressed more rapidly than females in all aspects of behavioral development, taking more flights per hour, making longer flights, and overall spending more time in the air than did females.

At the time the young birds were last seen in the area—usually at 10 to 15 days after fledging—males were spending more than 10 minutes per hour in the air, whereas as females were spending fewer than two minutes. Overall, females remained in the immediate vicinity of the nest for as many as 6 days longer than males. By the time fledglings left the vicinity of their nests (most departing flights occurred late in the day), all flew out of sight individually while engaged in the longest flights of their life. Prior to their departures, none of the birds had engaged in flights of more than two and a half minutes in duration, and the longest flights seen in most fledglings were less than two minutes. For me, the overall take-home message was that young Northern Harriers left the immediate vicinity of their nests (an area within a radius of about 400 m [1300 ft]), as well as parental care, without ever having caught prey larger than an insect, and without ever having flown for more than two and a half minutes straight, this despite the fact that this species is a small-bird and small-mammal specialist and aerial hunter that, as an adult or free-ranging juvenile, spends about a third of each day in flight, searching for prey.

Although I did not follow the birds on their outbound journeys using radio telemetry, then-graduate student Alan Beske, who also was studying harriers at the site, radio-tracked four other young birds after they had left their nests on the marsh. As was true of the birds I watched, all four of Beske's birds left the marsh on their own. The three that were tracked beyond the study site established temporary home ranges 71 to 171 km (44 to 110 mi) south of their nests, and maintained and hunted in them for 2 to 3 weeks, before continuing their southbound migrations.

Viewed in their entirety, the studies above suggest why the post-fledging period is the most dangerous time in a free-flighted raptor's life. At this stage of life, young raptors are truly inexperienced novices that are far from being well-trained by their parents in the whys and wherefores of being a bird of prey. As such, they are both easy prey for larger raptors and more likely to be incapable of finding sufficient food in difficult times. With little in the way of parental training, fledglings are, at best, "hard-wired" for adulthood and quite vulnerable. Indeed, it is small wonder that many fledglings starve or are eaten even before facing their first winter.

SYNTHESIS AND CONCLUSIONS

1. Reproduction is an essential aspect of raptor biology. Without it, there would be no next generation.
2. Most diurnal birds of prey are seasonally monogamous. Some larger species mate for life.
3. Polygamy occurs in a few species of raptors. Polyandry appears to be slightly more common than polygyny.

4. Because they often are vocal and sometimes perform elaborate aerial displays, including sky dances, breeding birds of prey can be conspicuous when courting.
5. Breeding raptors raise their young in a variety of nest settings. Some species build no nest at all, others nest in structures previously built or excavated by other species of birds, including other species of raptors. Still others build enormous stick nests or nest on cliffs. A few species nest in caves.
6. Some raptors usurp the active nests of other species of birds, including other raptors.
7. Many raptors nest in human-built structures, including nest boxes and nest platforms.
8. Many raptors place fresh greenery in their nests, to help sanitize them and protect nestlings from disease-causing bacteria and ectoparasites.
9. Several species of cliff-nesting raptors, including both Peregrine Falcons and Gyrfalcons, have been known to nest at the same site for hundreds of years or more.
10. Some raptors nest on the ground, particularly in treeless areas and on predator-free islands.
11. It typically takes both parents working together as a team to raise their young.
12. Bad weather, predation, and limited food availability can contribute to nesting failure.
13. In the past, some people purposely destroyed the nests and nest sites of diurnal birds of prey, including eagles.
14. Recently fledged raptors are relatively naïve and as such are easy prey to both avian and mammalian predators.
15. Many fledged young starve after their parents cease to provide food.
16. Mortality is particularly high during the first several weeks after young raptors leave the nest.

6 FEEDING BEHAVIOR

Considering that the way in which birds of prey kill is perhaps the most spectacular and interesting aspect of their lives, it is surprising how little we really know about how this is done.
Leslie Brown, 1976

MY GUESS IS THAT, like me, most raptor biologists were first attracted to birds of prey because they were fascinated with the birds' predatory behavior. Unfortunately, studying raptor predation itself is rarely straightforward, and is hardly ever easy. Because of this, many raptor biologists have come to focus on other aspects of the birds' biology, including breeding ecology, or on other, less-direct approaches to feeding behavior, including nest watches and diet studies. Unlike raptor breeding biology, which is closely tied to an active nest and therefore has a focal site for observation, raptor feeding behavior is, quite literally, all over the map. This is particularly true outside of the breeding season, when the home ranges of individual birds of prey can extend across hundreds, and sometimes thousands, of square kilometers. The predatory behavior of secretive woodland-dwelling species, for example, can be exceptionally difficult to study when there is no nest to watch while waiting for the hunting adult to return with prey, and the alternative of following birds traveling through the woods at 20 to 40 miles an hour simply is not possible. Open-habitat raptors make for easier targets, but they, too, can be challenging. As a result, this chapter's lead-in quote is almost as true today as when Leslie Brown penned it in 1976.

Although my own raptor career has wandered a bit—most notably into migration and movement ecology—I have managed to return to studying raptor feeding behavior many times; most recently, with Striated Caracaras in the Falkland Islands. This chapter focuses on many of those studies, together with those of other researchers who also have not been able to get rid of the "bug" of raptor feeding behavior, even if our studies have sometimes been indirect.

WATCHES AT NESTS

Because of difficulties associated with studying predatory behavior directly, raptor biologists often study it indirectly, including by making observations at active nests during the breeding season. Until the recent invention and proliferation of low-cost "camera traps," nest watches typically consisted of sitting in a hide with the clear view of a nest for long hours, day after day, waiting to see the kinds of prey parental raptors brought back for their nestlings and mates, and routinely scouring the areas around nest sites on regularly timed visits to pick up and identify prey remains. The advent of inexpensive camera traps has increased the efficacy of this task considerably, as it allows researchers to place arrays of these devices at nests and simultaneously collect data at a number of breeding attempts.

Unfortunately, although nest watches, camera-recorded or otherwise, give us a good idea of what parental raptors bring back to the nest and feed their young, they tell us little about how the birds actually catch their food, or whether the adults are feeding on the same prey. And, in fact, careful fieldwork demonstrates that the prey of adult and nestling raptors often differ considerably, with larger prey items being brought to the young and smaller items being consumed by the adults. This occurs at least in part because it is more efficient for parental birds to catch and ferry a few large items to the nest than it is to carry many smaller items, whereas it makes sense for parental birds to catch small as well as large prey for their own consumption at or near the point of capture. Thus, for example, American Kestrels breeding in nest boxes near Hawk Mountain Sanctuary in Pennsylvania frequently take insects while hunting for themselves, but are more likely to take mice, voles, and small passerines when securing prey for their nestlings.

Notwithstanding these limitations, observations of food brought to a nest can provide important clues about how parental birds seek out and capture prey. When studying breeding Northern Harriers in central Wisconsin in the 1970s, I frequently witnessed short runs of identical prey, including three or four similarly aged nestling Red-winged Blackbirds or larger numbers of hatchling Greater Prairie Chickens, being brought to a nest by both male and female parents, with a parent returning with prey at brief intervals and from the same compass direction. These observations led me to conclude that the parent had found a bird's nest with nestlings or a recently hatched brood of precocial young and was in the process of serially predating these food bonanzas.

Watches at nests also help identify differences in hunting behavior among individual birds that result in particular types of prey being more available to them than would otherwise be the case. The British ecologist Ian Newton found that, among the many radio-tagged females he followed during long-term studies of breeding Eurasian Sparrowhawks, one individual that hunted by stooping on flying prey took disproportionately more Eurasian Starlings that did other females that were breeding in the same forest.

Yet another piece of information that one can glean from patiently watching birds bringing back food to their nests is the identity of preferred prey. Frances Hamerstrom, who, as I mentioned earlier, studied fluctuations in the numbers of breeding harriers in central Wisconsin for three decades, also documented population fluctuations of the meadow voles at her site. In years when voles were plentiful, they and other small mammals made up more than 65% of all prey brought to nests by parental harriers, whereas in years when voles were at the low point in their four-year population cycles, small mammals made up only about 40% of the prey brought to nestlings, with small-bird prey becoming more important at such times. Clearly, the harriers preferred voles when they could get them, but fell back on avian prey when voles were scarce. That fewer harriers nested at the Buena Vista Marsh study site in years when vole populations were low also suggests voles were the preferred harrier prey.

The examples above demonstrate that it is possible to learn much about raptor feeding ecology and behavior while watching for parental birds to return to their nests with prey. Although this can be decidedly more difficult when the raptors in question are vultures or other species that swallow food for their young and regurgitate it to them when back at the nest, one can learn much about the timing of food deliveries and, in instances involving marked or otherwise individually recognizable parents, who, exactly, is feeding the young.

An excellent case in point regarding the timing of raptor predation involves a series of careful observations of Common Kestrels feeding principally on the common vole in recently reclaimed lands in the Netherlands in the 1970s. The work, which included dawn-to-dusk watches of Common Kestrels totaling more than 2900 hours, documented the daily patterns of individually color-marked and, sometimes, radio-tagged birds, whose foraging activities were studied for consecutive days while the birds were feeding on common voles in natural grasslands. Unlike those of many other rodents, which tend to be largely or completely nocturnal, the feeding activities of voles are spread out across both day and night. As a result, voles are vulnerable to both diurnal and nocturnal predators.

The difference in timing appears to be related to the voles' digestive physiology. Common voles feed mainly on nutritionally poor grasses, whose digestion requires near-continual gastrointestinal fermentation. This process involves the presence of symbiotic cellulolytic intestinal bacteria that live in a well-developed lower-digestive-tract caecum, or pouch, at the beginning of the large intestine, in which fermentation occurs. Digestion, which is similar to that in cows and other ruminant ungulates that also feed mainly on grasses, requires continual fermentation to work efficiently. Thus, voles must actively feed during the day, when kestrels and other raptors pose a threat of predation, as well as at night. And indeed, coincidental large-scale trapping of voles at the site demonstrated precisely synchronized patterns of feeding activities that peaked at approximately 2-hour intervals during each 24-hour cycle, a rhythm small-mammal biologists claim reflects the species' reliance on fermentation-based digestive physiology. Intriguingly, voles reduced the risk of diurnal aerial predation at the Dutch study site by timing their cyclic daytime feeds to begin at the end of a 2-hour inter-meal interval timed to begin at daybreak. This timing minimized the total number of feeding bouts occurring each day and, presumably, the risk of daytime predation by kestrels. Large-scale vole trapping at the site indicated that a postdawn 2-hour break in feeding did indeed occur and that most voles synchronized the timing of their feeding bouts accordingly, just as predicted.

Simultaneous dawn-to-dusk observations of six color-marked kestrels feeding at the site indicated that kestrels timed their flight-hunting activities so that they overlapped with periods of vole activity, with the birds catching most of the prey taken from flights during such periods. Moreover, the kestrels, which usually immediately consumed the first vole caught each day, "cached," or temporally stored, additional voles caught later in the day in hiding places, from which they subsequently retrieved and ate the voles episodically over the course of the day until shortly before nightfall, seemingly ensuring a pre-evening feed to carry them through the night.

As a result, individual kestrels exhibited remarkable consistency across days in the timing of their hunting behavior. In January 1978, for example, a female Common Kestrel at the site was recorded as being active an average of just over 9 hours daily, of which she spent approximately 34 minutes in hunting flight and the remainder of the time perch hunting, while catching an average of three voles daily. The bird visited the same hunting area each morning at approximately 10:00 A.M., during the first period of vole activity, and then returned to the same area in late afternoon to retrieve voles that she had cached there earlier in the day. When the researchers experimentally provided her with food in the morning in a different location that she rarely visited, the female immediately switched to hunting in that area each morning, but only in the morning, while shifting to other feeding sites in the afternoon. The researchers concluded that this bird's daily strategy of adjusting her hunts to coincide with vole activity peaks and the timing of experimentally provided prey reduced the time needed to flight-hunt daily by 10–22%, a seemingly small, but apparently significant, part of her daily energy budget. The researchers also concluded that the voles themselves also benefited in synchronizing their activity periods, in that being active simultaneously created a phenomenon of "safety in numbers," compared with voles that were out-of-sync and, therefore, unable to take advantage of what biologists refer to as a selfish-herd effect. Although similarly careful studies have yet to be undertaken in other species, almost certainly, other raptors that routinely feed on voles also time their hunts to take advantage of synchronous peaks in prey activity, and, for that matter, peaks in the activity of other rhythmically synchronous prey.

SEARCHING FOR AND CAPTURING PREY

In a PhD thesis on the New Zealand Falcon in 1977, the raptor researcher Nick Fox laid out seven distinct search techniques that birds of prey employ to look for, locate, and approach their prey. Although several of the tactics overlap, Fox's categories remain the most inclusionary I am aware of. They are (1) still-hunting, (2) fast contour-hugging flight, (3) soaring and hovering flight, (4) slow-quartering flight, (5) stalking, (6) listening, and (7) flushing from cover. Together with high-speed pursuit and stooping, they form the basis of this section of the chapter.

Still-Hunting

Used by many avian predators, including raptors, still-hunting consists of perching, typically at an elevated vantage point, while searching visually and often listening for prey. Having spotted a potential victim, still-hunting raptors drop from their perch and glide or flap unobtrusively toward their targets in an attempted surprise attack. Many species, including kestrels, frequently pump their tails and bob their head prior to takeoff, presumably to improve their depth perception. Although recently fed and reasonably well satiated raptors also still-hunt, hungry individuals do so more attentively and purposefully, assuming a stature that falconers refer to as "sharp-set." If human bird handlers can assess the status of their birds, it seems likely that potential prey can do so as well. A clever study by Frances Hamerstrom tested this possibility in Wisconsin in the summer of 1956, using a captive Red-tailed Hawk, a species that routinely still-hunts. Working in her backyard, Hamerstrom conducted a number of exposure tests on eight different days to see how neighborhood songbirds responded to her perched and tethered hawk. In four of the tests, the Red-tailed Hawk had been fed recently, and in four other tests, it had not been fed recently and was visibly sharp-set. The results were striking. When well fed, the hawk was "mobbed" (swooped on at close range or physically struck) by twelve songbirds, including Black-capped Chickadees, Gray Catbirds, and American Robins; when sharp-set, it was mobbed by more than 100 songbirds. Hamerstrom also noted that the hawk was mobbed more frequently when being flown sharp-set than when being flown well fed. Hamerstrom noted that when the bird was sharp-set, the feathers at the top of its head appeared to be more flattened, while those on the hind neck were held erect, and the eyes appeared less rounded, and that when well fed, the bird often perched with one foot tucked up under its belly feathers, something it never did when sharp-set. Clearly, potential prey on the lookout for still-hunting hawks are far less tolerant of them when the hunter is sharp-set and hungry than when it is well fed and satiated.

Fast Contour-Hugging Flight

Fast contour-hugging flight is the "ying" to still-hunting's "yang." This a highly active and far more energetically expensive form of hunting that occurs when a raptor flies at high speed one to several meters above the ground or vegetative canopy in an attempt to startle and secure potential prey before they know what hit them. Falcons that live in relatively flat, open, and treeless areas, including Gyrfalcons, Merlins, and Saker and Aplomado falcons, as well as many open-habitat accipiters, use this hunting tactic. The element of surprise and a capacity for highly agile flight are essential aspects of the technique, as is knowledge of local topography.

In the farmlands surrounding Hawk Mountain, Sharp-shinned Hawks and Cooper's Hawks use this fast contour-hugging flight when hunting songbirds at backyard bird feeders. Individual hawks establish trap lines in the neighborhoods they hunt, with individuals traveling low and fast among the houses involved and thereafter tightly circling individual homes while attempting to surprise and snare their prey. Unfortunately, many people place their bird feeders in the open and away from protective cover, making birds that feed at them more visible to their human fans, but also more vulnerable to avian predators.

Soaring Flight

In much the same way farmers now use drones to find livestock, raptors use high-altitude soaring flight to find ground-dwelling prey. The technique is simple enough. Weather conditions and topography permitting, the raptor launches into the air, gains altitude via low-cost soaring or kiting behavior, and then searches open areas for prey. Observations of the head movements of soar-hunting birds suggest that individuals of many species can scan the surrounding landscape while in flight as efficiently as if they were perch hunting, all the while manipulating and contorting their wings to pursue aerial hunting opportunities, as if they were held in place by invisible sky hooks.

Hovering Flight

Hovering flight, or "hover hunting," occurs when raptors maintain zero ground speed while flapping or kiting into the wind. Although doing so is energetically expensive when flapping is involved, creating an "aerial perch" in flat, treeless landscapes, where physical perches such as trees, fence posts, and power lines and stanchions are limited, expands the bird's hunting opportunities. Many species of kestrels and buteos hover-hunt when searching for both invertebrate and vertebrate prey. Hover hunting offers raptors with vision in the ultraviolet range (see Chapter 3) the chance to hunt directly above areas of high rodent densities, presumably increasing hunting success. In American Kestrels, females hover-hunt more than males, most likely because females hunt more often in open, treeless habitats with limited elevated perches than do males, or because females take small-mammal prey more frequently than do males, which often favor avian prey. Both American Kestrels and Common Kestrels, two species in which hover hunting has been well studied, typically hover-hunt at heights of between 10 and 25 m (30 to 80 ft), with a mean height of about 12 m (40 ft). Both are more likely to hover-hunt in moderate winds of from 10 to 30 km (6 to 18 mi) per hour, wind speeds that are close to the species' velocity of minimum power, and, as such, velocities at which flapping flight is least costly per unit time in the air. Both species also hover at lower heights as wind speed increases overall, most likely because doing so allows them to continue to fly at the velocity of minimum power. Although individual kestrels can hover-hunt continuously for a minute or more, most hovering bouts last from 10 to 30 seconds. When searching for small mammals, kestrels typically interrupt their descents several times while closing in on individual prey they have spotted, presumably to get a better look, or because small mammals stop moving, or "behaviorally freeze," when they sense they have been spotted by potential predators. Laboratory tests demonstrate that captive small mammals are more likely to move about when a nonmoving hawk silhouette is suspended above them than when it is moving, suggesting that within-descent stalls serve to induce movement in prey the raptors have spotted, making them easy to see and, therefore, easier to secure.

Species of raptors hover-hunt differently, most likely reflecting differences in their flight mechanics. In south-central Ohio, where I studied Roughlegs and American Kestrels, both hover-hunters, Roughlegs hovered at significantly greater heights (18 m compared with 12 m [60 ft compared with 40 ft]) and pounced less frequently (0.4 times per minute compared with 0.8 times per minute) than did the kestrels, although both caught small-mammal prey at similar rates. Many researchers have discussed the relative benefits of hover-hunting compared with still-hunting, with most agreeing that the former allows birds to hunt in areas where elevated perches are limited and that, while decidedly more expensive energetically than still-hunting, hover-hunting allows birds to catch prey more quickly overall. Presumably, the trade-off with high-cost hunting (hover hunting compared with perched still-hunting) is offset by the resulting higher rate at which prey are caught.

In Florida, where the American behavioral ecologist Tom Grubb studied the behavior of hover-hunting Ospreys, individuals hovered for only 3–4 seconds before flying to a different spot to hover-hunt again, less than half the length of time that overwintering kestrels and Roughlegs hovered in Ohio. Why this was so is unclear. It may be because Ospreys are more heavily wing-loaded than are Roughlegs and kestrels, and that the extra weight supported by their wings limits their ability to hover for longer periods, or because the Ospreys were searching for fish rather than small mammals, or because the Ospreys were hunting over water rather than over land. Whatever the reason, aspects of this hunting technique vary considerably among species.

Grubb also found that Ospreys were more successful per dive when diving from flights while hovering than in between hovers, and that cloudy weather and rough water surfaces reduced hunting success per unit of time, differences that he attributed to an increased difficulty in seeing fish. That Ospreys switched to catching more mullet and less crappie at such times, and that mullet were conspicuously silver-sided, moved more, and swam in schools, whereas crappie were darker overall, moved less, and were largely solitary during the course of the study, supports this idea. Finally, Osprey and other fish-eating birds have to spot and capture prey across two optical media, air and water. Because of this, they need to deal with the physical phenomenon of refraction (light bending) at the air-water interface, which may involve learning, and would explain why hunting success improves with age.

Slow Quartering

Sometimes referred to as "coursing," slow quartering is frequently employed by harriers, a group of long-winged, long-tailed, lightly wing-loaded raptors that are closely related to accipiters. Slow-flying harriers fly low, often at fewer than 5 m (15 ft) above vegetation, in flapping and gliding flight while turning frequently as they crisscross meadows, agricultural fields, and other grass- and scrubland habitat, searching, by both sight and sound, for small-mammal, avian, and invertebrate prey. All but one of the 601 pounces I observed in Northern Harriers overwintering in south-central Ohio farmlands occurred during slow-quartering flight, with adult males catching prey on 61% of their attempts, and females doing so 48% of the time. The only exception was a successful pounce on a small mammal by a perch-hunting male. A similar comparative study of quarter-hunting Northern Harriers in a South Carolina salt marsh and a Florida freshwater marsh by the raptor biologist Mike Collopy and me revealed significant differences in harrier behavior at the two sites. Although harriers in South Carolina captured prey in 15% of their pounces, those in Florida did so only 6% of the time, with the difference most likely linked to differences in prey taken at the two sites. In South Carolina, harriers took only birds, including songbirds and rails, whereas in Florida they took mainly cotton rats. In South Carolina, birds appeared relatively vulnerable to harrier predation, as most were flushed from "isolated, floating wracks of detrital vegetation" scattered across a tidal salt marsh and did not appear to be highly maneuverable in flight away from the harriers, with most of the flights tending to be rapid, straight-line attempts to reach protective cover. By comparison, in Florida, cotton rats were largely concealed in runways beneath tall stands of dense cover, and appeared to be difficult for the harriers to follow and capture. Intriguingly, although harriers captured prey on a per-pounce basis at higher rates in South Carolina than in Florida, harriers pounced significantly more frequently in Florida than in South Carolina and, consequently, captures per hour were about the same in both locations (0.27 captures per hour in Florida compared with 0.29 in South Carolina). The question of whether or not harriers at the two sites received the same amount of nutrition per unit of metabolic effort remains unanswered. Although this study focused on only a single species of bird of prey, the results suggest that both habitat differences and prey vulnerability can influence the hunting behavior and pouncing success of raptors.

High-Speed Pursuit and Stooping

The speed at which raptors approach prey that they have spotted depends not only on the flight abilities of the birds themselves, but also the maneuverability and escape tactics of the potential prey, as well as the ecological setting of the predation event. Cold-blooded, or poikilothermic, prey such as insects, amphibians, and reptiles are much easier to capture when temperatures are relatively low and their metabolic processes are not working at "full speed." Thus, snake eagles are more likely to capture and subdue prey on cool summer mornings than later in the day as temperatures rise, and the same is true for insectivorous birds of prey. It also is true for South American sloths and armadillos, whose characteristically lower metabolic rates make them particularly vulnerable to being preyed on by Harpy Eagles and Solitary Crowned Eagles. In addition to increased vulnerability to capture, prey with low metabolic rates also are more easily managed and manipulated once they are caught.

That said, the need to fly at high speeds is critical for successful predation in many raptors, and natural selection has produced some fast-flying birds of prey. Avivorous, or bird-eating, accipiters and falcons often are credited with high-speed maneuverable flight while pursing and subduing agile avian prey. The avian ecologist Thomas Alerstam used radar to track a Northern Goshawk and several Peregrine Falcons pursuing birds in his native Sweden. Alerstam found that peregrines began pursuits of songbirds from 0.8 to 1.3 km (0.5–0.8 mi) away in "vigorous flapping flight" while accelerating horizontally to 65–82 km (42–51 mi) per hour in level flight, and up to 138 km (86 mi) per hour in dives. The goshawk started its pursuit of a Western Honey Buzzard from 5 km (3 mi) off without accelerating flapping flight and dove at maximum speeds of approximately 110 km (66 mi) per hour, all of which are considerably lower than the terminal velocities (maximum speed in a vertical fall) of the two species involved. In a separate series of filmed observations, Peregrine Falcons and a goshawk were recorded striking prey at 65–70 km (40–44 mi) an hour, which, remarkably enough, is not much faster than the two species have been recorded migrating at Hawk Mountain Sanctuary, 50 km (30 mi) for peregrines and 60 km (38 mi) per hour for a single goshawk. Similarly, a southbound Peregrine Falcon migrating in tailwinds over the Gulf Stream between the coast of southern New Jersey and southern Florida traveled 1535 km (954 mi) in 25 hours at an average speed of 60 km (38 mi) per hour.

All of this is not to say that Peregrine Falcons don't fly fast when in a stoop. Professional photographers have recorded them stooping at speeds of up to 320 km (200 mi) per hour while diving in the company of humans jumping out of airplanes, and the claim that the species is the fastest flying bird seems secure. Adaptations for high-speed flight in Peregrine Falcons are many, including external nasal deflectors that enhance air intake during high-speed stoops after prey, strengthened flight feathers, and erectable back feathers that act as spoilers that reduce airflow separation and increase terminal velocity in high-speed stooping flight. Nevertheless, most prey caught by the species are ill prepared for this raptor's remarkable prowess in flight, and individuals rarely need to employ their maximum flight abilities while securing their prey. Overall, the Peregrine Falcon's ability to learn quickly how potential prey are likely to behave is as important as its speed in ensuring hunting success, and the same appears to be true of other birds of prey. Indeed, the fact that most raptors depend on their ability to catch several meals each day produces selection pressures that hone their skills in many ways to ensure success and survival.

GRASPING PREY

Ask anyone who routinely handles live birds of prey where the "business end" of a raptor is, and without hesitation they will point to the talons, or claws, not the beak. Falcons do have bilateral notches on their upper mandibles that create a "tomial tooth", which allows them to sever the spinal cord of their prey while swiftly administering a killing bite, but raptors use their talons to seize, contain, and, in most instances, kill their prey, with the beak serving largely to tear their food into bite-sized pieces for ingestion. Consequently, being pecked by a raptor's beak is one thing, but feeling the incredibly firm grip of a large diurnal bird of prey or owl through a thick leather glove is something else entirely. An important exception involves scavenging raptors, most of which have underdeveloped talons, and whose beaks can be quite dangerous. In such carrion eaters, the beak is used to rapidly and competitively remove meat from the carcasses of large animals during what might be called communal feeding frenzies. Thus, when biologists handle scavenging raptors, it is the beak that must be controlled carefully, as some of their most well-developed muscles are in the back of the neck, and their attack behavior while in the hand typically involves biting, tearing, and pulling of flesh rather than talon grasping.

Such cautions aside, the question then becomes: How do the talons of raptors work? Overwhelming strike capacity seems to apply in birds of prey, and enormous difference in relative size is important. The talons of a Harpy Eagle are 7.5 cm (3 in) long, or as large as those of a large-cat predator. In fact, raptor artwork often underrepresents the relative size of a raptor's talons, because depicting them life-size can make them appear out of proportion to the bird itself. But size alone does not a killing trophic appendage make, and a recent examination by two Australian biologists, Luke Einoder and Alastair Richardson, indicates that a "tendon locking mechanism," or TLM, in the toes of raptors has much to do with the efficiency with which birds of prey dispatch their quarry.

The two studied the toes and talons of nine species of diurnal birds of prey and three species of owls, as well as those of a cockatoo, a corella, and a raven. Their work reveals a distinct raptorial adaptation, including several unique features that separate the TLMs of diurnal birds of prey from those of owls.

Ornithologists have known for some time that many birds possess an automatic locking mechanism in their toes that allows them to maintain a grip force while perching, wading, hanging, and tree climbing. The mechanism enables elements of the tendons in the toes of birds to intermesh and physically lock the toes in place for long periods of time at low cost. Raptors, too, have a TLM, but theirs differ from those of other birds in two important ways. First, the TLM of raptors produces its "vise grip" within the toes themselves, rather from a point farther up in the leg, as occurs in most other birds. The difference means that raptors can grip their prey with their toes while simultaneously moving their legs, something the general locking mechanism of other birds does not allow. The TLM itself functions as a ratchet-like mechanism in which thousands of small, rigid, well-defined projections called tubercles on the ventral surfaces of each flexor tendon lock on transversely running folds on surrounding tendon sheaths, allowing the bird to clutch its prey without expending continuous musculature effort. Second, whereas the tubercles of non–birds of prey are largely rounded, those of hawks and owls are rectangular, with their longer sides running parallel to the surrounding folds, creating a surer and firmer grip.

Einoder and Richardson also found that tubercle size varied among raptor species. White-bellied Sea Eagles, Wedge-tailed Eagles, and Peregrine Falcons, for example, had the largest tubercles—controlling for the size of the birds themselves—whereas those of three accipiters, a harrier and two other falcons, were relatively small. This is not surprising, given that the first three species are known for their large talons, as well as for their penchant for taking and subduing seemingly oversized prey.

Many raptors capture prey of a particular size, and the size of the feet and toes of raptors often correlate closely with the size of their principal prey items. Size matching allows raptors to position their talons so that they most effectively pierce and penetrate their captures. The specialized, digitally centered TLM of raptors enables them to do this more efficiently than the standard TLM of other birds.

Owls differ from diurnal birds of prey in the arrangement of their digits in having two toes facing forward and two facing backwards; whereas all but a few diurnal birds of prey have three toes facing forward and only one, the hallux, or "killing toe" facing backward. (A notable exception, the Osprey, like owls, has two toes facing forward and two facing back, most likely to better hang on to slippery fish.) Not surprisingly, the hallux has a TLM that is substantially larger than that of the three other digits, whereas the TLMs of owls tend to be more similarly sized in all four toes.

The authors of this study are quick to point out that additional research involving a greater number of species and kinds of raptors is needed to fully assess how the ecological differences of birds of prey affect their TLMs. Indeed, given the importance of their toes and talons, it seems likely that the TLMs of raptors will be found to vary considerably among species, based on the diets and relative sizes of the prey they kill and eat. The piscivorous, or fish-eating, Osprey appears to be a case in point.

An analysis of Finnish Ospreys filmed while preying on trout suggests that Ospreys can catch prey with a single talon while striking the water surface at up to 70 kilometers (44 miles) per hour, in part because the talon snaps shut in 2/100 of a second, suggesting a tactile reflex rather than a voluntary action. In addition to an automatic grasping mechanism, the foot and toe pads of Ospreys are covered with short, sharp spines that help the bird keep its grip on slippery prey. Even so, Ospreys successfully capture prey on only one of every three attempts. Small fish are usually carried off immediately, whereas larger fish can keep the Osprey in the water for several seconds or more, while the bird attempts to rearrange and tighten its grip. Rarely are fish powerful enough to drag an Osprey further into the water, at which point the predator typically releases its grip and flies off with empty talons. Once airborne, fish usually are carried head first, most likely for aerodynamic streamlining.

Ospreys are unique among raptors in regularly carrying prey while migrating, possibly because their specialized toe pads, automatic TLM, and two forward-facing and two backward-facing toes combine to make it easy to do so. The distances that migrating Osprey travel while carrying a fish are unstudied, but I have seen northbound individuals arriving on the Spanish side of the 14-km (9-mi) wide Strait of Gibraltar clutching prey that presumably they had captured in shallow waters off the coast of Morocco in northwestern Africa.

The long-legged and extremely dexterous Crane Hawk of Middle and South America, and African and Madagascar harrier-hawks, or gymnogenes, have a flexible, double-jointed tarsal joint capable of bending 30° backwards and slightly sideways, physical adaptations that allow these species to reach into holes and crevasses to extract nestling birds.

Another raptor with an unusual method of talon-based prey capture is the sub-Saharan Secretarybird, a grassland species with long, storklike legs and short, stubby toes that spends much of the day walking about in search of both invertebrate and vertebrate prey. Unlike other raptors, Secretarybirds trounce on, rather than seize, potential victims, immobilizing their surprised prey in a rapid series of forceful kicks. Secretarybirds, which swallow most of their prey, including small tortoises and hares, whole, take venomous adders and cobras as well, attesting to the effectiveness of their unusual foot-stomping behavior.

Finally, keeping a firm grip on one's prey or disabling it with a hurried series of kicks serves not only to secure a meal, but also reduces the likelihood of potential prey injuring the raptor during defensive maneuvering in escape attempts, which is not the kind of food fight raptors want to engage in.

NOCTURNAL HUNTING

By definition, diurnal birds of prey feed mainly during the day. Nevertheless, many species of raptors routinely hunt during low-light conditions at dawn, dusk, or both. Crepuscular hunters (raptors that opportunistically or routinely feed at dawn, dusk, or both), comprise a number of falcons, including American, Common, and Lesser Kestrels, Northern Hobbies, Merlins, Aplomado Falcons, Orange-breasted Falcons, Peregrine Falcons, and, not surprisingly, Bat Falcons, along with harriers,

Broad-winged Hawks, and Red-tailed Hawks. Intriguingly, another crepuscular hunter, the Short-eared Owl, often roosts communally on the ground in the same field as harriers roost, with these day-night ecological counterparts routinely comingling while hunting crepuscularly near their roosts.

Raptors also hunt nocturnally, at least occasionally, and especially on moonlit nights. Species that do so routinely include Black-shouldered Kites, Bald Eagles, American, Common, and Lesser Kestrels, and Peregrine Falcons. Peregrines, in particular, do so routinely in well-lighted urban areas. In fact, this species has been reported hunting well into the night in many parts of its range, including North America, France, Germany, Poland, the Netherlands, Hong Kong, and Taiwan, suggesting the behavior is widespread. One of the more thorough studies of this phenomenon was conducted by the American raptor specialists Robert DeCandido and Deborah Allen, who studied Peregrine Falcons at the observation deck of the well-lit Empire State Building, more than 300 m (1000 ft) above the streets of densely populated Manhattan Island, in New York City. The researchers observed falcons hunting on more than half of the 77 nights they spent on the skyscraper during southbound raptor migration in 2004. Although Common Pigeons and bats also were seen Peregrines took only songbirds, ranging in size from Wood Warblers to a Yellow-billed Cuckoo. The falcons, which employed both perch hunting and aerial pursuit, chased 111 birds and caught 37 of them at a per-attempt success rate of 33%.

The Lesser Kestrel is another diurnal bird of prey that routinely hunts nocturnally under ornamental lights in urban areas. An Old World species that breeds colonially in the architectural nooks and crannies of chapels, churches, and cathedrals in European cities, Lesser Kestrels routinely take aerial insectivores feeding on insects attracted to the light at such sites. Detailed observations at two well-lit buildings, a cathedral and a smaller church about 500 m (1650 ft) from each other in Seville, Spain, revealed substantial nighttime hunting and nocturnal nestling provisioning in summer. Both historic buildings are illuminated for tourists, and the lighting attracts enormous numbers of flying insects, which, in turn, attract large numbers of aerial insectivores, including both bats and Lesser Kestrels. The study, conducted by the Spanish raptor biologist Juan Jose Negro and colleagues during the late 1990s, reported more than 400 prey deliveries to nests at the two buildings, 11% of which occurred at night.

The extent to which nocturnal hunting affects the nesting success of these and other kestrels breeding in urban areas has not been studied.

PRACTICE HUNTING AND PLAY BEHAVIOR

Play is a biologically significant attribute of many young birds. This is especially true for young raptors, which are less-successful hunters than are older birds, and typically suffer higher death rates during the first year of life. The degree to which differential mortality of particularly inept young raptors—versus learning—plays a role in this dichotomy remains unclear, but numerous anecdotal observations of play behavior in raptors suggest that purposeful learning helps reduce age-related differences in predation ability. The "drop-catch" play of nonfood objects, in which an individual raptor repeatedly releases and recaptures an object, has been reported in more than twenty-five species of predatory birds, including at least eleven species of raptors. One notable account reports an immature Scottish Golden Eagle dropping and subsequently regrasping a stick at least seventeen times in 10 minutes. Another involves an immature Martial Eagle manipulating a clump of elephant dung after initially "striking," and carrying it approximately 10 m (30 ft) off and thereafter ripping it apart and protectively mantling it—a sequence of events that the bird repeated four times during about 15 minutes, before flying off and perching in a tree. Other examples include in-flight Common Buzzards rapidly switching small sticks from one foot to the other, and back again.

At least three species of harriers seize and subsequently manipulate sticks, grass, corncobs, and other inanimate objects in their talons. Northern Harriers near winter communal roosts in south-central Ohio played extensively with mouse-sized corn cobs, tearing them apart and "sham eating" them. The exceptionally curious Striated Caracara of Tierra del Fuego and the Falkland Islands routinely approaches inanimate objects offered to them, manipulating such items while footing them, "mandibulating" them in their beaks, and attempting to fly off with them.

Although the actual function of such behavior remains somewhat controversial among animal behaviorists, both physical training and the practice of motor skills appear to be involved. My observations of the Northern Harriers support this, because the corncobs manipulated by the birds were more similar in size to the species' meadow vole prey than were a series of randomly collected corncobs that were available to them near the roost. At least one raptor researcher suggests that drop-catching most likely provides pleasurable positive feedback as well, since individuals repeatedly engage in this behavior. Although such behavior does not, in and of itself, involve the birds "consciously attempting to educate themselves or to physically train through play," it most likely acts through natural selection to produce that result.

TOOL USE

Tool use, a fundamental concept in assessing mammalian cognition, underwent something of a redefining moment when the noted primatologist Jane Goodall and the wildlife biologist and filmmaker Hugo van Lawick described it in Egyptian Vultures in 1966. The two observed a pair of vultures in Tanzania that were purposefully using small stones to crack the shells of Ostrich eggs so that they could eat the contents. Their initial observation was of two birds breaking into two eggs at an abandoned nest in the presence of both White-backed and Lappet-faced vultures "that were not, apparently, able to break open the eggs, themselves, although they repeated[ly] pecked at them and hurriedly fed on the contents when the Egyptians broke one open." A second observation involved an experiment in which four eggs were offered one at a time to three Egyptian Vultures. Two of the eggs were opened in less than 2 minutes after four to six hits, and the other two were opened after eleven and twelve hits. The first egg was open within 5 minutes, the second within 8 minutes. There also is a report from Israel of an Egyptian Vulture killing a monitor lizard after striking it with a stone, and one from Bulgaria in which two Egyptian Vultures were seen using a twig to gather recently shorn wool from the ground for use as nest lining. Striated Caracaras in the Falkland Islands routinely peck at, open, and consume the much smaller and more thinly shelled eggs of Upland Geese before consuming them; they do so, however, without the use of the stones even when the latter are available nearby.

Although not strictly tool use, drowning as a means of killing prey also has been reported. In North America, Sharp-shinned Hawks, Northern Harriers, and Cooper's Hawks all have been seen drowning avian prey ranging in size from small passerines to small waterfowl. And in the Netherlands, a Eurasian Sparrowhawk has been observed drowning a Common

Starling. The events occurred at shallow ponds, with the raptors forcefully submerging and sometimes kneading the prey for up to 10 minutes to drown it. Whether individuals learned this technique serendipitously and, thereafter, routinely employed it, or whether observational learning was involved, is not known. That purposeful drowning is rarely reported in the literature suggests that it may not be learned at all, but may only be used opportunistically and occasionally.

PREY TAKEN

Among the behavioral "rules" regarding how birds of prey pursue and secure food, versatility in predatory approach remains foremost in most species, as does enormous variability in prey types taken. Although some species of raptors are stenophagic, dietary specialists that feed on a narrow subset of prey types, most are euryphagic, dietary generalists that catch and consume dozens, and in some instances hundreds, of different kinds of prey. A potential prey item's relative size, abundance, and vulnerability rank as the three most important criteria for its inclusion in a raptor's diet. Because of their diversity in size, distribution, and flight ability, the world's birds of prey feed on an overwhelming variety of prey types, including tiny invertebrates, fishes, reptiles, amphibians, birds, and mammals, along with the large carcasses of elephants and whales. Overall, raptors consume tens of thousands of different prey species, including, in a few instances, plants.

Consider, for example, the diet of the Peregrine Falcon, arguably the world's most cosmopolitan bird of prey. This agile, high-speed-pursuit predator, which is capable of catching, subduing, and killing prey twice its body mass, feeds primarily on birds, but also takes insects, mammals, and, when hard-pressed, carrion. Peregrines also rob other raptors of prey. Although they appear to be particularly fond of Common Pigeons, Peregrine Falcons, as a species, take more than 500 kinds of birds globally, ranging in size from tiny North American wood warblers to large waterfowl, with one survey alone listing seventy-four different species of birds in the diet of a single regional population.

Individual falcons, however, can be quite specific in their diets. Sedentary Peregrine Falcons nesting on cliffs above water holes in the deserts of Western Australia can spend their entire adult life on a single territory, feeding on the relatively small number of species of birds that drink daily at the desert oasis below their rocky eyrie. Other individuals, such as the highly migratory birds breeding along the northern foothills of the Brooks Range in Arctic Alaska and overwintering in the tropics of South America, feed on decidedly different species on their Arctic breeding sites compared with their tropical wintering areas, and also take different prey while migrating between the two regions.

Although migrating Peregrine Falcons frequently soar on migration, particularly in tropical latitudes, they also engage in powered flight, and thus expend considerable energy while traveling. Many of their migratory pathways overlap those of their shorebird prey, creating what shorebird biologists have called a moving "landscape of fear" for the prey species involved. To say that a versatile diet has helped the Peregrine Falcon nest

Table 6.1 Diets of New World Peregrine Falcons. Prey in **bold** are taken as prey at two sites.

Site	Species included in diet
Breeding grounds in Arctic Alaska	Mammals Arctic ground squirrel, tundra vole Birds Black-throated Loon, Canada Goose, Northern Pintail, **Green-winged Teal**, American Widgeon, Greater Scaup, Red-breasted Merganser, scoter, Rough-legged Hawk, Peregrine Falcon, ptarmigan, Semipalmated Plover, **American Golden Plover**, Common Snipe, **Spotted Sandpiper**, Lesser Yellowleg, Pectoral Sandpiper, Long-billed Dowitcher, Semipalmated Sandpiper, Bar-tailed Godwit, Red-necked Phalarope, Parasitic Jaeger, Long-tailed Jaeger, Pomarine Skua, Sabine's Gull, Arctic Tern, Short-eared Owl, Say's Phoebe, Grey Jay, Grey-cheeked Thrush, Bluethroat, Arctic Warbler, Eastern Yellow Wagtail, Water Pipit, Great Grey Shrike, American Yellow Warbler, Redpoll, Tree Sparrow, Fox Sparrow, Lapland Longspur
Migration in coastal Texas	Mammals Not identified, but reported as present Birds Snowy Egret, Cattle Egret, Green Heron, Black-crowned Night-Heron, Northern Shoveler, **Green-winged Teal**, Redhead, Lesser Scaup, American Kestrel, King Rail, American Coot, **American Golden Plover**, Willet, Herring Gull, Laughing Gull, Royal Tern, Common Pigeon, White-winged Dove, Mourning Dove, Northern Flicker, Horned Lark, Grey Catbird, Eastern Meadowlark, Rusty Blackbird, Great-tailed Grackle, Brown-headed Cowbird, sparrow
Wintering grounds in South America (Brazil)	Mammals None reported Birds **Spotted Sandpiper**, Blue-and-white Swallow, White-collared Swift, Eared Dove, Band-tailed Pigeon, Common Ground Dove, Vermilion Flycatcher, finch

successfully across six of the world's seven continents (it is absent from Antarctica) may be a bit of an understatement.

Peregrine Falcons are not the only migrating raptors that follow their prey while migrating. Eurasian Sparrowhawks do the same across northern Europe as they shadow the movements of Chaffinches, Bramblings, and other abundant songbird migrants they prey on. In autumn, southbound Sharp-shinned Hawks in eastern North America do the same while tracking the movements of migratory songbirds, providing what the American raptor-migration specialist Paul Kerlinger calls "a conveyer belt" of food for bird-eating raptors.

At the other extreme, several raptors are dietary specialists, and are relatively inflexible in terms of their diet feeding on but a few species of prey. In tropical and subtropical regions of the Western Hemisphere, for example, wetland-habitat specialist Snail Kites survive almost exclusively on apple snails, a group of freshwater mollusks of the genus *Pomacea*. The kites find their gastropod prey while hovering above expansive freshwater marshes such as the Florida Everglades, Venezuelan Llanos, and Brazilian Pantanal. Kites snatch and carry exposed and extremely vulnerable snails in their talons to an elevated perch, where, while holding the mollusk in their feet, tear away the lidlike operculum that covers and protects the body of the snail, and extract the meat below by inserting their long and slender, hyper-curved bill into and along the curved shell of the snail while severing the muscle that attaches the body to the inner surface of its otherwise protective exoskeleton. The kite then consumes the soft body tissue either intact or by tearing it in pieces. Although the species also feeds on turtles, crayfishes, crabs, and even fishes, in most locations, apple snails make up the bulk of the habitat specialist's diet.

With more than 1200 recognized species, bats comprise the most speciose order (Chiroptera) in the class Mammalia. Although few vertebrate predators feed primarily on bats, two tropical birds of prey, the Bat Hawk of the Afrotropics and Oriental Region, and the Neotropical Bat Falcon, do specialize on them. That these two species are tropical birds of prey is not surprising, as both bat species' diversity and numerical abundance are highest in equatorial regions, and other raptors that opportunistically feed on them are most abundant in these regions as well. The two specialists are not closely related. The Bat Hawk's closest relatives include a number of Old and New World kites in the Accipitridae, and the Bat Falcon's include several New World falcons in the Falconidae. Bat Hawks, whose flight silhouettes resemble those of large falcons, weigh 600–650 g (21–23 oz), feed primarily on 20–75-g (0.7–2.5-oz) insectivorous bats and secondarily on small birds, including swifts. The species captures its prey in rapid aerial pursuit, and swallows them rapidly in flight, sometimes taking and eating as many as five bats in as few as 5 minutes. Bat Falcons, which resemble other small falcons, weigh less than half as much as Bat Hawks, and feed primarily on birds, including swifts, and large insects, and only secondarily on bats, with individuals living close to communal roosts typically specializing on them. Like Bat Hawks, Bat Falcons often consume smaller prey on the wing. Both species, which are largely crepuscular, appear to time their feeding effort to take advantage of the arrivals and departures of bats at large roosts, which sometimes number in the tens of thousands.

A Bat Hawk hunting bats at a roost in Zambia, for example, spent 15–20 minutes hunting each evening, during which it caught an average of seven bats, each of which weighed about 50 g (1.8 oz), remarkably taking only about 6 seconds to ingest each one. The hawk dropped about as many bats as it ate, possibly because the bats bit it, or because the hawk had trouble grasping them in its talons. Based on estimates of the respective body masses of predator and prey, the researchers conducting the study suggested that the Bat Hawk they were watching consumed approximately 8–9% of its body mass each day during its brief hunting efforts.

And indeed, rapid prey-handling and consumption appear to be an adaptation to the short time that the explosive, roost-evacuating "bat bonanza" is available to them. The anatomical and physiological consequences of rapid prey consumption in a fast-flying predatory raptor have yet to be studied in detail, at least as far as I know. Neither species is said to have distensible crops like those found in vultures, condors, and caracaras, all of which routinely gorge food, and the brief period during which they consume much of their daily prey must "stretch" both the flight ability and digestive capacity of these prey specialists. An examination of potential anatomical and physiological adaptations to this lifestyle is overdue.

Temporal and Geographic Changes in Prey Taken

Both weather and regional changes in prey availability influence the diets of individual raptors. American Kestrels overwintering in southern Ohio, for example, switch from hunting songbirds and small mammals to hunting insects when temperatures rise above 10 °C (50 °F), i.e., temperatures at which grasshoppers and other Orthopterans begin to move about. When temperatures drop below freezing and insects are largely inactive, kestrels switch back to perch-hunting and hover-hunting small mammals. The kestrels also switch to feeding on songbirds when heavy snows blanket the fields, protecting mice and voles from the aerial predators. Diet in the species also changes geographically. In Temperate Zone northeastern North America, nesting kestrels typically take small-mammal and insect prey, supplemented by a few birds. In tropical Venezuela, nesting kestrels take insects and lizards, and only rarely feed on birds and mammals. The take-home message is clear: for American Kestrels, both prey size and the ability to handle prey are important, but prey taxonomy and prey type play somewhat lesser roles. Field studies involving other common and widespread raptors show similar patterns.

Habitat Effects on Prey Taken

Feeding specializations among individual birds of prey often reflect habitat differences. In North America, Red-tailed Hawks living along interstate highways feed more on roadkills than do individuals living along roads in less trafficked areas. Red-tailed Hawks living in farmlands typically take rabbits, mice, and voles, whereas those living in forested areas and wooded suburban sites depend more heavily on squirrels and chipmunks. And in areas of North America where they co-occur with Northern Harriers, Red-tailed Hawks sometimes rob harriers of their prey, an opportunity that obviously is not available when harriers are not present.

Gender Differences

In species with significant size dimorphism, male and female raptors often specialize on different prey. In such situations, the larger, more powerful females take larger and slower prey than do their more agile male counterparts. As might be expected, gender differences in diet are particularly pronounced in raptors exhibiting extreme sex-related differences in body size. In some species, the differences are such that falconers flying these species hunt with one or the other sex when attempting to catch specific prey. A good example of sex-specific differences in both hunting behavior and diet occurs in North American populations of Northern Harriers, a species where males average about 350 g (12 oz) and females average about 525 g (19 oz), resulting in females being half again as large as their counterparts. Among harriers overwintering in south-central Ohio, males flew noticeably more rapidly while hunting fields than did females, and were far more likely to "overshoot" prey they had spotted than were their female counterparts, resulting in a significantly greater proportion of "hook pounces," in which they pirouetted over small-mammal prey before pouncing on them. On the other hand, slower-flying females were more likely to engage in slow-motion "soft pounces" when they tried to capture prey, and thereafter methodically probed the grassy, domed vole nests they extracted from the field in an attempt to secure nesting meadow voles suckling their young. Adult male harriers also caught and fed on significantly more songbird prey than did adult females (40% compared with 5%), whose diets consisted almost entirely of small mammals. First-winter plumage in harriers is identical in males and females, and the two sexes are difficult to distinguish in the field. Not surprisingly, the diets of unsexed juveniles were found to be intermediate between those of adult males and females.

Age Differences

In addition to sex differences in diet, many species also exhibit age differences, with juvenile diets, especially those of recently fledged young, differing significantly from those of adults. Fledglings specialize on smaller and less maneuverable prey than do their parents, presumably because such prey are easier to catch. Northern Harriers, for example, typically hunt and catch only insects during the first few weeks out of the nest. Within a month or so, however, and presumably after their flight skills have improved, young harriers begin taking both small-mammal and avian prey, as adults do. Even so, juveniles do so with less alacrity than older individuals, and at considerably lower success rates. Because of this, many juvenile birds of prey focus on taking less-maneuverable small-mammal prey during their first winter as opposed to taking birds. And, in at least one species of opportunistically scavenging birds of prey, the Striated Caracara, juveniles compensate for their youth by banding together to more capably compete with older individuals during encounters at large carrion.

Box 6.1 Cooperative "gang behavior" in young Striated Caracaras and other raptors

Striated Caracaras are relatively large, stocky, inquisitive, and opportunistically predatory and scavenging raptors. An island-dweller in southernmost South America, the species nests on the perimeters of penguin and seabird colonies, where individuals feed on both eggs and chicks, as well as on dead and dying adult birds. Clumsy predators but aggressive scavengers, caracaras also sometimes attack healthy nestling and fully grown seabirds and land birds. The species also competes with and sometimes displaces other birds, including Turkey Vultures, Variable Hawks, and Subantarctic Skuas, from carcasses. It also routinely feeds on the placentas, feces, and carcasses of seals, and scavenges the carcasses of sheep and other livestock. To say that the Striated Caracara is a dietary generalist is something of an understatement.

With a world population estimated at fewer than 2500 mature individuals, this regional endemic occurs only in Tierra del Fuego and the Falkland Islands, with the latter archipelago believed to be its stronghold. The Falklands population has been stable since the 1980s, even though breeding success through fledging indicates that it should have increased. Why this is so is not certain, although high juvenile winter mortality because of limited food resources has been suggested.

The large numbers of breeding seabirds that provide Striated Caracaras with abundant food in summer largely evacuate the islands in autumn for warmer waters to the north, and winter on the Falklands is no picnic for the nonmigratory caracaras that remain. My own research on 127-km^2 (49-mi^2) Saunders Island—a nursery island for 90 to 130 mainly juveniles and subadults that holds no nesting pairs—indicates that juveniles and subadults lose about 15% of their body mass in winter. Individuals feed feverishly at this time of year, and competition for limited food creates feeding frenzies among individuals. Indeed, in winter, caracaras rarely, if ever, exhibit signs of satiation or "slowing down" while eating, with most feeding attempts ending only once a bird has consumed all of the edible parts of an item, or is chased from it by another caracara. Most of the birds overwintering on the island do so at a farm settlement, where human activity provides a nutritional subsidy in the form of farm scraps.

Observations of feeding birds in and around the farming settlement, along with an analysis of regurgitation pellets collected at a night roost, indicate a winter diet of mainly native geese (killed either by Variable Hawks or by the farmers), beetles and other invertebrates, and the carcasses of domestic sheep.

The young birds that inhabit Saunders Island apparently do so for two reasons. First, a lack of breeding pairs on the island means that few competitively superior adults are there. Second, the family that farms the island does not harass the caracaras, permitting them access to leftovers and scraps in an around the farm without fear of being shot, something that is not necessarily true of other settlements on other islands in the archipelago. Unfortunately, there does not appear to be enough food for all of the birds that congregate at the farm settlement each winter. And this is where cooperative "gang behavior" comes into play.

Juveniles and subadult caracaras tend to hang out and feed in "avian gangs" of up to several dozen birds, especially when the food in question occurs in large quantities such as human-butchered sheep remains or the remains of 4-kg (9-lb) Upland Geese killed by 1–1.5 kg (3-lb) Variable Hawks. Screaming while in flight as they approach a carcass, relatively small gangs of young caracaras quickly build to mobs of upwards of several dozen individuals which, although far less predatory than the hawks, often dominate them by sheer numbers, and in so doing are able to make off with a sizeable share of the food. Importantly, adult caracaras arriving simultaneously also give way to the ravenous mass of younger birds, as the goose is picked clean in short order. Much the same occurs when pigs at the farm are fed their daily dose of shot Upland Geese, when a feeding frenzy at the pigpen negates any age advantage held by adult caracaras, as both young and old caracaras scramble for a share of the bonanza.

Juvenile and subadult Striated Caracaras are not the only diurnal birds of prey that exhibit gang behavior in nutritionally stressful times. Other scavenging raptors that might be disadvantaged by their smaller size also do so in order to compete for what might otherwise be unattainable food resources. Old World vultures in the genus *Gyps* engage in gang behavior to assert dominance at large carcasses. And in South America, the relatively small Black Vulture uses gang behavior to numerically dominate more massive Andean Condors at food sources.

It is not known if participants involved in gang behavior have "friends." Regardless, the strategy appears to be an effective way for otherwise less competitive raptors to secure food when feeding on "bonanza" nutritional resources.

Plant Eating

Not all birds of prey limit their diets to animals. The Palmnut Vulture of sub-Saharan Africa, for example, feeds mainly on the fleshy husks of oil-palm nuts, in addition to many species of invertebrates, including crabs, mollusks, snails, and locusts. The species depends so much on the nuts of oil-palm trees that its distribution in Africa is said to coincide almost exactly with that of several species of these trees, which some researchers have characterized as the vulture's most stable source of food. How Palmnut Vultures deal with a diet that is high in plant material is unclear, but researchers estimate that an individual feeding on as few as thirty-four oil-palm nuts a day would meet its daily caloric requirements. Perhaps not surprisingly, whereas the fruits amount to only 65% of gut contents in adults, they made up fully 92% in younger birds. In Colombia, Black Vultures, too, sometimes feed on oil-palm fruit, and the species reportedly feeds on the fruit of rubber trees as well. And in North America, both Black Vultures and Turkey Vultures sometimes feed on vegetables, including pumpkins, presumably when carrion is limited. In Costa Rica, I have seen Black Vultures pull unhusked coconuts onto roadways, and then wait for passing vehicles to crush them before feeding on the "meat." The species also feeds on dung of livestock, including pig manure. Swallow-tailed Kites also feed on the fruits of oil palms and black rubber trees. And in Tenerife in the Canary Islands, Ospreys have been reported gathering up, carrying in their talons, and subsequently deliberately feeding on algae at a reservoir. Why the birds did so is not clear.

Ingesting Non-prey Items

Most birds have a two-part stomach, an upper proventriculus for the chemical digestion of food and a lower gizzard for mechanical digestion. Many seed-eating birds, including species that range in size from small finches to ostriches, purposely ingest mineral grit and small stones to help the gizzard mill hard-to-digest seeds. Raptors also consume grit and small stones, with the latter referred to as "rangle" by falconers who feed it to captives. In birds of prey, as in cormorants, petrels, and fulmars, which also ingest stones, rangle appears to be associated with high-fat, meaty, and fishy diets. It is not known how frequently raptors ingest rangle in the wild, but captive individuals apparently do so weekly to monthly, and there are numerous reports of wild birds picking up pebble-size river stones while drinking at watering holes. The stones are believed to loosen and stir up mucus in the upper digestive tract, speeding its passage through the intestine and out the digestive tract.

Although the ingestion of such natural objects makes evolutionary sense, man-made objects and human trash can be mistakenly ingested as rangle, and when this happens, problems occur. Lead pellets from ammunition can cause lead toxicosis when ingested, and bottle caps, broken glass, and other relatively small sharp objects, widely known as "microtrash," can clog or otherwise physically damage the digestive tract. Large vultures are particularly vulnerable to this "junk food" for another reason. Many feed on specific portions of the large carcasses they consume, including soft tissues such as muscle and viscera. Bone is an important source of calcium in birds of prey, and many species of vultures search specifically for bone fragments to supplement their largely soft-body-tissue diet. This is particularly true during the breeding season, when female raptors require calcium for their eggshells, and developing nestlings need it to help grow and strengthen their skeletal systems.

But the need for calcium alone does not explain why large vultures, including the critically endangered California Condor, regularly take bottle caps, ammunition casings, and fragments of brightly colored glass, the consumption of which has been linked to the deaths of nestlings. Soft body tissue also does not contain keratin, which, like bone, is another typical dietary component of predatory species of raptors. This structural protein is found in the hair, hooves, and claws of mammalian prey, as well as the feathers, claws, and bills of avian prey. Because raptors cannot digest keratin, the protein typically makes up the bulk of the pellets that most raptors routinely regurgitate. Although regurgitation pellets function to expel nondigestible objects from the guts of birds of prey, the peristaltic contractions involved in their doing so also serve to scour the upper digestive tract of accumulating fat and mucus, a potentially significant cleansing function. And this latter function may help explain the bizarre variety of non-bone-like "junk food" in the diets of scavenging raptors. In fact, many scavenging birds of prey seek out and purposely ingest fibrous objects, including, grasses, mosses, leaves and twigs, for this important purpose. More than three-quarters of sixty-seven regurgitation pellets my colleagues and I collected from Striated Caracaras feeding at a farm site in the Falkland Islands contained significant amounts of lichens, mosses, ferns, grasses, and other plant material. Presumably, such material helped create the mass needed to form sufficiently sized pellets.

SYNTHESIS AND CONCLUSIONS

1. Feeding behavior can be difficult to study directly, and much of what we know about it has been gained indirectly via nest watches and diet studies.
2. Raptors catch prey in many ways, including still- or perch-hunting, fast contour-hugging flight, soaring and hovering flight, low and slow-quartering flight, stalking, listening, flushing from cover, and high-speed pursuit and stooping. Most species employ two or more of these tactics to catch prey.
3. In most raptors, the "business end" of the bird are its large talons. Beaks are used mainly to tear prey into bite-sized pieces for swallowing.
4. The toes and talons of raptors employ a highly specialized ratcheted talon-locking mechanism, or TLM, in which thousands of small, well-defined projections called tubercles allow individuals to securely grasp their prey without continuous muscular effort.
5. Many raptors hunt at night, assisted by human-made light or moonlight.
6. Egyptian Vultures use stones as tools to crush the shells of ostrich eggs, which they then eat.
7. At least several species of harriers and accipiters submerge and drown captured prey before eating them.
8. Predatory raptors feed on many types of prey, including small insects and other invertebrates, fishes, amphibians, reptiles, birds, and mammals.
9. Scavenging birds of prey feed on insect and other invertebrate carrion, as well as on vertebrate carrion ranging in size from songbirds and small mammals to elephants and whales.
10. Widely distributed dietary generalists like Peregrine Falcons take more than 500 bird species globally, whereas dietary specialists like Snail Kites feed almost exclusively on snails in the genus *Pomacea*.

11. Prey taken by individual raptors can vary geographically, seasonally, and by time of day.
12. Several species of birds of prey eat vegetable matter, including palm nuts, coconuts, and pumpkins.
13. Most raptors ingest mineral grit and small stones to enhance digestive processes, and to assist in the formation of regurgitation pellets.

7 MIGRATION

Birds are the greatest travelers in the world.
Frank M. Chapman, 1915

RAPTOR MIGRATION, the seasonal, to-and-fro long-distance movements of diurnal birds of prey between their breeding and wintering grounds, continues to amaze and inspire us in many ways. That an Amur Falcon weighing less than a pound manages to shuttle itself from its easternmost Siberian breeding grounds to an overwintering area in southern Africa, and back again, ranks among the most demanding of all raptor migrations. Traveling in large, multi-thousand bird flocks across more than 70 degrees of latitude *and* longitude, these kestrel-sized falcons routinely undertake a transoceanic outbound journey transecting up to 4000 km (2500 mi) of the Indian Ocean as part of an annual round-trip of more than 22,000 km (13,000 mi). And whereas Amur Falcons may claim the record for the most arduous migration of all birds of prey, other raptors, including the New World Swainson's Hawks, Old World Steppe Buzzards, and high-Arctic breeding Peregrine Falcons, make round trips almost as lengthy.

Like other birds, raptors migrate largely along north-south axes. Those breeding at high latitudes are more migratory than those that breed at lower latitudes, and, not surprisingly, the Northern Hemisphere, which has fifteen times as much land between 30 degrees and 80 degrees as does the Southern Hemisphere, has many more species of migrants. Asia alone has at least sixty-six species of migratory raptors, Europe at least thirty-eight, and North America at least thirty-three.

Historical references to raptor migration are legion. The oldest I can find is an Old Testament admonition to Job: "Doth the hawk fly by thy wisdom, and stretch her wings toward the south?" More recently, Holy Roman Emperor Friedrich II of Hohenstaufen (1194–1250), writing in *De arte venandi cum avibus* [The art of falconry], appears to be the first to address the question of how weather affects raptor migration, stating, "We notice also that when a favorable wind sweeps up, whether by day or by night, migrating birds tend to generally hasten to take advantage of it, and even neglect food and sleep for this important purpose."

Other observers, including the Spanish historian Gonzalo Fernández de Oviedo y Valdés, described the massive movements of birds of prey through the Caribbean as early as the sixteenth century. And the near-contemporary French zoologist Pierre Belon did much the same for the Old World in describing multitudes of Black Kites streaming across the Bosporus over Istanbul, Turkey. Clearly, people have known about the massive migratory movements of raptors for thousands of years.

Although migratory flight defines raptors behaviorally in much the same way as talons and keen eyesight define them physically, the phenomenon of raptor migration remained something of an ornithological backwater until high-quality prismatic binoculars became widely available in the latter part of the nineteenth century and ornithologists actually began to "see," and correctly identify, the species involved. Another major advance in the field occurred with the invention of bird banding at the beginning of the twentieth century. In the 1930s, reasonably priced binoculars, together with the appearance of modern field guides, helped fuel the new sport of "raptor watching" at traditional migration bottlenecks and other concentration points along major migration corridors in North America and Europe, an activity that today continues to provide information on both the sizes of regional populations and the migration strategies of birds of prey.

A transition to what many consider the "golden age" of raptor migration studies occurred in the last quarter of the twentieth century, when scientists began placing tiny VHF (very-high frequency) radio transmitters, and, shortly thereafter, UHF (ultrahigh frequency) satellite-based tracking devices, on individual birds of prey. The latter, which allow the birds themselves to tell us how and where they actually migrate—rather than our relying on theoretical models of their assumed movements based on watch-site observations and banding recoveries—has advanced the field noticeably in the last three decades. Consequently, we now have a much better understanding of the whys, wherefores, and complexities of raptor migration, one that, in many ways, presents a far more realistic "biological" picture of the phenomenon. This chapter reflects how these important new tools of raptor-migration study have enabled the growth of newly gained insights.

THE BASICS OF RAPTOR MIGRATION

Not all raptors migrate. In fact, a third of all birds of prey fail to undertake seasonal movements of any kind. Nearly half of this third are sedentary island-restricted species, which, having reached a tropical or near-tropical island or archipelago, became nonmigratory. Although many island raptors have close continental relatives from which they most likely evolved, others do not appear to, and the details of their colonizations remain, for the most part, unexplored. In any case, that large numbers of nonmigratory island-restricted raptors exist suggests that migratory behavior is a rapidly evolving and easily reversible trait in diurnal birds of prey.

Other primarily nonmigratory raptors include low-latitude continental species whose breeding populations lie mainly in the tropics, 23° north and south of the equator. Yet another group of largely nonmigratory birds of prey include nonmigratory low-latitude populations of species whose breeding ranges are spread across large latitudinal gradients, and whose higher-latitude populations are migratory. The Common Buzzard, an abundant and widespread species that breeds from 35° to 65° north across much of Eurasia, is a good example of a partial migrant whose southernmost populations are nonmigratory. Northernmost populations of Common Buzzards breeding in Scandinavia and northern Russia consist all-but entirely of long-distance migrants, many of which overwinter in southern Africa. At the other extreme, more southerly populations in Mediterranean Europe are overwhelmingly, if not entirely, sedentary. Populations of buzzards in midlatitudes exhibit a mixed migration pattern, with some individuals migrating and others not.

Seasonal shifts in prey availability play a major role in determining such latitudinal trends. Temperate Zone raptors that feed primarily on "warm-blooded" homeotherms, such as birds and mammals, all but a few of which are active year-round, tend to be less migratory than those that feed on "cold-blooded" poikilotherms, such as insects, fishes, reptiles, and amphibians, which tend to be inactive or otherwise unavailable in winter, necessitating migration in predator populations that depend on them.

Diversity in the migratory tendencies of birds of prey has led specialists to classify species as "complete," "partial," or "irruptive or local" migrants. In this scheme, complete migrants are species in which at least 90% of all individuals shuttle seasonally between separate breeding and nonbreeding areas; partial migrants are species in which fewer than 90% of all individuals migrate; and irruptive or local migrants are species whose movements occur more sporadically, and over shorter distances, than those of complete and partial migrants. As is true for most groups of birds, partial migrants are by far the most common type of migratory raptors, and complete migrants are the least common.

Table 7.1 Attributes of complete, partial, and irruptive and local migratory birds of prey

Species	Long-distance migrant[a]	Trans-equatorial migrant[b]	Rains migrant[c]	Altitudinal migrant[d]	Maximum flock size	Maximum distance traveled over water
Complete migrants						
Osprey	Yes	—	—	—	>50	>100 km
European Honey Buzzard	Yes	—	—	—	>1000	>100 km
Mississippi Kite	Yes	—	—	—	>100	—
Short-toed Snake Eagle	Yes	—	Yes	—	>10	<25 km
Pallid Harrier	Yes	—	—	—	<10	<25 km
Pied Harrier	Yes	—	—	—	<10	<100 km
Montagu's Harrier	Yes	Yes	—	—	>10	>100 km
Levant Sparrowhawk	Yes	—	—	—	>1000	<100 km
Chinese Sparrowhawk	Yes	Yes	—	—	>1000	>100 km
Grey-faced Buzzard	Yes	Yes	—	—	>1000	>100 km
Broad-winged Hawk	Yes	Yes	—	—	>1000	<25 km
Swainson's Hawk	Yes	Yes	—	—	>1000	<25 km
Roughleg	Yes	—	—	—	>10	<100 km
Lesser Spotted Eagle	Yes	Yes	—	—	<10	<100 km
Greater Spotted Eagle	Yes	—	—	—	<10	<100 km
Steppe Eagle	Yes	Yes	Yes	—	>10	—
Lesser Kestrel	Yes	Yes	—	—	>1000	>100 km
Red-footed Falcon	Yes	Yes	—	—	>1000	>100 km
Amur Falcon	Yes	Yes	—	—	>1000	>100 km
Eleonora's Falcon	Yes	Yes	—	—	>10	>100 km
Sooty Falcon	Yes	Yes	—	—	<10	>100 km
Northern Hobby	Yes	Yes	Yes	—	<100	<100 km
Partial migrants						
Turkey Vulture	Yes	Yes	—	—	>1000	<25 km
Black Vulture	—	—	—	Yes	>10	—
African Cuckoo Hawk	—	—	Yes	—	—	—
Jerdon's Baza	—	—	—	—	—	—
Pacific Baza	—	—	—	Yes	—	—
Black Baza	—	—	—	Yes	>100	—

Species	Long-distance migrant[a]	Trans-equatorial migrant[b]	Rains migrant[c]	Altitudinal migrant[d]	Maximum flock size	Maximum distance traveled over water
Crested Honey Buzzard	Yes	Yes	—	—	>100	<100 km
Swallow-tailed Kite	Yes	Yes	—	Yes	>100	<100 km
Black-shouldered Kite	—	—	Yes	—	—	—
White-tailed Kite	—	—	—	—	—	—
Scissor-tailed Kite	—	—	Yes	—	<50	—
Snail Kite	—	—	—	—	>10	—
Plumbeous Kite	—	—	—	—	>100	—
Red Kite	—	—	—	—	<10	<100 km
Black Kite	Yes	Yes	Yes	Yes	>1000	<100 km
Whistling Kite	—	—	Yes	—	>100	—
Brahminy Kite	—	—	Yes	Yes	>10	<25 km
Pallas's Fish Eagle	—	—	—	—	<10	—
White-tailed Eagle	—	—	—	—		<100 km
Bald Eagle	—	—	—	—	<10	<100 km
Steller's Sea Eagle	—	—	—	—	—	<100 km
Lesser Fishing Eagle	—	—	—	Yes	—	—
Palm-nut Vulture	—	—	Yes	—	—	—
Bearded Vulture	—	—	—	Yes	<10	—
Egyptian Vulture	—	—	—	Yes	<10	<25 km
White-backed Vulture	—	—	Yes	—	—	—
White-rumped Vulture	—	—	—	—	<100	—
Himalayan Vulture	—	—	Yes	—	<10	—
Griffon Vulture	—	—	—	—	—	—
Cinereous Vulture	—	—	—	Yes	—	<25 km
Lappet-faced Vulture	—	—	Yes	—	—	—
Beaudouin's Snake Eagle	—	—	—	—	—	—
Black-breasted Snake Eagle	—	—	—	—	—	—
Western Marsh Harrier	Yes	—	—	—	<10	>100 km
African Marsh Harrier	—	—	—	—	—	—
Eastern Marsh Harrier	—	—	—	—	—	—
Pacific Marsh Harrier	—	—	—	—	—	—
Long-winged Harrier	—	—	—	—	—	—
Spotted Harrier	—	—	Yes	—	—	<100 km
Black Harrier	—	—	Yes	Yes	—	—
Northern Harrier	Yes	—	—	—	<10	>100 km
Cinereous Harrier	—	—	—	Yes	—	<25 km
African Harrier Hawk	—	—	Yes	—	—	—
Dark Chanting Goshawk	—	—	—	—	—	—
Pale Chanting Goshawk	—	—	—	—	—	—
Gabar Goshawk	—	—	—	—	—	—
Grey-bellied Goshawk	—	—	—	—	—	—
Shikra	—	—	Yes	—	—	—
Brown Goshawk	—	—	—	Yes	—	<100 km
Japanese Sparrowhawk	Yes	Yes	—	—	>100	<100 km
Besra	—	—	—	Yes	—	—
Ovambo Sparrowhawk	—	—	Yes	—	—	—
Eurasian Sparrowhawk	Yes	—	—	Yes	<10	<100 km
Sharp-shinned Hawk	Yes	—	—	Yes	<10	<25 km
Cooper's Hawk	—	—	—	Yes	<10	<25 km

(*Continued*)

Table 7.1 (Continued)

Species	Long-distance migrant[a]	Trans-equatorial migrant[b]	Rains migrant[c]	Altitudinal migrant[d]	Maximum flock size	Maximum distance traveled over water
Bicolored Hawk	—	—	—	Yes	—	—
Chilean Hawk	—	—	—	—	—	—
Northern Goshawk	—	—	—	Yes		<25 km
Grasshopper Buzzard	—	—	Yes	—	—	—
White-eyed Buzzard	—	—	—	—	—	—
Common Black Hawk	—	—	—	—	—	—
Savanna Hawk	—	—	—	—	—	—
Harris' Hawk	—	—	—	—	—	—
Black-chested Buzzard Eagle	—	—	—	—	—	—
Grey Hawk	—	—	—	—	—	—
Red-shouldered Hawk	—	—	—	—	<10	<25 km
Short-tailed Hawk	—	—	—	—	<10	—
White-throated Hawk	—	—	—	Yes	—	—
White-tailed Hawk	—	—	—	—	<10	—
Variable Hawk	—	—	—	—		—
Zone-tailed Hawk	—	—	—	—	<10	—
Red-tailed Hawk	—	—	—	—	>10	<25 km
Common Buzzard	Yes	Yes	—	Yes	>1000	<25 km
Mountain Buzzard	—	—	—	—	—	—
Long-legged Buzzard	Yes	—	—	—	<10	—
Upland Buzzard	—	—	—	—	—	—
Ferruginous Hawk	—	—	—	—	<10	—
Red-necked Buzzard	—	—	Yes	—	<10	—
Indian Spotted Eagle	—	—	—	—	—	—
Eastern Imperial Eagle	—	—	—	—	—	—
Wahlberg's Eagle	—	—	—	—	<10	—
Golden Eagle	Yes	—	—	—	—	<100 km
Verreaux's Eagle	—	—	—	—	—	<25 km
Bonelli's Eagle	—	—	—	—	<10	<25 km
Booted Eagle	Yes	Yes	—	—	>10	<25 km
Rufous-bellied Eagle	—	—	—	—	—	—
Chimango Caracara	—	—	—	—	—	<25 km
Common Kestrel	—	—	Yes	—	<100	<100 km
Australian Kestrel	—	—	Yes	Yes	<10	<100 km
American Kestrel	—	—	—	—	<10	<100 km
Fox Kestrel	—	—	Yes	—	—	—
Grey Kestrel	—	—	Yes	—	—	—
Red-necked Falcon	—	—	Yes	—	—	—
Aplomado Falcon	—	—	—	Yes	—	<25 km
Merlin	Yes	—	—	Yes	<10	>100 km
Oriental Hobby	—	—	—	—	<10	<25 km
Australian Hobby	—	—	—	Yes	—	<100 km
New Zealand Falcon	—	—	—	Yes	—	<100 km
Brown Falcon	—	—	—	—	<10	<100 km
Lanner Falcon	—	—	Yes	Yes	—	—
Saker Falcon	Yes	—	—	—	—	<100 km
Gyrfalcon	—	—	—	—	—	>100 km

Species	Long-distance migrant[a]	Trans-equatorial migrant[b]	Rains migrant[c]	Altitudinal migrant[d]	Maximum flock size	Maximum distance traveled over water
Prairie Falcon	—	—	—	Yes	—	—
Peregrine Falcon	Yes	Yes	—	—	—	>100 km
Irruptive and local migrants						
Lesser Yellow-headed Vulture	—	—	—	—	—	—
Greater Yellow-headed Vulture	—	—	—	—	—	—
King Vulture	—	—	—	—	—	—
California Condor	—	—	—	—	<10	—
Andean Condor	—	—	—	—	—	—
Hook-billed Kite	—	—	—	Yes	—	—
Square-tailed Kite	—	—	—	—	—	—
Black-breasted Buzzard	—	—	—	—	—	—
Bat Hawk	—	—	—	—	—	—
Australian Black-shouldered Kite	—	—	—	—	<100	—
Letter-winged Kite	—	—	Yes	—	<100	—
Double-toothed Kite	—	—	—	—	—	—
Rufous-thighed Kite	—	—	—	—	—	—
White-bellied Sea Eagle	—	—	—	—	—	<25 km
African Fish Eagle	—	—	Yes	—	—	—
Hooded Vulture	—	—	Yes	—	—	—
Long-billed Vulture	—	—	—	Yes	—	—
Rüppell's Griffon	—	—	—	Yes	—	<25 km
Cape Griffon	—	—	—		—	—
Red-headed Vulture	—	—	—	Yes	—	—
Brown Snake Eagle	—	—	Yes	—	—	—
Southern Banded Snake Eagle	—	—	—	—	—	—
Western Banded Snake Eagle	—	—	Yes	—	—	—
Bateleur	—	—	Yes	—	—	—
Crested Serpent Eagle	—	—	—	—	—	—
Malagasy Marsh Harrier	—	—	—	—	—	—
Lizard Buzzard	—	—	—	—	—	—
Eastern Chanting Goshawk	—	—	—	—	—	—
Crested Goshawk	—	—	—	—	—	—
African Goshawk	—	—	Yes	—	—	—
Little Sparrowhawk	—	—	Yes	—	—	—
Collared Sparrowhawk	—	—	—	—	—	—
Rufous-breasted Sparrowhawk	—	—	—	Yes	—	—
Black Sparrowhawk	—	—	—	—	—	—
Red Goshawk	—	—	—	—	—	—
Crane Hawk	—	—	—	—	—	—
White-necked Hawk	—	—	—	—	—	—

(*Continued*)

Table 7.1 (Continued)

Species	Long-distance migrant[a]	Trans-equatorial migrant[b]	Rains migrant[c]	Altitudinal migrant[d]	Maximum flock size	Maximum distance traveled over water
Mantled Hawk	—	—	—	—	—	—
Great Black Hawk	—	—	—	—	—	—
Black-collared Hawk	—	—	—	—	—	—
Montane Solitary Eagle	—	—	—	—	—	—
Crowned Solitary Eagle	—	—	—	—	—	—
Roadside Hawk	—	—	—	—	—	—
White-rumped Hawk	—	—	—	—	—	—
Puna Hawk	—	—	—	Yes	—	—
Hawaiian Hawk	—	—	—	—	—	—
Rufous-tailed Hawk	—	—	—	—	—	—
Madagascar Buzzard	—	—	—	—	—	—
Augur Buzzard	—	—	—	—	<10	—
Jackal Buzzard	—	—	—	—	—	—
Harpy Eagle	—	—	—	—	—	—
Tawny Eagle	—	—	Yes	—	—	—
Gurney's Eagle	—	—	—	—	—	—
Wedge-tailed Eagle	—	—	—	—	—	—
Little Eagle	—	—	—	Yes	—	—
Ayres' Hawk-Eagle	—	—	Yes	—	—	—
Long-crested Eagle	—	—	—	—	—	—
Changeable Hawk-Eagle	—	—	—	—	—	—
Mountain Hawk-Eagle	—	—	—	Yes	—	—
Black Hawk-Eagle	—	—	—	—	—	—
Ornate Hawk-Eagle	—	—	—	Yes	—	—
Secretarybird	—	—	—	—	—	—
Carunculated Caracara	—	—	—	—	—	—
Mountain Caracara	—	—	—	—	—	—
Striated Caracara	—	—	—	—	—	—
Northern Crested Caracara	—	—	—	—	—	—
Yellow-headed Caracara	—	—	—	—	—	—
Collared Falconet	—	—	—	Yes	—	—
Pied Falconet	—	—	—	Yes	—	—
Greater Kestrel	—	—	—	—	—	—
Bat Falcon	—	—	—	—	—	—
Orange-breasted Falcon	—	—	—	—	—	—
African Hobby	—	—	—	—	<10	—
Grey Falcon	—	—	—	—	—	—
Black Falcon	—	—	Yes	—	—	—
Laggar Falcon	—	—	—	—	—	—

Note: A dash indicates no data.

[a] *Long-distance migrants* are species in which at least 20% of all individuals migrate >1500 km.

[b] *Transequatorial migrants* are species in which at least 20% of all individuals migrate across the equator.

[c] *Rains migrants* are species that routinely migrate in response to seasonal rains.

[d] *Altitudinal migrants* are species in which some individuals migrate from high-altitude breeding areas to lower areas during the nonbreeding season.

With the exception of seabirds, swifts, and swallows, raptors spend more time flying both while searching for prey and while migrating than do most other birds. As a result, energy-efficient flight is important to them. Because raptors have oversized wings for their body mass and are lightly wing-loaded, many are capable of extended, energy-efficient, nonflapping soaring flight. Soaring is an especially efficient form of long-distance travel, both during and outside of migration, and many birds of prey are obligate soaring migrants, species that require soaring flight to complete their long-distance journeys. It is believed that soaring flight initially developed in raptors because it reduced the costs of searching for prey and carrion, however, many species of diurnal birds of prey actually soar more often and for longer periods when migrating than at other times of the year. Almost all truly long-distance migrants are obligate soaring migrants, including New World Broad-winged Hawks and Swainson's Hawks, and Old World Eurasian Honey Buzzards, Lesser Spotted Eagles, and Steppe Eagles.

ORIGINS AND EVOLUTION OF MIGRATION

Birds, including raptors, have been migrating for a long time. Indirect evidence of avian migration dates at least as far back as the Cretaceous period, more than 65 MYA. The occurrence of recognizably "modern" raptors in Europe in the early Eocene epoch, coupled with the fact that many living species of birds of prey regularly undertake long-distance seasonal movements, suggest the likelihood of raptor migration for at least 40 million years. One can only infer a behavior from the fossil record and, even then, only when information about paleoclimates and the ancient juxtapositions of continental landmasses is readily available. It is now clear that many "behavioral patterns", including migration—"behavioral tools" may be a better term—can appear and disappear rapidly in a species, making migration behavior in a particular species difficult to track over evolutionary time. Nevertheless, a growing knowledge of the timing and geography of existing systems of raptor migration, together with careful laboratory studies of the genetics of migration in European songbirds, provide a useful framework for understanding the origins of the phenomenon.

A growing consensus among avian biologists now recognizes that raptor migration is a genetically based, remarkably flexible behavioral response to a combination of both intrinsic and extrinsic factors, including intraspecies competition, flight mechanics, breeding location, continental geography, climate, and the seasonal availability of food resources, together with the population dynamics of the birds themselves.

A genetic basis for migratory behavior suggests that birds, including raptors, have genes for both sedentary and migratory behavior and that these genes, when expressed at the population level, can produce sedentary populations, as well as partial and completely migratory populations. A hypothetical initial sedentary population of raptors living in a nonseasonal or slightly seasonal environment provides the starting point for how this can come about. As long as sedentary young in the population survive and reproduce, the population will remain sedentary. Furthermore, if the number of young that survive to reproductive age equals the number of adults that die each year, the population will remain stable. On the other hand, if the number of young that survive exceeds the number of adults that die, the population will grow. If a population increases to the point that resources in the environment become limiting and individuals increasingly compete for those resources, selection will favor dispersal by some individuals if dispersants are more likely to survive and reproduce than nondispersants. Because intraspecies competition is likely to favor experienced adults over inexperienced young in such situations, juveniles rather than adults are more likely to adopt a dispersal strategy. Thus, not surprisingly, a number of empirical studies indicate that post-breeding-season dispersal is far more common in young raptors than in adults.

At least initially, selection should favor juvenile dispersal into similar, reasonably nonseasonal habitats close to the species' breeding range. If the population continues to grow and spread, however, dispersing juveniles may be forced to occupy more distant and, presumably, less-similar habitats, some of which may be more seasonal (suitable at certain times of the year for food finding, but not at others). At this point, natural selection will favor a strategy in which dispersants into seasonal habitats will remain there for part of the year, but move to other locations at other times of the year (they will become migratory), in response to regional seasonal differences in food availability. Once this occurs, the former sedentary population becomes a partial migratory one.

When such a series of events plays out against latitudinal gradients in habitat seasonality, leapfrog migration, in which migratory individuals breed at higher latitudes than sedentary ones and overwinter at lower latitudes, often results. Overall, this line of reasoning, which draws heavily on the American ornithologist George Cox's earlier models for the evolution of migration behavior, places the origins of raptor migration squarely on the phenomenon of shifting competition for limited resources.

Migration also can evolve in sedentary populations when environmental conditions "extrinsic" to the population shift from nonseasonal to seasonal in all or part of a species' range. On the other hand, partial migration would be expected when both sedentary and migratory individuals in the population are reasonably successful, and complete migration would be expected when sedentary individuals were consistently less successful than migratory individuals.

Large numbers of raptors are difficult to maintain in captivity and, not surprisingly, most experimental evidence for the genetic basis of their migration is inferred from work on smaller birds, particularly songbirds. There is one exception, however. In the 1930s, the German ornithologist Rudolf Drost conducted a series of "displacement" experiments involving migrating adult and juvenile Eurasian Sparrowhawks. The work provides one of the best examples of how genetics and learning together play important roles in determining migration geography in raptors. Eurasian Sparrowhawks are bird-eating accipiters, similar to North American Sharp-shinned Hawks. Northern populations of Sparrowhawks are migratory, and large numbers of Scandinavian breeders migrate along the shorelines of the Baltic and North Seas each autumn en route to wintering areas in southwestern Europe. Drost trapped and banded several hundred of these migrants on an island in the North Sea off the coast of Germany. The birds were captured at a songbird banding station in gigantic funnel traps as they sought cover and prey in trees that had been planted there to attract migrating passerines. The captured birds were aged by plumage and banded. Two hundred and nine of them were transported 600 km (410 mi) east-southeast to eastern Poland, where they were released. A second, control, group was held and subsequently released at the capture site.

Recaptures of several dozen control birds indicated that sparrowhawks released at the capture site migrated southwest and spent the winter in the Netherlands, Belgium, northwestern France, and, in one instance, Portugal. Thirty-six of the displaced individuals also were recaptured in winter. Displaced juveniles had continued to move southwest, and most were recaptured in central Europe, east-southeast of the population's traditional

Box 7.1 Changes in migration behavior in partial migrants

Merlins. Merlins are compact, circumboreal falcons that breed in forested and open habitats across much of the Northern Hemisphere. There are ten subspecies, some of which are more migratory than others and, overall, the species is considered a partial migrant. In the early part of the twentieth century, one of three Northern American subspecies, a pale grey prairie form called the Richardson's Merlin, began to expand the northern limits of its wintering range from Colorado and Wyoming into southwestern Canada. Reports of the expansion placed overwintering Merlins in Saskatchewan in 1922 and in Alberta in 1948. The expansion, which continued well into the latter half of the twentieth century, was especially apparent in urban areas. By 1970, Richardson's Merlins not only were overwintering in cities, but were breeding there as well. Since then, nonmigratory populations of "city" Merlins have appeared in numerous urban areas throughout southern Canada and the northern United States.

Several factors appear to have played a role in this shift from migratory to nonmigratory behavior in this population. The initial northward expansion of winter areas coincided with the regional expansion of the species' predominant urban prey, the House Sparrow, an Old World exotic introduced into North America in the 1850s that had spread into the American West thereafter. It seems likely that increased prey availability, including both House Sparrows and Bohemian Waxwings—the latter being attracted to urban areas by ornamental trees—contributed substantially to the Merlin's wintering farther north. A second factor, declining human persecution throughout the period—in the 1930s, Richardson's Merlin was still characterized as "never allowing [humans] to come within gunshot range"—also may have played a role.

Sharp-shinned Hawks. Sharp-shinned Hawks are small, partially migratory, woodland raptors that breed across most of forested Canada and the United States. Like Merlins, the hawks were once described as vicious bird killers, and the species was heavily persecuted throughout much of the eastern part of its range into the 1960s. Members of the species were particularly vulnerable to shooting during migration, when large numbers concentrated along traditional flyways. This, together with the misuse of organochlorine pesticides in the 1940s–1960s, reduced sharpshin numbers throughout much of mid–twentieth century eastern North America. Fortunately, the widespread use of the organochlorine pesticide DDT was banned in Canada and the United States in the early 1970s, and the sharpshin received full protection under the Migratory Bird Treaty Act in 1972. Not surprisingly, it began to rebound in eastern North America shortly thereafter.

The hawk's secretive nature makes it particularly difficult to find and study during the breeding season, and populations of the species are best monitored during autumn and spring, when large numbers of migrants concentrate along major migration corridors and at migration bottlenecks. When counts of sharpshins reported at migration watch sites in the northeastern United States began to decline in the 1980s, concerns were raised about the species' status. Initial reports of the decline suggested somewhat benign explanations, including the possibility that the species had reached its natural carrying capacity in the region following a period of explosive population growth in the 1970s. As counts continued to decline into the late 1980s and early 1990s, however, more ominous explanations were advanced, including pesticides, acid precipitation (and its impact on the sharpshin's songbird prey base), and the loss of forested nesting habitat. Because the reduction was particularly acute at coastal count sites where juveniles dominated the flight numerically, several observers suggested that reproductive success had declined.

By the early 1990s, conservationists were suggesting that eastern populations of Sharp-shinned Hawks were in "free fall." Oddly enough, declines were not reported at count sites west of the Great Lakes. In the end, it turned out that none of the explanations above was correct. Rather than reflecting a shift in population numbers overall, declines in the counts of sharpshins at eastern migration count sites reflected a shift in the species' migration behavior, for at the same time that numbers of migrating sharpshins were declining at hawk counts throughout much of southern New England and the Middle Atlantic States, their numbers were increasing in early-winter Christmas Bird Counts north of the count sites.

The northward shift in the species' wintering areas appears to be due to a phenomenon migration biologists refer to as migration short stopping. First-year migrants appear particularly inclined to migration short stopping, which occurs when food availability increases along the migratory route and individuals stop or slow their migratory movements to take advantage of it. For sharpshins, the food in question was increased numbers of backyard birds. Observations indicate that sharpshins rank above domestic cats as the most frequently seen predator at backyard bird feeders. Increases in bird feeding throughout the northeastern United States and eastern Canada during the last quarter of the twentieth century and a series of especially warm winters beginning in the 1980s and continuing into the 1990s, coupled with reduced human persecution, appear to have induced the shift. Intriguingly, observations suggest that numbers of Sharp-shinned Hawks at bird feeders in the northeastern United States decline after December. Thus, it remains possible that some sharpshins may simply delay their movements until after the first of the year. Either way, this species' rapid response to changing environmental conditions provides another example of how quickly migration habits can shift when conditions merit.

When my colleagues and I first reported that migratory short stopping rather than population declines were likely responsible for declining numbers of sharpshins at migration watch sites in the northeastern United States, many conservationists challenged our findings. That said, our explanation has stood the test of time. Indeed, after our findings were published in the mid-1990s and migration short stopping was placed on the "radar screens" of other raptor migration scientists, the phenomenon, which is typically linked to climate amelioration or increased food availability in winter in the form of human subsidies, or both, has been described in American Kestrels in the American West, as well as in Red Kites, Marsh Harriers, Short-toed Snake Eagles, and Booted Eagles in Europe.

wintering areas. By contrast, displaced adults directed their movements west upon being released in Poland, and most of their recoveries were substantially closer to the population's "targeted" wintering ground. Drost's results, which are similar to those of displacement experiments involving Common Starlings and Hooded Crows, suggest that naïve, first-autumn raptors possess genetic instructions regarding directional orientation, but lack the navigational skills needed to correct their course if displaced by weather, or scientists. Experienced adults, on the other hand, apparently are able to make midcourse corrections and reestablish themselves relative to their intended destinations.

Displacement experiments similar to Drost's have yet to be performed on other species of raptors, but observations of wind-drifted and storm-deflected migrants suggest that natural displacements such as these also are more likely to affect juvenile than adult raptors, suggesting that learning—as well as genetics—plays an important role in the development and maintenance of migration patterns.

The theory of the evolution and control of bird migration outlined above posits that partial migration maintains sufficient genetic variability in most species of migratory raptors and as conditions change, natural selection can track the changes, and individuals can respond appropriately. Examples of such changes involving Merlins, Sharp-shinned Hawks, and other raptors support this idea.

MIGRATION STRATEGIES

Migrating birds of prey need to accomplish three things: (1) locate appropriate wintering areas on their outbound journeys and locate appropriate breeding areas on their return migrations, (2) time their migrations to coincide with appropriate climatic and ecological conditions both en route and at their intended destinations, and (3) complete their outbound and return migrations in a physical condition that leaves them fit and capable of successfully overwintering and reproducing. Although different species face these challenges differently, there are common threads to successful "flight strategies" during migration, including how atmospheric considerations affect their migrations and how flight mechanics affect their flight behavior.

Atmospheric Considerations

Earth's atmosphere provides raptors with three ingredients necessary for successful long-distance migration: oxygen for metabolism, a gaseous medium for generating lift during flapping and soaring flight, and vertical updrafts to generate lift for soaring flight. A fourth atmospheric component, horizontal wind, can help or hinder migrants.

Although the atmosphere extends approximately 120 km (75 mi) above the surface of the earth, most of its mass is within 5 km (3 mi) of the surface. This narrow layer of dense air is where most "weather," including thermals, deflection updrafts, and horizontal winds, occurs, and it is here that temperatures and oxygen pressures are most similar to those at the earth's surface. It also is where most migration occurs.

Visual observations of migrants at watch sites indicate that many raptors migrate within several hundred meters of the surrounding landscape. Radar studies and studies in which migrants were followed in fixed-wing aircraft, however, suggest that many others travel at heights that make them impossible for observers to see from the ground. More recent studies involving satellite-tracked birds confirm this. The American raptor migration specialist Paul Kerlinger used radar to assess the altitudes at which raptors migrate at several sites in the United States. Radar tracks of spring migrants in New York State indicate that most typically travel between treetop level and about 1500 m (one mi) above the ground, and that more than 80% of migrants travel within 1000 m (0.6 mi) of the surface. A study of migrants on the Gulf Coast of Texas produced similar results, as did one in southern Israel, where the upper limit of the flight was 2000 m (1.2 mi), and where >90% of the migrants flew within 1000 m (0.6 mi) of the ground. In the latter study, Levant Sparrowhawks—one of only a handful of raptors known to migrate at night as well as by day—flew at up to 3000 m (1.8 mi) at night in the same area. On the other hand, all but one of ten species of raptors migrating above forested Alaska, near the beginnings and ends of their migrations, flew at mean altitudes <150 m (500 ft)—the single exception being the Golden Eagle, which averaged 230 m (750 ft)—probably because they were searching for food, which is typical of migrants near the starts and finishes of their journeys.

Raptors also are known to travel at significantly greater altitudes. Turkey Vultures, Broad-winged Hawks, and Swainson's Hawks followed in fixed-winged aircraft in tropical Panama migrated at up to 4000 to 5000 m (13,000–16,000 ft) above the ground. And Bald Eagles migrating across Canada and the western United States regularly reach altitudes of up to 4500 m (14,500 ft). Old World vultures sometimes travel at even greater heights. Radio-tracked Griffon Vultures are known to soar at up to 10,000 m (6 mi), and a Rüppell's Vulture was struck by an airliner over

West Africa, traveling at an altitude of 11,300 m, or about 7 mi high, at the time of the collision.

High-altitude flight in frigid, thin air is both physiologically and aerodynamically challenging for raptors. One problem is that "normal," low-altitude, hemoglobin cannot capture, transport, and release oxygen effectively in thin air. Unlike most birds, Rüppell's Vultures have at least two adaptations to exceptionally high flight. The first is a conspicuous thick layer of dense insulating down covering the lower surface of the patagium, mentioned in Chapter 2. The second is their ability to produce four types of hemoglobin, two of which function effectively at low partial pressures of oxygen encountered at high altitudes, and two of which function best at lower altitudes. It is not known whether other high-flying migratory raptors, such as Himalayan Vultures, possess this significant metabolic adaptation.

Thermals and Deflection Updrafts

Because raptors migrate where most weather occurs, meteorological phenomena are of considerable significance to migrating birds of prey. Updrafts, or vertical winds, provide numerous opportunities for low-cost soaring flight and almost always assist migrants during their travels.

Thermals

Thermals are isolated pockets of warm, rising air that form when different surfaces in landscapes receive and absorb different amounts of sunlight. Dark surfaces have lower reflectivity indexes, or surface albedos, than do light-colored surfaces, and absorb more solar radiation as well. Dry surfaces, including rocky outcrops and parched fields, heat more quickly than do "evaporatively cooled" wet surfaces, such as water and living vegetation. In hilly terrain, surfaces that are oriented perpendicular to incoming solar radiation—east-facing mountain slopes in the morning and west-facing slopes in the afternoon, for example—warm more quickly than do shaded and less-perpendicular surfaces.

Land-based thermals typically require sunlight to sustain them. Because of this, the strongest and largest thermals typically occur on bright, sunny days between midmorning and midday, after the sun is high enough to differentially warm the landscape and create thermals, but before strong afternoon horizontal winds begin to tear them apart. Because land-based thermals are fueled by sunlight, thermals are stronger in summer than in winter in the Temperate Zone, and are stronger overall in the tropics. Also, because the sun rises more vertically and, thus, more quickly in the tropics, thermals form earlier in the morning and extend to later in the afternoon there.

Thermals are most useful to migrants when they are wide enough for raptors to circle within them and strong enough to enable individuals to reach heights sufficient so that they can glide in the intended direction of migration to the next thermal.

Land-based thermals routinely reach heights of more than 2000 m (6600 ft) outside of the tropics, and higher still within them. Although the diameters of some thermals exceed 1000 m (3300 ft), in the Temperate Zone they typically range in size from several meters in diameter to several hundred meters. Stronger thermals tend to be taller and wider than weaker thermals. Thermal strength (the speed at which air rises within the thermal) varies predictably within each individual cell, and decreases significantly both with height above and with distance from the center of the cell. Not surprisingly, raptors that soar within thermals typically seek out the cores of the vortices and circle within them, and typically leave thermals before reaching the top.

Although most thermals are isolated from one another, thermal streets, or linear arrays of thermals aligned with the prevailing wind or with shorelines, are sometimes created during periods of intense solar radiation, particularly in the tropics. The locations of thermal streets often are apparent by the linear arrays of clouds that form above them. Under certain circumstances, particularly large and powerful thermal updrafts can extend for hundreds of kilometers. When such formations are aligned along the principal axis of migration, migrants are able to soar linearly within them in much the same way as they slope-soar along similarly aligned deflection updrafts that form above mountain ridges.

Thermals typically form over land, but not always. Indeed, the strongest thermals on earth occur over the Gulf Stream off the eastern seaboard of the United States, where warm surface water from the Gulf of Mexico, surrounded by cooler water, creates thermal updrafts that sometimes extend more than 10 km (33,000 ft) above the Atlantic Ocean. A satellite-tracked Peregrine Falcon used these thermals, together with a tailwind created by a passing hurricane, to complete a more than 1500-km (930-mi), 25-hour overwater journey from southern New Jersey to south Florida in mid-October 2008, a record that stands as the longest and most rapid overwater journey of any known migrating peregrine.

An adult Barred Forest-Falcon, an accipiter-like, but falcon-related, Neotropical species that demonstrates considerable regional variation in appearance. Note the whitish iris of this individual, which is typical of Barred Forest-Falcons in the southern Andean portion of its range. (Bolivia; photo by Sergio Seipke)

An immature Crowned Solitary Eagle. This seldom-seen large raptor is scarce to rare throughout its South American range. See Chapter 4 for details on rarity. (Argentina; photo by Sergio Seipke)

An adult Black Hawk-Eagle, a slender, medium-sized Neotropical "eagle." Note the forward facing eyes that serve to enhance binocular vision. See Chapter 1 for details regarding the names of raptors, and Chapter 3 for details on vision. (Ecuador; photo by Sergio Seipke)

An adult female Snail Kite carrying an apple snail in its talons. Unlike many raptors, which are dietary generalists, Snail Kites are dietary specialists that feed almost entirely on apple snails. See Chapter 6 for details on diet. (Argentina; photo by Sergio Seipke)

A juvenile Snail Kite. Note the bird's slender, curved, upper mandible, which has evolved to allow this raptor to slice off and extract apple snails from their shells. (Argentina; photo by Sergio Seipke)

An adult Grey-faced Buzzard on autumn migration in southern Thailand. When photographed, this bird was traveling southbound along the Malay Archipelago en route to an unknown wintering area, most likely in Malaysia or Indonesia. See Chapter 7 for details on migration geography. (Thailand; photo by Sergio Seipke)

An adult Black Baza migrating with an insect in its talons. Many insectivorous diurnal birds of prey catch insects in the air and feed on them while migrating. See Chapter 7 for details on feeding during migration. (Thailand; photo by Sergio Seipke)

Two adult Black Bazas roosting near the Chumphon raptor-migration watch site in southern Thailand. These curious-looking, largely insectivorous birds of prey, which migrate through the region in large flocks and roost communally in the region's forests at night, are believed to mimic cuckoos. See Chapter 2 for details on plumage mimicry in birds of prey. (Thailand; photo by Sergio Seipke)

An adult Red-throated Caracara looking at a wild bee. This species feeds mainly on bees, wasps, and hornets, whose nests it crashes into at high speeds to break open before feeding on the inhabitants. See Chapter 6 for details on feeding behavior. (Brazil; photo by Sergio Seipke)

An adult Crane Hawk picking up a pebble. Many raptors ingest small stones, or "rangle," to help them digest food mechanically in their digestive tracts. The Crane Hawk's long feet and hyperflexible, double-jointed legs allow it to probe deeply into the nesting holes and hiding places of small birds and animals. (Brazil; photo by Sergio Seipke)

A captive, recently fledged Common (aka Rock) Kestrel. The postfledging period is considered by many to be the most dangerous period in the bird's life. Note the tomial tooth on the bird's upper mandible. See Chapter 2 for details on fledglings. (South Africa; photo by Sergio Seipke)

An adult male Amur Falcon on the wintering grounds in Africa. Amur Falcons undertake what many consider to be the most arduous migratory journeys of all raptors. See Chapter 7 for details. (South Africa; photo by Sergio Seipke)

An adult Jackal Buzzard with abnormal plumage. Note the unusual white feathers on the top of the head, upper back, and sides of its neck and chest. See Chapter 2 for details on abnormal plumage. (South Africa; photo by Sergio Seipke)

An Osprey carrying a fish while migrating. The forward-facing orientation of the fish helps the Osprey aerodynamically. (Florida; photo by Shawn P. Carey)

Three adult Black Vultures. Although most raptors are not social, many obligate scavenging birds of prey are, both within and outside of the breeding season. (Florida; photo by Shawn P. Carey)

An adult Black Harrier. Like other harriers, this bird has owl-like facial disks that funnel sounds to its external ears. Field tests involving Northern Harriers support the hypothesis that harriers locate their prey by sound as well as by sight. See Chapter 3 for details. (South Africa; photo by Robert E. Simmons)

A wing-tagged juvenile White-headed Vulture photographed the day after the wing tag and a satellite-tracking device had been placed on the bird in Kruger National Park. The brown, as opposed to white, feathering on the head identifies the bird as a juvenile. (South Africa; photo by André Botha)

A parental adult Hooded Vulture (lower right) and a recently fledged juvenile Hooded Vulture. The shorter (and still growing) upper bill on the juvenile is typical of many recently fledged birds of prey. Note the large external ear opening behind the eye of both birds. Although the supposition has yet to be tested, some believe that Hooded Vultures locate carcasses by listening for hyenas and other predators bone cracking and squabbling over their kills. (South Africa; photo by André Botha)

A recently fledged Striated Caracara flying overhead at close range while "inspecting" the photographer. Although almost all birds of prey are relatively shy, Striated Caracaras routinely approach humans to within several meters, making them perfect subjects for photographs. Why they do so is unclear, although some believe that it may be searching for food the humans may be carrying. See Chapter 6 for details. (Falkland Islands [Malvinas]; photo by Scott Weidensaul, © 2011)

A photo collage of an adult male (right) and an adult female (left) Bateleur soaring above Kruger National Park. Bateleurs display plumage dimorphism in adults, with the white on the underwing of the female extending farther on the wing than on the male. (South Africa; photo by Sergio Seipke)

In addition to Gulf Stream thermals, oceanic (or sea) thermals also occur, and they, like land-based thermals, play a critical role in shaping the geography of long-distance raptor migration over water. Sea thermals regularly form in tropical and subtropical oceans and seas 5° to 30° north and south of the equator in the Trade Wind Zone. Within these latitudinal belts, predominant northeasterly winds in the Northern Hemisphere and southeasterly winds in the Southern Hemisphere blow relatively cool, subtropical surface air toward the equator. As they do, the increasingly warmer surface waters of the Equatorial Zone heat the cooler air, producing bands of sea thermals. Because the temperature differential between the cooler trade winds and the warmer sea exists both day and night, sea thermals occur 24 hours a day, meaning that migratory raptors can use them to subsidize their flights both day and night. The existence of these predictable oceanic updrafts allows several species of seabirds, including Magnificent Frigatebirds, to travel long distances during multiday foraging flights. Sea thermals also allow Chinese Sparrowhawks, Grey-faced Buzzards, and other migratory raptors to soar hundreds of kilometers over water along the East Asian Oceanic Flyway (Map 7.1). They also make it possible for the Amur Falcon to undertake the longest overwater passage of any migratory raptor, a seemingly implausible 4000-km (2500-mi)

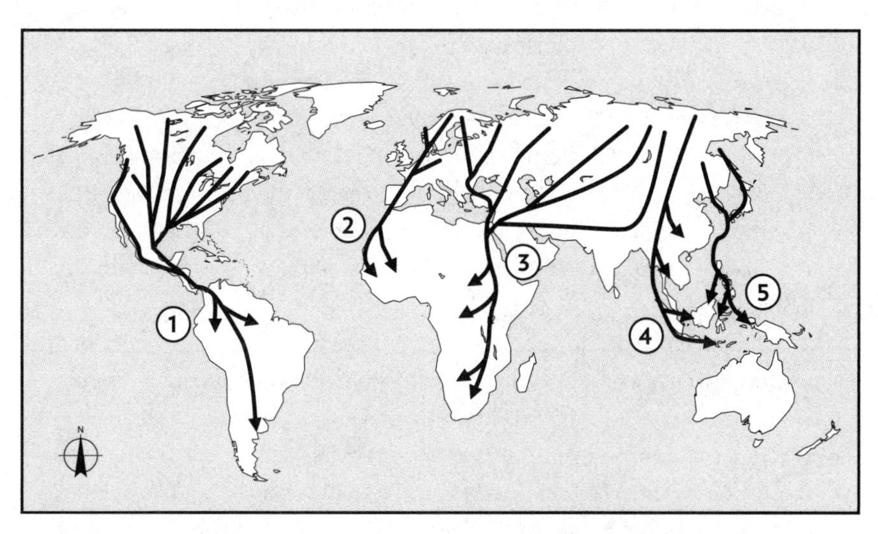

Map 7.1. The world's principal raptor-migration flyways. (1) Trans-American Flyway (2) Western European–West African Flyway (3) Eurasian–East African Flyway (4) East-Asian Continental Flyway (5) East-Asian Oceanic Flyway (Updated from Bildstein 2006.)

Box 7.2 Thermal soaring in the Middle East

One of the most thorough studies of thermal soaring in raptors was conducted by the Swiss ornithologist Reto Spaar and others, who "radar-tracked" the individual flight paths of more than 2000 migrants traveling above the Arava Valley and the Negev Desert of Israel during several springs and autumns in the early 1990s. Soaring conditions and flight behavior varied predictably over the course of each day. Most of the migrants, which included several species of harriers, Eurasian Honey Buzzards, Steppe Buzzards, and Steppe Eagles, began each day migrating in relatively weak thermals that usually formed within several hours of sunrise. On most days, climbing rates increased through the morning, before peaking at a little over 3 meters per second (7 mph) by late morning, and began to decrease by late afternoon. Overall, climbing rates for individual species averaged from 1.5 to 2.1 meters per second (3.5 to 5.0 mph). Some of the birds climbed at rates of up to 5 meters per second (11 mph) during strong midday updrafts.

Migrants ascended to greater heights when their climbing rates were faster, indicating that they were able to determine the strengths of thermals they were in, remaining in stronger thermals longer than in weaker ones. Observations of individuals circling as few as one or two times in weak thermals before gliding in the direction of another updraft suggest that the birds were able to determine the strength of a thermal quickly, and that they actively searched for stronger ones.

Migrants soaring over the valley often climbed to 2000 m (6600 ft), whereas those traveling across the Negev Highlands rarely exceeded 1000 m (3300 ft). On particularly sunny days, the characteristic pattern of individuals alternating between circle soaring in thermals and directed, straight-line gliding between thermals, was replaced with directed, straight-line soaring as the migrants made use of larger-scale thermals and thermal streets.

Within- and between-species comparisons of two of the most numerous soaring migrants at the site, Steppe Eagles and Eurasian Honey Buzzards, led Spaar to conclude that at least for these two migrants, the behavior of the two species was more similar under similar weather conditions than was the behavior of each species under different weather conditions. Even so, Spaar recorded consistent species differences in the flight behavior of several of the raptors he watched. Steppe Eagles, for example, whose wingspans are 40% greater than those of Eurasian Honey Buzzards, rarely soared in the weak early-morning and late-afternoon thermals, whereas Eurasian Honey Buzzards regularly did. Larger raptors tend to have larger minimal turning radiuses than do smaller birds, and weaker early- and late-day thermals have smaller diameters than stronger midday thermals, and the difference in flight behavior between these two species most likely is related to the difference in their size.

My own observations of migrants both in the region and elsewhere reveal similar body-size affects. In southern Israel, small Levant Sparrowhawks circling in thermals together with larger Steppe Buzzards pivoted more tightly and climbed at greater rates. And the same relationship repeats itself among migrating Broad-winged Hawks, Swainson's Hawks, and Turkey Vultures soaring together in thermals in Costa Rica and Panama. The smallest of the three, the Broad-winged Hawk, turns in tighter circles and climbs more rapidly than the intermediate-sized Swainson's Hawk, which, in turn, circles more tightly and climbs more rapidly than the larger Turkey Vulture. That smaller migrants outperform larger migrants in their use of weaker thermals has important implications for the daily distances covered by different species.

outbound passage across the Indian Ocean between India and East Africa, mentioned earlier in this chapter. Finally, these thermals, together with northeasterly trade winds, also allow Ospreys to cross the Caribbean from eastern Cuba and western Hispaniola to the South American mainland of northwestern Venezuela and northeastern Colombia.

In sum, it is difficult to overstate the extent to which both land- and sea-based thermals affect long-distance raptor migration. Every one of the world's twenty-two species of complete migrants soars within thermals at least occasionally when migrating. And more than half of these species, including the Eurasian Honey Buzzard, Mississippi Kite, Grey-faced Buzzard, Chinese Sparrowhawk, Broad-winged Hawk, Swainson's Hawk, Lesser Spotted Eagle, and Greater Spotted Eagle, depend on this type of flight to complete their long-distance journeys.

Deflection Updrafts

In addition to soaring in thermal updrafts, raptors also soar in deflection updrafts, which occur when horizontal winds strike surface discontinuities, including mountains, buildings, and tall vegetation, and are deflected up and over them. As is true of thermals, deflection updrafts can be too small or too weak for migrants to soar within them. They also can be too turbulent for migrants. Many deflection updrafts, however, create ideal soaring conditions, and at such times, large numbers of migrating raptors are attracted to them. Deflection updrafts rarely extend more than 300 to 400 m (1000 to 1300 ft) above the surrounding landscape, and migrants that soar within them often are more visible to observers on the ground than are migrants soaring in thermals. Mountain ridges, gorges, cliffs, and elevated coastlines often provide the best sources of deflection updrafts.

New World migrants use deflection updrafts in the Appalachians of eastern North America, the Rockies of western North America, the Sierra Madre Oriental of eastern Mexico, the Talamanca Mountains of Costa Rica, and the Andes of northwestern South America. Old World migrants use them in the Alps of Switzerland and France, the Tien Shan and Hindu Kush of central Asia, and the southern Himalayas of Nepal and India, as well as along the escarpments of the Great Rift Valley of East Africa and the Middle East.

Migrants use deflection updrafts in three ways: first, by alternately circle soaring within them and then gliding to the next updraft along the region's principal axis of migration, second, by straight-line, slope soaring when mountain ridges produce linear arrays of updrafts that parallel the principal axis of migration, and third, by soaring at great altitudes in lee waves created by deflection updrafts. The latter are standing, or stationary, waves that form on the lee, or downwind, sides of mountain ranges, under specific conditions of wind and temperature. Although lee waves sometimes extend 10,000 m (33,000 ft) above the mountains that create them, most form at 2000 to 3000 m (6600 to 9900 ft) above the landscape. Griffon Vultures and other soaring migrants use lee waves to help cross the 14-km (9-mi) wide Strait of Gibraltar between southern Spain and Morocco.

In some situations, including near the beginnings of their migrations when migrants are more likely to be deflected from a preferred direction of travel by local conditions, individuals sometimes slope soar within deflection updrafts for relatively short distances, even when doing so may take them off course. When this happens, migrants are said to be following a leading line.

Leading lines differ from diversion lines, such as land-water interfaces and lowlands adjacent to high mountain ranges, along which migrants concentrate not because they are attracted to them, but because they are trying to avoid what lies beyond. Migrants will follow diversion lines for hundreds of kilometers or more, far longer than most leading lines, even to the point of reversing their direction of travel along peninsulas and erratic forest-desert edges. Because of their substantial regional and sometimes continent-wide effect in diverting and concentrating large numbers of migrants along narrow flight lines, diversion lines frequently are associated with migration bottlenecks, such as the Strait of Gibraltar and the Bosporus, at the western and eastern ends of the Mediterranean Sea, respectively. By comparison, leading lines typically divert migrants for much shorter distances, sometimes for as few as several kilometers (Box 7.3).

Box 7.3 The Central Appalachian Mountains and leading lines

The northeast-to-southwest oriented Central Appalachian Mountains of eastern Pennsylvania form one of the better-known leading lines for migrating raptors. Autumn cold fronts, which typically pass through the region at 4- to 5-day intervals, approach the Appalachian's even-topped "corduroy ridges" at right angles, and their northwest winds are deflected up and over them. The result is a series of parallel leading lines for outbound migrants. The Kittatinny Ridge, the southeastern-most mountain in the region, creates the strongest leading line in eastern Pennsylvania, most likely because it offers the last opportunity for southbound migrants to slope-soar there. Ridge adherence along this leading line appears to be considerable, with as many as 95% of all migrants traveling along it continuing to do so after reaching a 1.3-km (0.8-mi) wide "water gap," or break in the ridgeline. Even so, simultaneous counts at multiple sites along the Kittatinny suggest that many migrants seen along the ridge remain on it for several kilometers to several dozen kilometers, rather than for most of its 400-km (250-mi) length, and that this is true even on days when excellent opportunities for slope soaring prevail.

Several factors, including the species involved, the strength of the deflection updrafts being produced, and the coincidental occurrence of off-ridge thermals, are known to affect the extent to which mountain ridges such as the Kittatinny function as leading lines. Migrants that pass through eastern Pennsylvania in August and September, for example, thermal-soar more and slope-soar less than do later migrants, apparently because of the availability of strong late-summer thermals then. And whereas watch sites in the region typically record their best late-season flights during periods of strong northwest winds, strong early-season flights are most common on days when strong thermals form and winds are light or variable, or on days when northeasterly tailwinds dominate. In fact, the largest single-day count in the 80-year history of Hawk Mountain Sanctuary, the oldest watch site on the Kittatinny, occurred on a somewhat cloudy day with east-northeasterly winds, during which more than 20,000 migrants were counted crossing the ridge on a broad front on weak tailwinds—apparently in pursuit of the drifting thermals—rather than slope soaring along it.

A good example of seasonal shifts in the use of leading lines by raptors involves the flight behavior of Red-tailed Hawks at Hawk Mountain Sanctuary. A long-term analysis of the species' flight at the site indicates that passage rates peak during the two days following the passage of a cold front. At such times, clear skies and moderating temperatures create ideal conditions for the formation of thermal updrafts, and coincidental northwest winds produce abundant deflection updrafts, which suggest that Red-tailed Hawks are more likely to migrate when opportunities for soaring flight increase. But the story does not end there. Southbound Red-tailed Hawks migrate through eastern Pennsylvania from late summer through early winter each year, and August and September cold fronts bear little resemblance to those that pass through the region in November and December. And a 1997 study of Red-tailed Hawk migration at the site revealed that early-season cold fronts enhance redtail movements principally by creating opportunities for cross-ridge thermal soaring, whereas late-season fronts do so by creating opportunities for ridgeline slope soaring.

Taken as a whole, these and other observations suggest that leading lines are better thought of as linear patches of favorable habitats and atmospheric conditions along which migrants concentrate for brief periods while taking advantage of local flight, feeding, or roosting conditions, and not long-distance "highways in the sky."

Horizontal Winds

The extent to which horizontal winds are aligned with the regional principal axis of migration determines whether they help or hinder the passage of migrating raptors. Tailwinds, or winds that are aligned with the preferred direction of travel, help migrants by "pushing" them forward. Head winds, or winds aligned against the preferred direction of travel, hinder migrants by "pushing" them backward. Crosswinds, or winds that intersect the preferred direction of travel at perpendicular and near perpendicular angles, can hinder migrants by shifting the intended direction of travel via a phenomenon known as wind drift. Field evidence suggests that migrating raptors often alter their flight behavior to take advantage of favorable winds, as well as to reduce the costs of unfavorable winds.

In general, wind speed increases with height. Because of this, raptors migrating into strong head winds fly lower and those migrating with strong tailwinds fly higher than those migrating in light or variable winds. Similarly, raptors migrating in coastal areas tend to fly lower than those migrating along inland corridors, presumably to reduce the risk of being blown out to sea. This is particularly true of raptors about to undertake water crossings in crosswinds.

The direction, force, and regularity of a region's prevailing winds change with latitude. Overall, prevailing winds are westerly between 30° and 60° latitude north and south of the equator, and easterly north and south of these two belts. The predictable nature of these large-scale latitudinal belts of easterly and westerly winds is brought about by two phenomena: differential heating of earth's surfaces by the sun, and the Coriolis effect (aka Coriolis force), which is caused by inertia and the fact that the earth rotates on a north-south axis.

The differential heating of the earth's surface means that temperatures are higher and thermals are stronger near the equator than they are poleward. As warm air rises near the equator, it is replaced by relatively cooler surface air from the north and south. The warm rising air, in turn, cools as it ascends through the atmosphere and begins to spread poleward. Eventually, the twin streams of high-level air cool sufficiently and descend to the surface at approximately 30° north and south of the equator. They then turn toward the equator as surface airstreams. As they do, the earth's daily rotation "deflects" these winds to the right in the Northern Hemisphere and to the left in the Southern Hemisphere, via the Coriolis effect.

Because Temperate Zone air is southbound in the Northern Hemisphere and northbound in the Southern Hemisphere, the deflection is westward in both hemispheres, resulting in parallel bands of easterly trade winds 5° to 30° north and south of the equator. The Coriolis effect again "deflects" these near-surface parcels of air to the right in the Northern Hemisphere and to the left in the Southern Hemisphere, which, because the air masses are moving poleward, means eastward in both hemispheres. This, in turn, creates the midlatitude westerlies, two regions of predominant winds 30° to 60° degrees north and south of the equator.

The breeding and wintering areas of most migratory raptors are aligned along north-south axes, and the existence of latitudinal bands of alternating predominant westerly and easterly winds presents a migration challenge for species prone to wind drift. This is particularly true for long-distance, transequatorial migrants, whose outbound and return movements traverse several of these bands. Although wind drift can occur at any stage of migration, field observations suggest it is more likely to occur earlier rather than later in migration, and that migrants are most likely to compensate for wind-drift displacement toward the end of their migrations. Some researchers have suggested that migrants purposefully allow themselves to be wind drifted when doing so allows them to complete their journeys more quickly and at a lower cost than by delaying flights during periods of crosswinds, or by correcting for them continually en route.

Such behavior, called adaptive wind drift, likely explains the circuitous "elliptical" round-trip journeys of several long-distance migrants whose outbound journeys in autumn initially are influenced by Temperate Zone westerlies, and whose return journeys in spring initially are influenced by trade wind easterlies. In North America, for example, large numbers of Broad-winged Hawks breeding in eastern Canada and the northeastern United States are drifted southeast in autumn along flight lines that sometimes take them east of the Appalachian Mountains, whereas in spring, the same individuals, which are drifted west by easterlies in the southern United States, pass northwest of the Appalachians along the southern shorelines of Lakes Erie and Ontario. In the Old World, Pallid Harriers from eastern Europe migrate south to their sub-Saharan wintering grounds through Turkey and the Middle East while circumventing the eastern Mediterranean, but return to their breeding grounds along a more western central Mediterranean route via Sicily, southern Italy, and the Balkans. And in the Middle East, outbound Steppe Eagles fly south along the Arabian Peninsula east of the Red Sea before crossing into Africa, whereas return migrants in spring fly north through Egypt west of the Red Sea, before turning east at Sinai.

Observations suggest that juveniles are more prone to wind drift than are experienced adults. Satellite-tracked adult and juvenile Ospreys and Eurasian Honey Buzzards migrating between nesting areas in northern Europe and wintering grounds in Africa, for example, indicate that first-year migrants are wind drifted in strong crosswinds, whereas adults compensate and are drifted only one-third as much. Similarly, coastal flights of Sharp-shinned Hawks in eastern North America are dominated overwhelmingly by juveniles, whereas inland flights are not, presumably because juveniles are more likely to be drifted by the region's prevailing northwest winds. Large numbers of juvenile Broad-winged Hawks overwinter in peninsular Florida, far north of the species' principal overwintering areas in Central and South America and substantially east of the species' typical North American migration route, suggesting that first-year migrants in this species, too, are more likely to be blown off course than are more experienced adults.

Intriguingly, first-year Sharp-shinned Hawks banded on migration along a coastal flyway in New Jersey often are recaptured migrating along inland routes in later years, and first-year Broad-winged Hawks banded while overwintering in Florida have been recovered as adults in southern Mexico and Guatemala along the species' principal migration corridor. This suggests that juveniles learn from their first-year mistakes and correct for them in later years.

Flocking

Almost all species of migratory raptors flock at least occasionally while migrating. Most flocks are small single- and mixed-species groups of a couple to a dozen or so birds, most occur along major flyways, migration bottlenecks, and other points of concentration, most last for several minutes to less than an hour, and most appear to be little more than avian traffic jams that result when numbers of raptors travel in the same direction at the same place at the same time. On the other hand, several species flock on a more regular basis and on a much larger scale. In such species, flocking appears to be an essential aspect of successful migration. A few

Box 7.4 Superflocking in Broad-winged Hawks

Broad-winged Hawks begin coalescing into small groups of several birds to several hundreds of birds almost as soon as they initiate their migrations across eastern Canada and the northeastern United States. A few of these high-latitude flocks—particularly those diverted by the northern shorelines of the Great Lakes—number in the tens of thousands. Flocks of broadwings are so predictable that hawk-watchers have a special term for them: kettles. Most of North America's approximately 2 million Broad-winged Hawks overwinter in the Neotropics, and by the time the broadwing flight reaches southern Mexico, flocks have swelled substantially, with some exceeding 50,000 birds. Except for several minutes of flight in early morning and late afternoon when birds are departing from and returning to their roosts, all but a few broadwings migrate in large flocks for at least several weeks. Although such flocks are no longer surprising, early observations of similar events appear to have baffled early observers. The late nineteenth century–early twentieth century ornithologist Frank Chapman, for example, once described a springtime passage of what were probably Broad-winged Hawks in the Sierra Madre Oriental of Veracruz, Mexico, as resembling "a swarm of bees circl[ing] about and among each other in a most remarkable and confusing fashion" that "in spite of their wheeling . . . passed rapidly northward and were soon out of sight."

Researchers have attempted to explain these enormous flocks in several ways. One hypothesis suggests that traveling in large flocks allows migrants to find thermals and other updrafts more quickly than when traveling alone, thereby reducing the cost of migration by increasing the time spent in soaring flight. A second suggests that large flocks form because inexperienced juvenile migrants cluster around and follow more experienced adults. A third suggests that social attractions are not involved at all, and that large groups are simply the consequence of crowded flyways and the patchy distribution of thermals (birds aggregate in large flocks simply because that is where the updrafts are). The first two hypotheses, which are not mutually exclusive, suggest that flocking improves the chances of birds successfully completing their migrations; the third does not. Below, I discuss evidence for each hypothesis.

The idea that migrants might be attracted to one another because traveling in flocks helps individuals find and use updrafts more efficiently has been around for some time. Powered, as opposed to soaring, flight is expensive. Most estimates place the metabolic cost of flapping flight at well over ten times that of resting metabolism, whereas soaring flight has been estimated to be no more than 2.5 times as costly as resting metabolism, making it less than 25% as expensive as flapping flight. (Actually, in Turkey Vultures, soaring flight appears to be no more costly than resting metabolism.)

I have calculated the energetic consequences of soaring on migration in Broad-winged Hawks, and they are considerable. By my calculations, broadwings burn 65 to 75 calories a day, both during the breeding season and on the wintering grounds. Assuming that an individual broadwing flies all the way from southern Canada to central Brazil at its most efficient air speed of approximately 50 to 55 km per hour (30 to 35 mph), it would need to travel 2 to 3 hours a day to complete its journey of 6500 to 8000 km (4000 to 5000 mi) in two months. Migration using powered, or flapping, flight almost doubles the bird's metabolic needs each and every day of the flight. A broadwing using this method needs to find more food in less time, and in unfamiliar territory, than it does either when breeding or overwintering. On the other hand, assuming that the bird uses soaring rather than powered flight to complete 80% of its migration—which seems reasonable given what is known about the species' flight behavior en route—its metabolic needs increase by only about 20%, which, while not inconsequential, are more likely to be met.

North American hawk-watchers have known for some time that the sizes of flocks that broadwings travel in are related to the magnitude of the day's flight. At Hawk Mountain Sanctuary, for example, on days in which a hundred or so broadwings are seen, flocks range in size from several birds to several dozens of birds, whereas on days in which a thousand or so broadwings are seen, flocks range in size from several dozens of birds to several hundred birds. Observations in the Neotropics suggest that the same thing also happens there, albeit at a much larger magnitude. Observations in Caribbean-slope Costa Rica at the Kéköldi watch site, where the broadwing flight exceeds 500,000 birds annually, indicate that flocks swell enormously on high-magnitude flight days, and that flocks of more than 10,000 birds are far more common on days when 25,000 or more broadwings are counted than on days when fewer of these migrants pass.

A recent study of the flight behavior of broadwings at the Kéköldi hawk watch tested the extent to which flock size influences soaring behavior in the species. Then–Sherbrooke University students Vincent Careau and Jean-François Therrien tracked more than 2000 soaring broadwings for 30 seconds each while recording the number of times the bird flapped, as well as whether it was traveling within a flock and, if so, flock size. On average, broadwings flapped less than five times during each 30-second observation compared with the seventy to eighty flaps that would be expected during full powered flight. Moreover, individuals in flocks flapped significantly less than those that traveled alone, and individuals in larger flocks flapped significantly less that those traveling in smaller flocks even when flock size exceeded 1000 birds. Overall, these observations suggest that Broad-winged Hawks soar more efficiently when flying in flocks than when flying alone and that soaring efficiency increases with flock size. The apparent benefits of flocking behavior for broadwings, however, do not end there.

A two-year study of broadwing flocking behavior at Hawk Mountain Sanctuary indicated that adult Broad-winged Hawks are significantly more likely to fly in flocks than are juveniles (78% of all individuals versus 62%), and that adult Broad-winged Hawks are more likely to fly in larger flocks. The study also showed that when adults and juveniles glided between thermals in mixed-age flocks, adults were more likely than juveniles to be in the leading half of the flock. These findings suggest that juveniles have fewer flocking skills than do adults, and that juveniles are more likely to follow adults into flocks than vice versa. Following experienced adults probably assists juveniles in two ways: first, by helping them in finding thermals faster, and second, by helping them find appropriate wintering areas.

Taken as a whole, it appears that superflocking helps migrating Broad-winged Hawks two ways: first, by enhancing their ability to find good soaring space and, second, by helping inexperienced migrants locate appropriate wintering areas.

superflocking migrants predictably coalesce into groups of hundreds to tens of thousands of birds that, once formed, can remain together for days or even weeks. All but one of the eighteen known superflocking migrants are long-distance, transequatorial travelers, and most superflocks form in and around the tropics. North America's Broad-winged Hawk is a well-studied example of this type of obligate flocking.

Flocking appears to help migrants by enhancing their ability to find good soaring space, and by helping inexperienced migrants locate appropriate wintering areas. There is no evidence to suggest that superflocking is simply the result of crowding along major migration corridors. And indeed, there is substantial evidence to the contrary. Most superflocking migrants, including Eurasian Honey Buzzards, Chinese Sparrowhawks, Levant Sparrowhawks, and Broad-winged Hawks, have acutely timed migrations in which most of each year's flight passes within a week or two, and, sometimes, within a single day, whereas nonflocking migrants have far-more protracted flights. This, together with the fact that superflocking migrants begin assembling into flocks almost as soon as they begin to migrate, and that they travel in flocks of increasing size long before they reach significant bottlenecks, suggests that the birds themselves, and not particularly narrow migration passageways, create the supersized flocks in which they migrate.

Superflocking can have a downside, however, particularly when birds travel in mixed-species groupings. Although individuals in single-species flocks rarely collide, species differences in flight mechanics create conditions for numerous "near-misses," as well as collisions within mixed-species flocks. The problem is especially obvious at watch sites along major bottlenecks in Latin America and the Middle East.

In Mexico and Central America, migrants traveling in enormous mixed-species flocks dominated by Turkey Vultures, Swainson's Hawks, and Broad-winged Hawks sort themselves by species when streaming between thermals or when flying unidirectionally within large thermal streets. In such situations, agile and buoyant Broad-winged Hawks stream above the somewhat less buoyant Swainson's Hawks, which, in turn, stream above the relatively more ponderous Turkey Vultures, while decidedly less numerous Ospreys, Mississippi Kites, and Peregrine Falcons also pass in well-formed, single-species subgroups, typically at the perimeters of the main flight line. Even so, whenever large, mixed-species streams of migrants spiral together in multidirectional kettles, near-misses and collisions occur at higher frequencies than in single-species flocks. The same phenomenon repeats itself in southern Israel when migrating Levant Sparrowhawks and Steppe Buzzards comingle in mixed-species flocks. When soaring individuals do collide, most use flapping flight to realign themselves and regain their position within the flock.

Other potential downsides of flocking behavior include its potential to enhance the spread of contagious diseases and interspecies predation, as well as prey robbery within and between species that feed aerially while migrating. Traveling in large flocks also may affect where raptors roost at night and, thus, their habitat needs. On the other hand, roosting in large numbers can reduce the potential impact of nighttime predation on individual birds. Migrating in large flocks also restricts the timing of migration, which, in turn, may result in individuals migrating earlier than their body condition merits.

Nevertheless, flying together in groups allows many species of birds of prey to achieve what they otherwise would not be able to do, migrate long distances between appropriate breeding and wintering areas. Turning such social behavior on and off seasonally allows raptors to do what many Temperate Zone organisms cannot do, successfully migrate long distances across a seasonally changing world.

Box 7.5 Orientation and navigation research on migration

Because of both finances and logistics—big birds tend to be both more expensive and more difficult to keep in captivity in large numbers—most orientation and navigation research has been done with relatively small and easy-to-keep seed-eating songbirds and domestic pigeons. As a result, except for the single displacement experiment involving Eurasian Sparrowhawks mentioned earlier, experimental approaches to orientation and navigation are lacking for raptors, and most ideas regarding their way-finding behavior during migration has been inferred from the results of experiments involving other species of birds. That said, observations at migration watch sites, together with banding returns and satellite tracking, suggest that the navigational skills of raptors are similar to those of other long-distant migrants, and it seems reasonable to assume that the factors underlying their abilities are similar as well.

To successfully complete long-distance, goal-oriented journeys, migratory raptors need to be able both to orient and to navigate themselves. Orientation is the ability to maintain a specific geographic direction across an external grid on a curved surface, such as the longitudinal and latitudinal network that cartographers use to plot locations on the surface of the earth. Simply put, to be able to orient themselves, migrants need to know north from south and east from west. Navigation, on the other hand, is the ability to move toward a goal while reorienting en route and, equally important, to know the distance between one's current location and the appointed goal.

An easy way to distinguish the significance of these two essential aspects of successful long-distance migration is to imagine a situation in which someone is lost in the woods and wants to get out as quickly as possible. The use of a magnetic compass or, for that matter, any orientation device that allows a person to travel in a constant direction is of little value by itself, particularly if one uses it to travel deeper into the woods and away from the nearest way out. Only when a person also has an accurate map of the area and knows their location on it, as well as that of their intended goal, does the compass become a useful tool for navigating oneself out of the woods.

A growing body of both field and laboratory evidence suggests that birds use one or more of three environmental cues to orient themselves while traveling long distances. The cues include the sun, the stars, and the earth's magnetic field. The "solar" compass was the first to be discovered. Working in the late 1940s, the German behaviorist Gustav Kramer noticed that in the months when they would otherwise be migrating, captive Common Starlings hopped and flitted about their cages more often than at other times of the year, a phenomenon that one of Kramer's colleagues had called *Zugunruhe* (German for *migratory restlessness*) several years earlier. Kramer noted that the migratory restlessness he observed was far more frequent on sunny than on overcast days and that most of the hopping and fluttering was aimed in the direction the birds would normally migrate. Kramer's observations led him to propose that the birds were using the sun's location in the sky to orient themselves. To test this idea, Kramer performed a series of experiments in which he used mirrors to shift the "direction" of the sun that the birds saw, and confirmed his suspicion. The results of additional experiments, in which the starlings' normal day-night cycles were "time-shifted" 6 or 12 hours with artificial light, so that the birds could no longer use the sun's relative position in the sky to orient correctly, indicated that the birds were using an internal 24-hour clock, together with the sun's horizontal position in the sky, or azimuth, to orient themselves.

Many species of birds migrate at night, however, and within a few years of Kramer's discoveries, other scientists looked to see if these nocturnal migrants were using individual stars or constellations of stars as cues for orientation and navigation. The American behavioral ecologist Steven Emlen conducted a series of unusual "laboratory" experiments to test this possibility. Emlen placed captive Indigo Buntings in a planetarium in which he shifted and reconfigured nighttime skies in ways that demonstrated that the birds used stars to orient themselves during migration. He also discovered that the birds were able to do so only if they had been exposed to the natural movements of the constellations in question as juveniles.

Earth's magnetic field provides a reliable reference for both direction and location, and by the 1960s, experiments with caged individuals of several species added this planetary feature to the list of compass cues. Subsequent attempts to determine how birds actually detect and correctly interpret the earth's magnetic fields suggest that at least some species use magnetically sensitive photo-pigments for directional orientation, and that precise arrangements of tiny magnetite crystals (Fe_3O_4) within individual cells enable birds to sense magnetic anomalies and declinational differences in the earth's magnetic field, as well as its polarity, thereby allowing them to create "magnetic maps" of the landscapes they migrate across.

Most recently, many researchers have suggested that migrants use multiple sources of information, including celestial and magnetic cues, along with polarized light and, possibly, predominant wind patterns, to help orient during long-distance travel. Explanations for navigational maps, however, remain more elusive. Work involving pigeons in Italy suggests that under certain conditions, olfaction can play a major role in helping birds map their current locations relative to their geographic goal. Whether the same holds for long-distance avian migrants remains unclear. Indeed, the point to which birds use true navigation, as opposed to simple compass orientation, to locate their wintering areas and breeding grounds is still an open question, with some researchers suggesting that the former may come into play only near the end of the long-distance travel, when its value in pinpointing an individual's exact breeding and wintering areas might become critical.

Most of the scant evidence we have for navigational ability in raptors comes in the form of band-recovery data and resightings of individually marked birds. For species that have been banded in large numbers, there is evidence that many individuals, particularly males, return to the same nest site, or within several nesting territories of where they have bred in previous years. The phenomenon, which is well studied in several populations of Ospreys and Peregrine Falcons, apparently occurs in many species of raptors, including long-distance, transequatorial migrants such as Broad-winged Hawks, Swainson's Hawks, Northern Hobbies, and Lesser Kestrels.

Although winter-site fidelity is decidedly less well studied than breeding-site fidelity, it, too, appears to occur. Examples include Roughlegs wintering on the same winter home ranges in California, migratory American Kestrels doing the same in southern Florida, and migratory North American Turkey Vultures doing the same in Central and South America.

Additional support for multidirectional navigation in raptors comes from direct observations of thousands of migrants traveling along some of the world's most heavily used migration corridors. North America's most abundant complete migrant, the Broad-winged Hawk, provides the archetypal example. Each year, millions of Broad-winged Hawks migrate from northern Temperate Zone breeding grounds, centered in southern Ontario, to wintering areas stretching from southern Mesoamerica into northern and central South America. The 8000 to 13,000 km (5000 to 8000 mi) one-way journey is far from unidirectional. The eastern populations of Broad-winged Hawks begin their outbound migrations along a roughly 240° southwesterly arc that transports them from southeastern Canada and the northeastern United States to the Gulf Coast of southern Texas. Once there, the flight is redirected approximately 60° to the east, thereafter following a roughly 180° southerly course toward coastal Veracruz, Mexico, before shifting another 60° east and following a southeasterly course into northern Central America. By the time the flight has reached western Panama, outbound Broad-winged Hawks are flying directly east toward Colombia, in northwestern South America. Once the flight reaches Colombia, the birds disperse across an arc of 180° en route to individual wintering areas across Colombia, Ecuador, Venezuela, and elsewhere in South America. This decidedly circuitous land-based journey, which circumnavigates the Gulf of Mexico and Caribbean Sea, enabling the birds to soar on land-based thermals en route, presumably presents a fair share of navigational challenges, particularly for juveniles attempting their first long-distance migration "south."

In addition to these and other direct observations of the principal flight lines of populations of long-distance migrants, the movements of single satellite-tracked birds reveal that individuals sometimes use all-but identical paths on successive outbound and return migrations, as well as on successive outbound routes in sequential years. (See Outbound and Return Migration, below.)

Although such observations provide little evidence for the mechanisms involved, taken as a whole, they strongly suggest that raptors are capable of navigating and orienting during their long-distance migrations.

WING SHAPE AND FLIGHT BEHAVIOR

A noted student of animal movements, Klaus Schmidt-Nielson, once remarked, "flight has the potential of being the most energetically costly form of animal locomotion." Although theoretically true, natural selection has shaped both the flight mechanics and flight behavior of birds over the course of millions of years, and today, modern birds, including raptors, are remarkably efficient flying machines.

As mentioned above, cross-country thermal soaring substantially reduces the energetic costs of long-distance flight. One might expect, therefore, that most, if not all, species of raptors would soar extensively on migration. But this is not so. Although superflocking migrants regularly soar while migrating and although all species of raptors soar at least occasionally on migration, several species frequently employ powered flapping flight en route, and most migrants typically alternate between the two flight types. At least three phenomena contribute to the less-than-consistent use of soaring flight. First, parts of many migration routes lack dependable soaring opportunities, and birds have little choice but to use more expensive flapping flight at such times. Second, smaller-bodied migrants often need to travel at higher speeds than soaring allows. The latter is because smaller birds, which have higher metabolic rates than larger birds, are living life in the "metabolic fast lane." A higher metabolic rate allows them to build large fat reserves quickly, but higher metabolic rates also require them to migrate quickly before their reserves run out. As a result, smaller migratory raptors tend to be high-energy "time minimizers" en route, whereas larger species tend to be lower-speed "energy minimizers." Third, a raptor's wings are shaped by many selective forces, including nonmigratory foraging flight and predator avoidance, as well as energy efficiency in migratory flight, and not all raptors have wings that are particularly well suited for soaring. Most soaring migrants, for example, are species whose foraging behavior depends heavily on soaring flight and whose risk of predation is little compromised by their inability to engage in high-speed flapping flight.

The amount of time a migrant spends soaring on migration, as well as the type of soaring it employs, is reflected in several aerodynamic aspects of its wings. Species that depend heavily on thermal soaring, for example, tend to have proportionately larger wings than other migrants. Proportionately larger wings result in lighter wing loading, which allows a bird to fly more slowly without stalling. This, in turn, enables a migrant to circle more tightly in thermals, thereby increasing the efficiency of its cross-country flight. Another characteristic feature of thermal-soaring migrants is wing slotting, a condition in which the tips of the outermost flight feathers are separated both horizontally and vertically when the wings are outstretched, creating the appearance of "fingered" wings. The roughly parallel series of aerodynamic surfaces that results from this configuration substantially reduces drag near the wingtips of slow-flying, soaring migrants, thereby increasing their flight efficiency. Although both of these characteristics are valuable during low-speed soaring flight, large wing surfaces and wing slotting often compromise the energetic efficiency of other types of flight, including high-speed soaring and gliding flight, high-speed prey-pursuit flight, and flapping flight. As a result, the wings of nonsoaring migrants differ considerably from those of soaring migrants.

Raptor migration specialists recognize five generalized "groupings" of wing shape within the anatomical spectrum: (1) kites, (2) harriers and osprey, (3) accipiters, (4) buteos and eagles, and (5) falcons. Kites and falcons have relatively long and slender high-aspect-ratio wings that feature pointed tips. Harriers and ospreys also have relatively long and slender high-aspect-ratio wings, but the tips of their wings are rounder than those of kites and falcons, and usually feature wing slotting. Most accipiters have relatively short rounded wings with some wing slotting. Buteos and eagles tend to have large and somewhat oversized, broad wings, which often feature wing slotting.

Efficient, high-speed linear soaring and gliding, including slope soaring, soaring in large thermal streets, and soaring over water, requires wings with relatively high-aspect ratios and high wing loading. Migrants such as Chinese Sparrowhawks, Grey-faced Buzzards, and Amur Falcons that depend on this type of soaring flight to complete their journeys tend to have relatively smaller and more pointed wings than do thermal-soaring migrants.

Raptors that depend on powered flight to complete much of their journey, such as Ospreys, harriers, most accipiters, and most falcons, employ a variety of flapping styles while doing so. Osprey and harriers, for example, typically intersperse relatively lengthy bouts of slope and thermal soaring with flapping flight. Accipiters usually employ undulating flapping flight, which also is called intermittent gliding and flapping flight, during which relatively short bursts of flapping flight alternate with similarly brief periods of linear soaring or, more typically, gliding. Falcons often engage in prolonged bouts of high-speed flapping flight en route, particularly in weather that grounds most other migrants.

Like many birds, raptors are able to reconfigure the shape and attitude of their wings depending on local flight conditions. Peregrine Falcons, for example, usually hyperextend their wings to the point of wing slotting when circling in small thermals, and harriers tend to "flatten" their otherwise dihedral wings under similar circumstances. Ospreys, which also circle-soar in thermals on relatively flattened and fully outstretched and slotted wings, quickly reassume their characteristic "M-shaped" profile when flex gliding and linear soaring at higher speeds. Other migrants, too, reconfigure their wings as conditions merit, and although many bird-watchers think of the flight profiles of raptors as being set in stone, careful observations indicate that most migrants constantly refine their flight configurations while traveling, presumably to reduce the energetic costs of their travels.

NOCTURNAL MIGRATION

Although birds of prey often are characterized as diurnal migrants, observations suggest that at least some migrate by both night and day. Species that are most likely to migrate nocturnally include (1) those that engage in considerable flapping flight while migrating, (2) those that migrate long distances over water, and (3) small-bodied, high-energy time minimizers that complete their journeys as quickly as possible. Obligate nocturnal travelers include Chinese Sparrowhawks and Grey-faced Buzzards migrating between Japan and the Philippines, Amur Falcons crossing the Indian Ocean from India to East Africa, and Merlins traveling across the North Atlantic between Iceland and Great Britain, all of which undertake overwater journeys of at least 450 kilometers (270 miles). Other species that migrate at night over water include radio- and satellite-tracked Peregrine Falcons that shortcut over the Atlantic Ocean along the Eastern Seaboard of North America and cross the Caribbean Sea between the Greater Antilles and South America, Ospreys that fly out over the Atlantic Ocean between Scotland and southern Europe, as well as cross the Mediterranean Sea between southern Europe and North Africa, and Northern Harriers, seen crossing the less than 30-km (18-mi) wide mouth of Delaware Bay between Cape May Point, New Jersey, and Cape Henlopen, Delaware.

Many overwater migrants time their flights to coincide with sea thermals (see above), tailwinds, or both. Both conditions reduce energy metabolic costs en route, and the latter reduces the amount of time spent over water where "landing" is not possible, which appears to be particularly important. Raptors also sometimes migrate over land at night. Evening flights of migrating Turkey Vultures have been reported in both Texas and Colombia, and Levant Sparrowhawks routinely migrate well into the night in the Middle East.

EXTREME WEATHER

Raptors typically time their migrations to take advantage of reasonably benign weather conditions in the regions they travel through. Many species migrate through major hurricane and typhoon regions, however, and cyclonic storms and other forms of extreme weather, including heavy rains, sometimes disrupt the regional geography, timing, and ultimate success of both short- and long-distance migratory movements. Hurricanes passing through eastern North America, for example, are known both to delay migrations and to shift longitudinally the principal passageways of migrants. Five days of rainy weather on 17–21 September 1938, coinciding with the coastal passage of the "Great Hurricane" of that year, stalled and compressed the southbound movements of Broad-winged Hawks at Hawk Mountain Sanctuary for almost a week, after which an unprecedented 5-day passage of more than ten thousand individuals—more than 95% of that season's total flight—was counted at the sanctuary. A record-breaking flight of 70,000 broadwings at Hawk Cliff, Ontario, on 16 September 1961, followed Hurricane Carla's movement through the midsection of the Great Lakes on 13–14 September. And a record-shattering flight of 500,000 broadwings on 17 September 1999 at Lake Erie Metropark, at the lake's western end, followed the passages of Hurricanes Dennis and Floyd to the east, earlier in September. Finally, a satellite-tracked Peregrine Falcon flying from Haiti toward Venezuela in late October 1998 returned to the island, having completed more than 80% of its 650-km (390-mi) overwater passage, after encountering southerly head winds associated with Category 5 Hurricane Mitch.

ENERGY AND WATER RESOURCES FOR MIGRATION

Raptors fuel their migratory movements in three ways: by building fat reserves prior to migration and burning fat en route, by feeding regularly on migration, and by soaring on thermals and mountain updrafts. Although most species use all three sources of energy, many depend disproportionately on one or two of these "fuels" to complete their migrations.

Fat Reserves

Fat is the fuel of choice for most migratory birds. Numerous small songbirds, for example, are known to double their lean body mass prior to migration, and many others are known to regularly interrupt their flights and replenish fat stores en route.

The phenomenon of premigratory fattening has not been studied in detail in raptors, and the extent to which it occurs has been inferred almost entirely from examinations of individuals captured while migrating. Data suggest that the fat stores of migratory raptors rarely exceed 20% of their lean body mass. Overall, researchers have found that juveniles have lower fat stores than adults, and that females tend to carry more fat on their return migrations than on their outbound flights. Most likely, the former reflects the fact that juveniles are less efficient predators and migrants than are adults, and the latter reflects the likelihood that natural selection favors females who arrive on their breeding grounds with sufficient fat to produce eggs.

One problem with studies involving trapped migrants is the likelihood of a sampling bias. Raptors trapped during migration almost always are lured with food, and it seems reasonable to assume that food-stressed birds are more likely to be captured than are their less-hungry counterparts. As a result, the stored-fat values reported in these studies almost certainly represent the lower end of the amounts that are carried by most migrants. And indeed, one could argue that the fact that many investigators have had difficulties capturing long-distance transequatorial migrants near the midpoints of their migrations suggests that the most of these birds are carrying sufficient fat at this point in their journey and are not "fuel stressed."

Routine Feeding en Route

Many raptors, particularly those that travel alone and that migrate short distances, regularly feed en route. Some individuals feed daily, others do episodically and opportunistically when weather conditions preclude efficient migration, and others still do so only when they encounter areas of abundant and easily available prey en route. Studies involving satellite-tracked raptors indicate that many begin their migrations slowly, increase their speeds as they approach and pass the midpoints of their journeys, and then slow down again as they approach their destinations. Presumably, such differences reflect differences in the likelihood of feeding en route, with individuals hunting more when starting out in an effort to build fat reserves for the journey, and again toward the ends of their migrations, either to avoid metabolizing protein as fat stores are depleted, or to sample prey availability on potential wintering areas.

Some species apparently time their migrations to take advantage of migratory prey en route. Peregrine Falcons, which fuel their migrations by feeding en route, typically interrupt their journeys for a week or more to feed at traditional stopover sites that predictably host seasonal concentrations of migrating shorebirds, waterfowl, and other waterbirds. The impact of falcon predation in these situations is such that researchers have suggested that prey species sometimes shift their migrations in an attempt to decouple them from that of the falcons. Other largely bird-eating accipiters, including the Sharp-shinned Hawk in North America and the Eurasian Sparrowhawk in northern Europe, apparently time and position their migrations to take advantage of the concentrated movements of their songbird prey.

Many raptors, including Black Kites, Broad-winged Hawks, Merlins, and American Kestrels, routinely hunt swarming insects while migrating, some of which are migrating themselves. Others regularly prey on birds and mammals, either before or after their daily movements, or throughout the day during periods of poor weather. In Israel, Egyptian Vultures interrupt their migrations to feed at garbage dumps and on roadkills. Ospreys, which often fuel their migrations by feeding en route, frequently carry prey while migrating, usually with the fish held tightly in both feet and oriented headfirst in an aerodynamic fashion.

Feeding rates can vary considerably within species, with individuals traveling longer distances being more likely to stop and refuel en route than those migrating shorter distances, and individuals with greater fat stores being less likely to feed en route than those with smaller reserves. Overall, feeding appears to be far more common immediately before and after long-distance travel across inhospitable landscapes, such as deserts and large bodies of water, than at other times.

Soaring Flight

As mentioned above, many raptors, particularly those that migrate long distances, depend heavily on energy-efficient soaring flight to complete their migrations. Almost all transequatorial and most complete long-distance migrants are superflocking, obligate soaring migrants that time their migrations so that the overwhelming majority of the flight passes any one site in fewer than two weeks. Several of these calendar migrants, including the Amur Falcon and the Swainson's Hawk, stopover in relatively small groups and feed for up to a month or more early in their migrations before massing into truly enormous flocks and flying, presumably without feeding, for most of the remainder of their journeys. Others either lay down considerable fat stores prior to migration or feed episodically and opportunistically on superabundant insects en route.

Raptors that depend heavily on this source of energy to complete their migrations often reach their destinations in a weakened state. For example, many of the Turkey Vultures and Swainson's Hawks that migrate between breeding sites in North America and wintering areas in South America arrive at the latter in late autumn fat-depleted and at barely 70% of their typical body mass. Most, however, quickly regain body condition and lay down fat needed for the return journey well in advance of their departures the following spring.

Water Needs

It is usually assumed that migrants, including raptors, secure sufficient water when traveling long distances via fat metabolism, which results in the by-production of metabolic water. Even so, many raptors, including Eurasian Honey Buzzards, Black Kites, Steppe Buzzards, Short-toed Snake Eagles, and Lesser Spotted Eagles, have been observed drinking at migratory stopover sites, most notably in Africa and the Middle East. Some species appear to be especially dependent on drinking water while migrating, even to the point of using brackish and saltwater sources. Individuals avoid salt poisoning by excreting excess amounts via paired salt glands located above their eyes. The glands, which occur in many raptors and serve mainly to reduce salt loads resulting from the consumption of desiccated prey, are capable of concentrating and dumping salt at rates that enable migrants to drink seawater-strength saltwater without ill effect.

TIMING OF MIGRATION

Many factors, including age, gender, and flocking, as well as the extent of feeding en route, affect the timing of raptor migration.

Flocking and Nonflocking Migrants

The timing of peak migration is decidedly more acute and temporally more predictable in super-flocking migrants than in species that migrate alone or that flock episodically. At Hawk Mountain Sanctuary, half of all Broad-winged Hawks—the only obligate flocking migrant at the site—typically pass within eight days each autumn, whereas the site's nonflocking migrants take 16 to 39 days to do so. Presumably, flocking species, which benefit by traveling in large groups, time their flights to increase the likelihood of doing so, whereas nonflocking species do not. And indeed, the latter, many of which feed regularly on migration, most likely "space" their movements, both temporally and spatially, to reduce competition while traveling.

Outbound and Return Migration

Numerous factors affect the timing of outbound migration from the breeding grounds. Many adults replace some or all of their flight feathers prior to autumn migration and then suspend molt during their movements, presumably to increase flight performance en route, and molt behavior—males and females tend to time their postbreeding molts differently—typically affects the departure schedules of individual birds. Bird-eating raptors that depend on migrating birds to fuel their migrations typically time their movements to coincide with those of migrating prey. And species that depend heavily on soaring flight, particularly those that depend on late-summer and early-autumn thermals in the Temperate Zone, need to time their departures to take advantage of this phenomenon. Finally, almost all raptors need to attain at least modest fat stores prior to migration regardless of how they fuel their journeys overall, and an individual's ability to lay down this fuel can affect its departure schedule.

One difference between outbound and return migration is that the former coincides with seasonal peaks in prey availability whereas the latter does not. As a result, outbound migrants are more likely to fuel their journeys by feeding en route than are return migrants, possibly slowing their rate of travel. Outbound migration also differs from return migration in that every migrant involved in the latter has at least some experience migrating long distances, whereas as about half of those involved in outbound migration are inexperienced young-of-the-year. Taken together, these two factors suggest that migrants might travel faster on return migration than outbound migration, and, indeed, there is evidence for this in many species of birds, including many raptors.

One factor that almost certainly does not affect the faster rate of travel in spring is the "urge to reproduce." It makes little sense for raptors to migrate back to the breeding grounds at anything other their optimal rate of travel given their physical state, weather conditions, and prey availability en route, and there is no indication that reproductive "urges" induce them to do otherwise. A forced march back to the breeding grounds in spring is counterproductive, particularly when arrival there in appropriate body condition is so important for territorial establishment and breeding, and when, in most circumstances, an earlier departure from the winter grounds is likely to allow migrants to arrive earlier on the breeding grounds in appropriate condition. And indeed, whereas satellite tracking indicates that individuals in many species complete their return journeys more quickly than their outbound migrations, the reverse is true in many others.

Age and Gender Differences

Ornithologists have known for some time that within species, individuals of different ages and genders often differ in their migration schedules. The phenomenon has been widely studied in raptors, particularly in species whose plumages allow individuals to be aged or sexed in the field. Unfortunately, few universal patterns have emerged. Some researchers have argued that adult males, which do most of the hunting on the breeding grounds, and therefore, have a better knowledge of the prey base there than do females, should extend their stays there and depart from these areas later in autumn and return to them earlier in spring than their mates. Others have suggested that males should return to the breeding grounds earlier than females because they are under intense selection pressure to acquire nesting territories before females arrive.

In terms of age-related differences, many researchers have suggested that inexperienced juveniles should follow rather than precede adults on their first outbound migrations. But although this makes some sense in terms of the relative navigational skills of the two ages classes, and although in many species juveniles do follow adults during the formers' initial outbound movements, in many other species, where adults remain on the breeding grounds and molt prior to migration, the reverse is true. Another potential reason for age-related differences in migration timing is that juveniles and adults often follow different routes on autumn migration.

Another factor complicating age-related differences in migration timing is that adults often over-winter closer to their breeding grounds than do juveniles, possibly because doing so allows them to return to these areas earlier in spring. In some species, including the Northern Goshawk and the Roughleg, males overwinter farther from their breeding areas than do females, whereas in many others the reverse is true. The first situation is often attributed to behavioral dominance or climate, with smaller males being out-competed by larger females for the closer and presumably preferred wintering areas, or being less able to tolerate the ferocity of northern winters; whereas the second situation is believed to result from males attempting to remain as close as possible to the breeding territories so that they are better able to reestablish them in the spring.

Unlike outbound migration in autumn, there is considerable evidence for adults preceding juveniles during return migration in spring, and no evidence, whatsoever, of the reverse. Almost certainly, this is driven by the fact that whereas adults are under strong selective pressure to return as early as possible to their breeding grounds, individuals born during the previous season, which in many species do not breed until their third calendar year of life, are not under the same selective pressure. Even so, the extent to which adult precedence in spring is due to age-related differences in navigational and flight efficiencies remains largely unstudied, and it too, may play a significant role.

BREEDING AND WINTERING ECOLOGY OF MIGRANTS

Breeding Ecology

There are many indications that the breeding biology of migratory raptors differs from that of sedentary species. The percent of migratory raptors increases with latitude and, overall, migratory raptors tend to breed in more seasonal habitats than do sedentary species. One result of this is that migratory populations encounter more temporally acute pulses of prey availability than do sedentary populations, which provides them with a greater peak in prey abundance, but at the same time restricts prey abundance to a much shorter period. Thus migrants tend to have briefer breeding seasons than sedentary species, and often breed at higher densities. Seasonally abundant prey, particularly small mammals and small birds, are usually more important in the breeding-season diets of migrants than in those of sedentary species, and, overall, migrants are more likely to breed in open habitats than in forests, possibly because the former tend to offer greater flushes of seasonal prey than the latter. And when migrants do breed in forests, they are more likely to do so in deciduous than coniferous woodlands, either because the former offer more suitable seasonal prey than the latter, or because the former are less likely to be inhabited by sedentary raptors than are the latter.

Wintering Ecology

Relative to the breeding season, migratory raptors have been little studied in winter, both because they are not “attached” to a nest site and are far more mobile while overwintering, and because many overwinter in the tropics where field ornithologists are few and where most fieldwork is focused on local species. Although the increased mobility that raptors gain in winter by not being tied to a nest sometimes brings them together at large roosting assemblages and at superabundant food resources, the increased freedom of movement also makes them more difficult to study for long periods. The extent of increased raptor mobility in winter has become evident as a result of satellite telemetry.

Three Lesser Spotted Eagles that bred in Germany and that overwintered in East Africa, for example, ranged across regions of between 11,000 to 25,000 km^2 (4,250 to 9,700 mi^2)—areas about the size of the state of New Jersey—while feeding on swarming termites during several winters in the mid-1990s. Even more impressive is the nomadic overwintering behavior of a Wahlberg’s Eagle that bred in Namibia in southwest Africa, and subsequently overwintered 3500 km (2200 mi) north in west-central Africa. The bird spent the first 6 weeks on its “wintering grounds” ranging across an enormous 60,000-km^2 (23,000-mi^2) West Virginia sized area that included parts of Cameroon, Nigeria, and Chad, before “settling down” into a far smaller 5000-km^2 (1900-mi^2) area of savanna in northern Nigeria during the austral winter of 1994. Other species that have been tracked by satellite, including Ospreys, are far less nomadic in winter, and show considerable inter-year site fidelity to individual home ranges. The same appears to be true for American Kestrels overwintering in Florida, and for Turkey Vultures from North America overwintering in the Neotropics (see below).

One thing that is certain about overwintering raptors is that local distributions are largely determined by food availability. Most northern breeders overwintering within or south of the tropics do so either in coastal and inland wetlands (e.g., Turkey Vulture, Osprey, Western Marsh Harrier, Peregrine Falcon, etc.) or in open grasslands and savannas (e.g., Mississippi Kite, Swallow-tailed Kite, Levant Sparrowhawk, Swainson's Hawk, etc.), seasonal habitats in which the benefits of year-round site fidelity are few and, consequently, densities of sedentary species are often quite low; and in which northern-winter flushes of productivity provide ideal conditions for northern migrants. Examples of such habitats include savannas and grasslands in East Africa, where overwintering populations of more than a dozen migrants feed on rainy-season-related insect and rodent outbreaks, and those of West Africa, where migrants congregate along dry-season fire lines to feed on locusts and grasshoppers. Migratory raptors that overwinter in forested habitats tend to do so along forest edges and in second-growth woodlands, rather than in large, undisturbed forested tracts.

There is no evidence that nonmigratory species competitively exclude migrants from undisturbed forests, or that predation risks are greater in large wooded areas than in other habitats, and it seems likely that migratory raptors choose more open and disturbed habitats either because the latter are more similar to those they inhabit during the breeding season, or because flushes of seasonally abundant prey are more readily available there, or both. Whatever the reason, the arrival of large numbers of migratory species in open and semi-open habitats can significantly affect regional patterns of species diversity. The expansive grasslands and savannas of the Great Rift Valley of East Africa, for example, host larger numbers of raptors than do adjacent tropical rainforests, a phenomenon that has important consequences for raptor conservation in the region. East Africa's wide-open spaces, and in particular some of its more productive farmland areas, also host nonbreeding populations of short-distance, intra-African migrants. And, perhaps not surprisingly, both types of migrants arrive at or near the beginning of the region's short rainy season, at the peak of prey availability. Remarkably enough, when the timing of the rainy season shifts among years, so does the arrival of the migrants, in spite of the fact that the latter includes an array of ecologically diverse species.

On a species basis, most migratory raptors have decidedly smaller wintering ranges than breeding ranges, and in some species the difference is considerable. The African wintering ranges of at least three species of Eurasian migrants, the Short-toed Snake Eagle, Red-footed Falcon, and Northern Hobby are less than one-third the size of their breeding ranges. The same appears to be true of the South American wintering areas of at least two species of North American migrants, the Mississippi Kite and Swainson's Hawk. This, together with the fact that many of the areas used by overwintering raptors also host large numbers of intracontinental, short-distance migrants, as well as many resident species, has led some to suggest that overwintering migrants might compete both among and within species for food and other resources (Box 7.6).

Box 7.6 Competition among raptors overwintering in farmlands in south-central Ohio

Competition among and within overwintering species of raptors was much in evidence in the farmlands of south-central Ohio when I studied them for four winters in the 1970s. In winter, the region's open-habitat raptor community is dominated by four species: Northern Harriers, Red-tailed Hawks, Roughlegs, and American Kestrels. Northern Harriers and Roughlegs are exclusively winter visitors in the region, whereas populations of Red-tailed Hawks and American Kestrels include both year-round residents and winter visitors. My estimates indicate that raptor numbers more than doubled on the site in winter, at the same time that periodic snow cover reduced the availability of the small mammal prey base that all four principal species feed heavily on. All four species differ significantly in habitat use and hunting behavior, a fact that had led previous researchers to conclude that niche segregation reduced or even eliminated interspecies competition within the feeding guild, something that was definitely not true in south-central Ohio.

Individuals of all four species frequently hunted close together, sometimes to the extent that representatives of each species hunted the same small field at the same time. Although many of these instances passed without obvious interaction, many of these juxtapositions resulted in increasingly predictable interspecies interactions. Most of the interactions involved an individual of one species displacing a member of another species or its own, either from a hunting perch or from the airspace over the field. Many of the interactions were subtle, with one bird leaving its perch or the airspace when the second bird was still more than 200 m (660 ft) away. And, indeed, I often did not see a second bird until after the first bird had departed. Although in many of these instances the departure of the first bird may have precipitated the arrival of the second, the two individuals were "interacting" nevertheless.

I recorded what I interpreted to be 172 overt instances of aggressive behavior during the course of the four-winter study, or a rate of about one encounter for each 4 hours in the field. In 80% of these interactions, neither of the participants possessed prey. Most of the "without-prey" encounters were directed at perched birds, and in most instances the perched bird was displaced. Smaller species were more likely to be the aggressor in interspecies interactions that did not involve prey. When such interactions were successful, the approaching bird typically hunted in the area after the perched bird left. When such interactions were unsuccessful, the approaching bird almost always left the area. Harriers and Roughlegs most frequently initiated such interactions.

Interactions begun by Roughlegs differed from those initiated by harriers in being brief high-speed aerial pursuits, rather than the prolonged bouts of repeated diving. Every time a harrier successfully displaced a Roughleg, the harrier subsequently hunted in the area. And every time the harrier was unsuccessful, it left the area, presumably to hunt elsewhere. Roughlegs always successfully displaced the harriers they approached, and always remained in the area after interacting with them. Overall, these differences appear to reflect the different motivations of the two species. Roughlegs, which are approximately twice as massive as harriers, were attempting to determine if the harriers they approached possessed prey. Harriers, on the other hand, were attempting to chase Roughlegs from areas that they, themselves, wanted to hunt in; the reason for this became apparent when

I watched Roughlegs and harriers interact when the harriers were feeding on or carrying prey.

Interactions in which prey was involved included thirty attempts at prey robbery in which a bird had just captured it and eight attempts at carcass displacement. Harriers were approached and chased by other raptors twenty-eight times—or fully one-third of the time that I saw them with prey—and were by far the most frequent victims. Nineteen of these attempted piracies were made by Roughlegs or by Red-tailed Hawks, the remainder by other harriers. Whether individual harriers suffered substantially as a result of piratical behavior is not known. That they actively avoided hunting near Roughlegs, the most common interspecies pirate, and may have been forced into suboptimal feeding areas as a result, suggests they did. Indeed, it seems likely that the successful perpetration of several robberies in rapid succession, especially during periods of severe weather, could significantly compromise a harrier's energy reserves. Harrier densities were relatively high in the area, but whether or not competition with larger raptors limited their numbers regionally is not clear. Nor is it clear if the numbers of Roughlegs, a species that was regularly pursued and sometimes pirated by larger Red-tailed Hawks, were, in turn, limited by the redtails.

American Kestrels, which by far were the smallest open-habitat raptors in the area, typically were not harassed by the three larger species, most likely because they fed on insects rather than vertebrates as soon as temperatures rose above freezing and insects became active, and because they tended to hunt close to active farm buildings and other human-dominated landscapes, something the three larger species tended to avoid.

Consistent and strategic piracy, such as reported above, is known in only a few species of raptors, including, perhaps most notably, Bald Eagles, which routinely take prey from Ospreys, Ferruginous Hawks, Red-tailed Hawks, and Roughlegs, as well as from conspecifics. Nevertheless, prey robbery has been observed on numerous occasions among migrants, both at stopover sites and en route, as well as on the wintering grounds. Overall, it appears to be most likely in areas where several or more species of migrants aggregate around abundant or superabundant prey, particularly if the site also hosts large numbers of resident raptors. Although the ecological significance of such interactions remains largely unstudied, evidence from Central and South America indicates that it can be significant (see text).

Each year, more than two million migratory Turkey Vultures from western North America pour into wintering areas that stretch from southern Central America into northern South America. The migrants involved are members of the *meridionalis* subspecies of the Turkey Vulture; the local residents belong to the smaller-bodied *ruficollis* race. At close range and in good light, *ruficollis* is easily distinguished from *meridionalis* by their conspicuous yellow to whitish-green nape bands and crown patches, which the former lack. In the Llanos wetlands of central Venezuela, where interactions between migrant and resident Turkey Vultures were studied for three winters in the early 1990s, migrants dominated and bullied residents at carcasses, so much so that many of the latter shifted from feeding in seemingly preferred open areas to feeding in gallery forests along rivers when migrants were present. Significantly, whereas the body condition of migrants increased during the winter dry season of their co-occurrence, the condition of the residents declined. And even though carrion appeared to be abundant throughout the period, residents did not begin breeding until the migrants had left. These observations suggest substantial competition between migrants and residents, at least in this part of Venezuela. And, the same appears to be true along the Caribbean coast of northern Colombia, where the annual arrival of migratory Turkey Vultures boosts local populations sevenfold, and where *meridionalis* similarly dominates and displaces *ruficollis.* That the migrants also are able to hold their own against local populations of Black Vultures, something that the smaller nonmigratory Turkey Vultures apparently are not capable of, allows the former to exploit a wider range of food resources, including leftovers at garbage dumps, which are dominated by Black Vultures at other times of the year.

WATER-CROSSING BEHAVIOR

For most raptors, traveling over water is energetically more costly and potentially more dangerous than traveling over land. This is particularly so for facultative and obligate soaring migrants. As a result, many raptors avoid even short water crossings whenever possible. Indeed, 55% of all complete migrants and 96% of all partial migrants rarely, if ever, undertake water crossings of 100 km (60 mi) or more. Even so, several raptor migrants regularly undertake long-distance, overwater journeys during their long-distance movements.

Raptors flying over large expanses of open water risk being blown out to sea by strong crosswinds. Consequently, most raptors that migrate long distances over water do so within the confines of the world's great seas, including the Mediterranean and Caribbean, or along oceanic archipelagos, rather than across open oceans. Exceptions include Amur Falcons that travel up to 2000 km (1200 mi) across the Indian Ocean between peninsular India and East Africa on their outbound migrations, Merlins that shuttle 800 km (480 mi) one-way across the North Atlantic between breeding grounds in Iceland and wintering areas in Great Britain and Europe, and New World Peregrine Falcons that regularly "shortcut" hundreds of kilometers across the temperate and tropical Atlantic between North and South America each autumn.

Notwithstanding such movements, there are numerous examples of raptors avoiding even relatively short overwater passages. Many migratory raptors use peninsulas to reduce the amount of overwater travel. Well reported New World examples of this include Cape May Point in southern New Jersey at the mouth of Delaware Bay, 30 km (18 mi) north of Cape Henlopen, Delaware; Kiptopeke, Virginia, at the southern tip of the Delmarva Peninsula at the 22-km (14-mi) mouth of the Chesapeake Bay; and southernmost peninsular Florida and the Florida Keys along the 100-km (62-mi) wide Straits of Florida. Old World examples include Falsterbo, southwestern Sweden, 25 km (15 mi) from eastern Denmark; Tarifa and the Rock of Gibraltar in southernmost Iberia, 14 km (9 mi) across the Strait of Gibraltar from northern Morocco; and Cape Rachado, Malaysia, 22 km (14 mi) via the Strait of Malacca from Sumatra. In all such instances, migrants travel the length of a peninsula before making an overwater crossing.

In addition to using peninsulas to shorten overwater travel, many migrants, including those traveling across broad-fronts, island-hop across oceanic archipelagos and continental islands in apparent attempts to reduce the costs and risks of overwater travel. New World examples of island hopping include flights of accipiters and falcons along the Florida Keys at the southern tip of Florida, north of the Florida Straits, and along the Sabana-Camaguey Archipelago in Cuba, south of the straits; and flights of Ospreys, Merlins, and Peregrine Falcons, along the Greater and Lesser Antilles of the West Indies. Old World examples include Rough-legs island hopping across the Bay of Bothnia, a northern extension of the Baltic Sea, between eastern Sweden and western Finland; and Chinese Sparrowhawks and Grey-faced Buzzards island hopping from Korea through southern archipelago Japan, the Ryu Kyu Islands, and Taiwan to the Philippines, a mostly overwater journey spanning 25° of latitude, or more than 2500 km (1500 mi).

Observations at Cape May Point, along the Atlantic coast of southern New Jersey, and the Strait of Gibraltar, between southern Spain and northern Morocco, provide examples of the extent to which migrating raptors modify their behavior to reduce the energetic costs and physical hazards of overwater travel.

Southbound raptors reaching Cape May Point in autumn face several options: they can continue south over water across 30-km-wide Delaware Bay to Cape Henlopen, Delaware; they can retreat over land more than 100 km (62 mi) northwest along the northern edge of the bay and make a less than 2-km (1.2-mi) river crossing at its headwaters on the Delaware River; or they can hedge their bets, backtrack part of the way and cross a narrower portion of the bay short of the river. Obligate soaring species, for example, are far less likely to shortcut across water than are species that regularly engage in powered flight on migration. Thus, whereas Turkey Vultures and Broad-winged Hawks crossed the body of water on fewer than one-third of their approaches, Sharp-shinned Hawks and American Kestrels did so most of the time. Not surprisingly, widely distributed, broad-frontal migrants, including Ospreys, Northern Harriers, and Peregrine Falcons, were most likely to continue over water, and crossed on more than 90% of their approaches. Species that were the least likely to attempt overwater passage also were more likely to abort crossings and return to the shoreline than were species that regularly attempted crossings.

During most overwater flights, birds flapped continually or intermittently flapped and glided en route. Three of fourteen Peregrine Falcons appeared to engage in energetically efficient dynamic soaring while crossing. Although some individuals soared to great heights before crossing, most flew relatively low over the water, often within 5 m (16 ft) of the surface.

Observations of more than 7000 Sharp-shinned Hawks demonstrated that this species was more likely to undertake water crossings in light rather than strong winds, and at higher rather than lower altitudes, something that appears typical of most species of raptors, and also that it was less likely to cross during crosswinds than during headwinds. Finally, individuals rarely crossed during periods of low visibility, regardless of the wind.

Studies at the 14-km-wide Strait of Gibraltar corroborate these findings. Almost all of the 150,000 or so raptors that routinely cross this major migration bottleneck between Western Europe and northwestern Africa each autumn do so along an approximately 20-km (12-mi) long windswept stretch of Spanish coastline between Punta Marroqui, Tarifa, and Punta del Carnero, Algeciras, or from the "Rock of Gibraltar," several kilometers east (Box 7.7).

Box 7.7 Water crossings at the Strait of Gibraltar

In general, easterly, or "Levante," crosswinds tend to forestall migrant passage at the Strait of Gibraltar, most likely because they increase the energetic cost of the trip as well as the likelihood of an individual being blown out to sea. Although migrants do cross in light rain, individuals rarely depart in heavy fog, and birds encountering fog over the Strait almost always abort their passage and return to shore.

Eurasian Honey Buzzards—the most powerful flapping flyers among soaring migrants at the site—are less likely to be held back by crosswinds than are more lightly wing-loaded and far more buoyant Black Kites, which are typically grounded at such times. And indeed, whereas many Eurasian Honey Buzzards cross the Strait immediately upon reaching it, regardless of the wind, Black Kites that reach the Strait after midmorning on windy days typically spend the rest of the day and night roosting in Spain, and cross the next morning before the winds build. During periods of particularly persistent Levante winds, thousands of these soaring migrants have been known to spend a week or more alternately "kiting" into the wind by day, and roosting in agricultural fields by night, waiting for the winds to abate. Although a few of the kites opportunistically hunt and scavenge for food at such times, most apparently feed little during such delays. Short-toed Snake Eagles and Booted Eagles also appear reluctant to cross at such times, and their numbers, too, tend to build in the area during adverse winds, as do the numbers of Griffon Vultures, the largest obligate soaring migrant at the site.

Griffon Vultures are ideally suited to short- and intermediate-distance soaring flight and, indeed, are masters of the technique. The species, however, is incapable of sustained flapping flight, and their apparent reluctance to cross the Strait of Gibraltar under all but ideal soaring conditions is not without reason. Seemingly loath to cross during strong east-to-west Levante crosswinds, Griffon Vultures are willing to wait in Spain for weeks for better weather, and, indeed, some of the individuals that mass at the Strait never cross. When they do cross, most do so within two hours of noon, typically on warm days with light winds and, almost always, after having soared to great heights along the coast. Many of the vultures, most of which are dispersing juveniles, cross in flocks of dozens to hundreds of individuals. Presumably, the vast majority successfully complete the overwater passages. There are many records of griffons drowning in the Strait while attempting to cross in autumn and spring. And in spring, individuals can be seen resting for hours after making landfall on the beaches near Tarifa, having reached the shoreline in apparently exhausting flapping flight within a few meters of the surface. Some even "land" in the surf and wade to shore.

At the other extreme, migrants that routinely employ flapping flight rarely hesitate to cross the Strait. Marsh, Hen, and Montagu's Harriers readily cross except during all but the strongest Levante winds, typically as individuals. And the same appears to be true of the Ospreys, Eurasian Sparrowhawks, Common Kestrels, Lesser Kestrels, and Northern Hobbies.

Viewed in their entirety, the observations above suggest that a raptor's reluctance to fly over water has at least as much to do with the chance of being blown out to sea, or being physically incapable of sustained flapping flight, as with any attempt to minimize the energetic costs of its flight.

LEAPFROG AND CHAIN MIGRATION

Leapfrog migration occurs within species when populations breeding at high latitudes migrate substantially farther than, and geographically "leap over," nonmigratory or less migratory populations breeding at lower latitudes, thereby reversing their latitudinal relationships between seasons. Chain migration occurs when migratory populations that breed at higher latitudes migrate approximately the same distance as those that breed at lower latitudes, thereby maintaining their latitudinal relationship between seasons.

The Common Buzzard, a widespread species that breeds across most of Europe and much of Asia, from 35° to 65° N, is a classic example of a leapfrog migrant. Common Buzzards are long-distance complete migrants at and near the northern limits of their range, short- or intermediate-distance partial migrants immediately to the south, and largely sedentary in southern portions of their range. Additional examples of leapfrog migration involve Turkey Vultures in western North America, Egyptian Vultures in southern Europe and northern Africa, Northern Harriers in North America, Western Marsh Harriers in Europe, Eurasian Sparrowhawks in Europe, Wahlberg's Eagles in southern Africa, Common Kestrels in Western Europe, and American Kestrels in North America.

ELLIPTICAL MIGRATION

Raptors often follow different routes during their outbound and return migrations. In many instances, the differences reflect climatic or other external events, including prey availability, that favor different passages at different times of the year; in others cases the differences apparently result from adaptive wind drift, a situation in which birds allow themselves to be wind-drifted early in their journeys to increase their rates of passage then, and later compensate for such drift near the end of their migrations. In some instances, outbound and subsequent return migrations are clockwise; in other instances, they are counterclockwise.

One well-studied example of elliptical migration involves the outbound and return migrations of Broad-winged Hawks in North America. Enormous, multi-thousand bird flocks of this species regularly funnel north and south along the Mesoamerican Land Corridor from southernmost Texas to northwestern Colombia each spring and autumn, as they shuttle back and forth between wintering areas in Central and South America, and breeding areas in North America. North of southern Texas, however, the Broad-winged Hawk flight traces an elliptical path, in which outbound migrants in autumn fly south considerably farther east than returning migrants fly north in spring. As a result, many of the Broad-winged Hawks that pass Hawk Mountain Sanctuary and other hawk watches in the Central Appalachian Mountains of eastern Pennsylvania on outbound migration in autumn find themselves migrating hundreds of miles west of these sites during return migration in spring. Thus, whereas Hawk Mountain Sanctuary records autumn counts of more than 5000 birds in most autumns, but typically only several hundred birds in spring, watch sites along the southern shorelines of Lake Erie and Lake Ontario northwest of the sanctuary record large broadwing flights in spring, but only small flights in autumn.

The decidedly elliptical route taken by broadwings results from the region's prevailing winds, which are easterly below 35° N and westerly above it, together with the species' propensity for adaptive wind drift. Simply put, autumn outbound migrants from the core of the species breeding grounds in eastern Canada allow themselves to be pushed eastward through New England and the Mid-Atlantic States by the region's prevailing westerly winds, only to redirect their flight westward to circumnavigate the Gulf of Mexico farther south. In spring, the same birds allow themselves to drift westward through northern Mexico by the region's prevailing easterly winds, only to redirect their flight eastward once they have reached the largely insurmountable water barriers of Great Lakes Erie and Ontario farther north.

Another notable elliptical migration involves Chinese Sparrowhawks and Grey-faced Buzzards, both of which island-hop in large numbers on outbound migration across Japan, Taiwan, and the Philippines each autumn, but return to their breeding grounds in spring largely via mainland China north of Taiwan.

TEMPERATE ZONE MIGRATION AND TROPICAL MIGRATION

Most of what we know about raptor migration is based on observations of migrants in the northern Temperate Zone. More than half of all long-distance migrants, as well as numerous short- and intermediate-distance migrants, are transequatorial travelers that spend substantial amounts of time migrating to, within, and, in a few instances, across the tropical regions of the world, and understanding these movements is the new frontier for raptor-migration science.

Several circumstances suggest that raptors might behave differently in tropical regions than in temperate regions. Most of the world's principal migration flyways include tropical bottlenecks, points at which long-distance transequatorial migrants assemble in numbers far greater than in temperate areas. Solar energy is more available in the tropics than in the Temperate Zone, and thermals are decidedly stronger and more readily available to soaring migrants there. Finally, the Trade Wind Zone, stretching from 5° to 30° north and south of the equator, permits migrants to travel long distances over water on sea thermals there.

It has been known for some time that flock size increases appreciably towards the equator, particularly among soaring migrants. Only recently, however, have biologists come to recognize a relationship between increased flock size and decreased wing flapping, and thus a link between flock size and energy expenditure en route. The impact of this relationship on migrants has yet to be studied in detail, but one obvious consequence, a substantial increase in flight efficiency, may explain how many long-distance migrants manage to complete their transequatorial journeys while feeding little, if at all, en route.

The sun, too, contributes significantly to increased migration efficiency in the tropics. Thermal streets are far more common in tropical regions, and their relative abundance increases the likelihood of linear soaring in the region. Increased linear soaring, together with the confluence of regional corridors into continental and intercontinental flyways, produce enormous rivers of raptors that characterize many tropical hawk watches. Another factor enhancing raptor migration in the tropics is that the sun rises and sets more vertically there. This means that thermals form earlier and last later in the day in the tropics, which, in turn, means that soaring migrants can begin migrating earlier in the morning and continue to migrate later in the afternoon than at higher latitudes.

Intratropical Migration

Most tropical raptors are sedentary, stay-at-home species that time their breeding to take advantage of seasonal fluctuations in prey availability locally. Others, however, migrate within the tropics, typically among regions that exhibit particularly pronounced wet-dry cycles. The latter comprise a seemingly small group of several dozen raptors known as intratropical migrants. Distances traveled within the tropics often are modest, and in many instances the species involved appear to be partial rather than complete migrants. Consequently, for many intratropical migrants, the movements produce a shift in the center of species abundance without substantially changing the perimeter of their range. This, together with the fact that intratropical migration has been little studied, results in a frustratingly incomplete picture of these movements.

Several patterns are beginning to emerge, however, particularly among a group of African migrants, whose movements appear to be closely tied to those of the Intertropical Convergence Zone, or ITCZ, a latitudinal band of heavy rains that wavers seasonally north and south across the world's tropical regions. Convective activity within the ITCZ is intensive and extensive, with moisture-laden clouds forming continually to create all-but-daily midday rains, making the doldrums one of wettest places on earth. This band of profound precipitation, which extends farther north and south over land than over water because of greater day-night temperature variations in the former, straddles the tropics year-round. This "meteorological equator" shifts northward in boreal summer and southward in the winter, creating a series of rainy seasons across more than 20° of tropical latitudes. The cell's effect on intratropical migrants is best known in the Afrotropics, where many of the world's thirty-nine species of "rains migrants" occur.

Rains migrants in the region include the African Cuckoo-Hawk, a small and secretive kite that moves north in July and September on the heels of the retreating ITCZ in West Africa, and the Black-shouldered Kite, which migrates to avoid the rains.

ALTITUDINAL MIGRATION

Altitudinal migration in raptors consists of relatively short-distance movements between areas of high and low altitudes, usually along mountain slopes. Such movements occur because temperatures decrease at a rate of approximately 0.7 °C (1.3 °F) per 100 m up to about 10 km (6 mi) above sea level. Raptors inhabiting exceptionally high-mountain plateaus and mountain ranges, including Egyptian Vultures, White-rumped Vultures, Indian Vultures, and Lesser Fish Eagles in the Himalayas of central Asia, and Cinereous Harriers and Aplomado Falcons in the Andes of South America, are particularly prone to this type of migratory movement. Because the distances involved tend to be short, and because the movements themselves often are coupled with longer latitudinal migrations, altitudinal migration is decidedly less well understood than its latitudinal counterpart. Nevertheless, at least forty-five species of partial and irruptive or irregular migrants are thought to migrate altitudinally.

A GOLDEN AGE OF RAPTOR-MIGRATION SCIENCE

Scientific understanding often advances rapidly on the heels of innovative technology. This certainly has been true in raptor-migration science, with new satellite-tracking technology advancing the field in leaps and bounds. Developed at the US Army's Applied Physics Laboratory in Maryland in the 1980s, satellite-tracking technology now includes the use of 5- and 10-g, solar-powered, platform transmitter terminals (PTTs) capable of transmitting a 200–350-g raptor's location for several years. In a few instances, larger units affixed to larger raptors have been used to track individual birds of prey for more than a decade.

Initially tested on Bald Eagles in the autumn of 1984, satellite tracking has been used to follow the movements of one or more individuals of at least forty species of relatively large-bodied raptors. Used in conjunction with the French-US Argos-Tiros satellites, the system can provide locations of ± 250 m across the surface of the earth, although many locations for moving targets such as raptors typically fall within the range of ± 2 km. Recently, PTTs with GPS receivers have increased locational accuracy to ± 20 m.

One drawback to satellite tracking is its expense. The general rule of thumb is that it costs somewhere on the order of $1,500–4,000 for each unit, and about $200–800 per-year, per-unit for data downloads, with the magnitude of the expense depending on the manufacturer and whether the data are sent via Argos-Tiros satellites or GSM cell-phone towers. A new system, operating off the International Space Station and expected to lower costs significantly, is scheduled to be deployed in 2017 or 2018.

Most early studies using this technology managed to track only a handful of birds. But that has changed over the years. In the United States, a study involving Ospreys tracked 117 adults over a total of 164 "Osprey years," and a study of Alaskan Golden Eagles tracked more than seventy birds. An ongoing study of North American and South America Turkey Vultures had tracked the movements of fifty-nine birds as of mid 2016, including one individual that has been tracked for more than a decade. An ongoing study of Spanish Black Kites has tracked ninety-two individuals across 364 migration "episodes" as of 2014. All of these have added considerably to our knowledge of raptor migration in ways that were unimaginable two decades ago.

For the first time in history of raptor migration science, the birds "themselves" are now telling *where* and *when* and, to an extent, *how* they migrate and this, in turn, is allowing us to better understand the phenomenon of raptor migration. No longer are we simply generalizing or "modeling" how raptors migrate. Satellite tracking is now teaching us precisely how raptors migrate on individual-bird and individual-year bases, as well as changing the way we think about raptor migration. Until recently, many students of raptor migration have been so enamored with the birds' ability to do what they do that we have tended to characterize the migratory movements of birds of prey as something of a fait acompli, rather than as a flexible and modifiable behavioral work in progress.

My own and colleagues' studies of the migratory movements of more than fifty satellite-tracked Turkey Vultures in North and South America, for example, paint a far more intriguing picture of the movement ecology of this partial migrant than I initially imagined. Assuming that the movements of the birds we have studied are representative of other partial migrants—and there is evidence that they are—birds of prey, while astute navigators, are also quite flexible in their migratory behavior, both within and across geographic populations. But first some background.

We began studies of Turkey Vultures in September 2003, when Cornell University graduate student Jamie Mandel and Hawk Mountain Sanctuary research biologist David Barber caught and tagged an adult Turkey Vulture in the Central Appalachians of eastern Pennsylvania. Since then, numerous collaborators and I have caught and tagged vultures not only in Pennsylvania, but in Saskatchewan, southern California, southwestern Arizona, and the states of both La Pampa and Rio Negro, Argentina. Our long-term goal has been to study differences and similarities in the movement ecology of geographically distinct populations of this abundant partial migrant. As anticipated, some of the birds tagged in Pennsylvania traveled fewer than 100 kilometers (62 miles), into southern New Jersey, whereas others traveled well over 1600 km (950 mi), into southern Florida, and still other did not migrate at all, but rather stayed in Pennsylvania year-round. Most of the vultures tagged in Arizona migrated, but some only as far south as Mexico, whereas others went as far as central Panama. As expected, all of the birds in Saskatchewan migrated, most as far as Venezuela or Colombia, although several that were tagged as nestlings spent their first winter in central America. The birds from southern California overwintered in Mexico and Guatemala, and those from La Pampa and Rio Negro, Argentina, migrated north into tropical Bolivia, Brazil, and Colombia in austral winter.

On a population level, vultures that migrated the farthest traveled the fastest. For example, the vultures from western Canada that migrated more than 6000 km (3750 mi) to South America completed their outbound journeys to the wintering grounds in about 52 days, while averaging 246 km

(153 mi) per day, whereas the Pennsylvania vultures, traveling less than 1300 km (810 mi) to the southeastern United States, took an average 47 days to compete their outbound journeys, while averaging only 61 km (38 mi) per day. Vultures from the other four geographic locations were intermediate in distances and speeds of travel. In all six populations, individuals returned to their breeding areas more quickly on their return migrations in spring than they took to travel to their wintering areas in autumn. Although none of this was necessarily surprising, following the migrations of individuals from six geographically distinct populations affirms the remarkably flexible nature of migratory behavior in the species, which in part probably explains why this species, the most migratory of all scavenging birds of prey, is also the most widespread and, arguably, the most numerous of all vultures.

What was surprising was the amount of migratory flexibility that the satellite tracking revealed, both *within* geographic populations of Turkey Vultures, as well as *within* individual birds.

Consider Irma, for example. We caught and tagged Irma on 6 August 2004 at a rubbish dump in Penn Argyl, Pennsylvania, 15 km (9 mi) west of the Delaware River, which forms the Pennsylvania-New Jersey border. Named after the wife of Hawk Mountain's first ornithologist, Maurice Broun, Irma (whose sex we did not determine) was an adult when tagged (at least a three-year-old bird, based on bill color). "Her" satellite tracking device remained active into late 2015, more than a decade after it was deployed. For the first two winters after she was tagged, Irma migrated approximately 100 km (62 mi) south into southern New Jersey. Since then, except for one brief 140-km (87-mi) round-trip to coastal New Jersey in May 2013, she has remained sedentary and nonmigratory in a relatively small, <2000 km^2 (<1240 mi^2) home range that straddles the Pennsylvania–New Jersey state line, centered on the small city of Easton, Pennsylvania (Map 7.2). Why Irma switched from being a short-distance migrant to being sedentary is not clear. The two winters in which she migrated were not notably different, weather-wise, from those that followed, and we know of no new food resources that might have been available to her in later winters. Nor do we know why she made the round-trip to coastal New Jersey, although it may have been due to nest predation, as it occurred when breeding vultures would have been tending eggs in the area. But whatever the reason, Irma is not alone in this type of shift in migration behavior.

Jennie, one of the adult female vultures we tagged in the Sonoran Desert of southwestern Arizona in late May of 2013, also has showed both migratory and sedentary tendencies, but unlike Irma, both occurred in the same year.

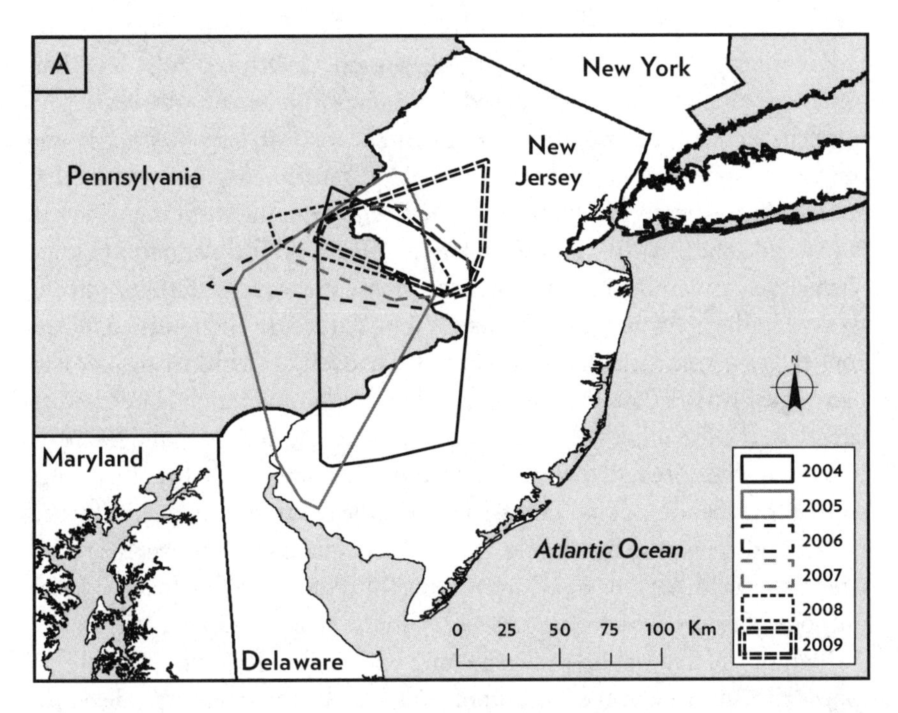

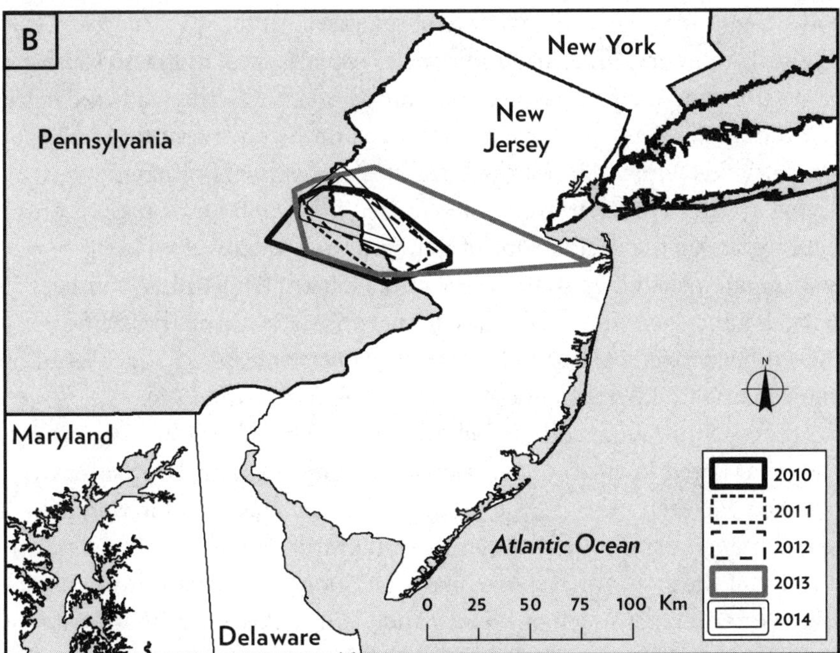

Map 7.2. Home ranges of Irma, an adult Turkey Vulture, 2004–2014. (A) The extended home ranges in 2004 and 2005 represent seasonal migrations of about 100 km (62 mi). (B) The 2012 home range, which includes easternmost New Jersey, represents a 2-day flight in May of that year, and may have followed disturbance at an active nest. Note the lack of seasonal migrations in 2006–2014.

Jennie migrated south into Sonora, Mexico, on 12 October 2013, less than 5 months after she had been tagged. Four days into her trip, however, and after she had traveled 466 km (290 mi) into northwestern Mexico, Jennie reversed course and backtracked for 6 days, returning to her summering area north of Tucson, Arizona, on 22 October, where she spent the winter of 2013–2014. The next two autumns, Jennie showed no inclination to migrate whatsoever, remaining in much the same home range through that winter as she had in the previous year. Why Jennie reversed course and returned to Arizona in the autumn of 2013 is not clear. She had left her summering area and had begun what I thought would be her outbound migration about a week later than the five other vultures we had tagged earlier that year, and her body mass at the time of capture was actually somewhat higher than the others, suggesting that she was not food stressed at the time of capture. Whatever turned her around appears to have worked, as she has managed to survive three winters in southern Arizona as of this writing, in spite of the fact that she is the only one of twelve Turkey Vultures tagged in the area that did not migrate.

Two of the five vultures tagged along with Jennie in May 2013 were no longer transmitting in the autumn of 2014, but the three that were also provide insight into the flexible nature of migration in the species. Desert Rat, Edward Abby, and Julie, all adult females when tagged, migrated southeast along the same Pacific Coast corridor into Mesoamerica they had used in the autumn of 2013, and all three overwintered in the same spot they had used the pervious winter, but all three traveled to distinctly different overwintering areas. Desert Rat overwintered 3350 km (2080 mi) southeast of the capture site on the Pacific slope of southernmost Mexico close to the border with Guatemala. Edward Abbey traveled 3860 km (2400 mi) to southern El Salvador and overwintered on the Gulf of Fonseca, again on the Pacific slope. And Julie traveled 5650 km (3510 mi) and overwintered on Lake Gatun on the Panama Canal in Atlantic slope central Panama (Map 7.3).

In many migratory birds, including many migratory raptors, individuals that breed in the same part of the species breeding range also overwinter near each other in the same part of the species wintering area, a phenomenon ornithologists refer to as migration connectivity. Obviously this is not the case with the Arizona population of Turkey Vultures, which overwinters across much of Mexico and Central America. But despite the lack of migration connectivity at the population level, individual connectivity was very much in evidence, with all three migrants overwintering in the same place in successive years: Desert Rat in southernmost Mexico,

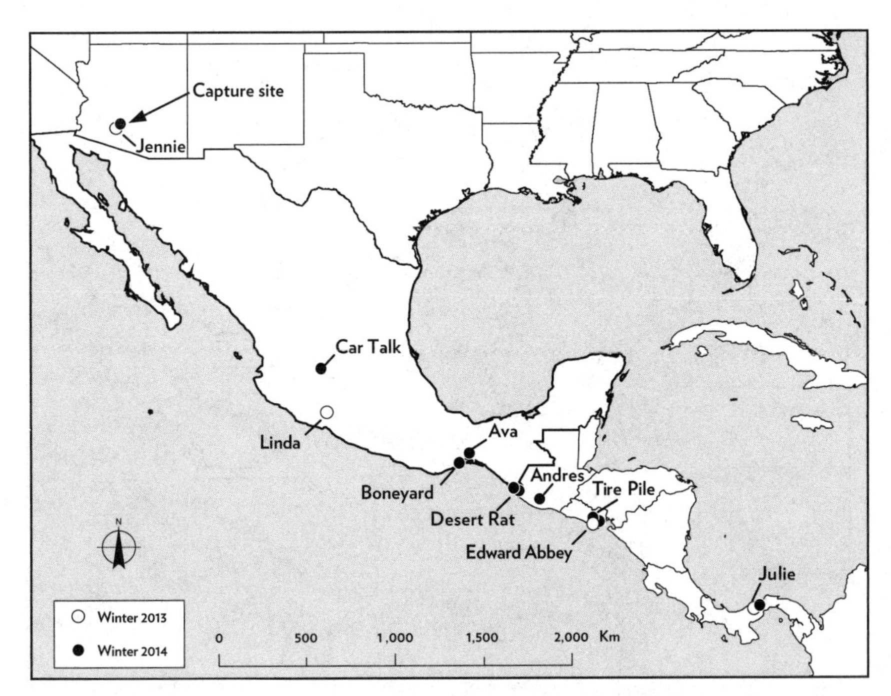

Map 7.3. Migratory movements of Turkey Vultures satellite tracked from the Sonoran Desert to overwintering areas in Mexico and Central America, 2013–2014. Note considerable variation in distances traveled among individuals and among-year consistency within individuals.

Edward Abbey on the Gulf of Fonseca in southern El Salvador, and Julie on Lake Gatun, in central Panama. That all three spent their winters in moist, coastal tropical areas, after having summered in the Sonoran Desert of southwestern Arizona, is also something of a mystery.

Yet another example of migration flexibility involves Leo, an adult female Turkey Vulture trapped and tagged as a breeding adult at her nest scrape in an abandoned farmhouse, at 53° N latitude in Leoville, central Saskatchewan, on 16 June 2007. We have tracked Leo on eight round-trips to her wintering range in the Llanos wetlands of western Venezuela at 7.5° N latitude. Leo has maintained very much the same winter range each winter and returns to precisely the same breeding site each spring. For the most part, she also migrates along the same migration flyway each spring and autumn, a more than 6000-km (3730-mi) one-way journey that takes her from Saskatchewan, Canada, through North Dakota, South Dakota, Nebraska, Kansas, Oklahoma, and Texas in the United States, and thereafter through eastern Mexico, Guatemala, Nicaragua, Costa Rica, Panama, and Colombia, into Venezuela (Map 7.4).

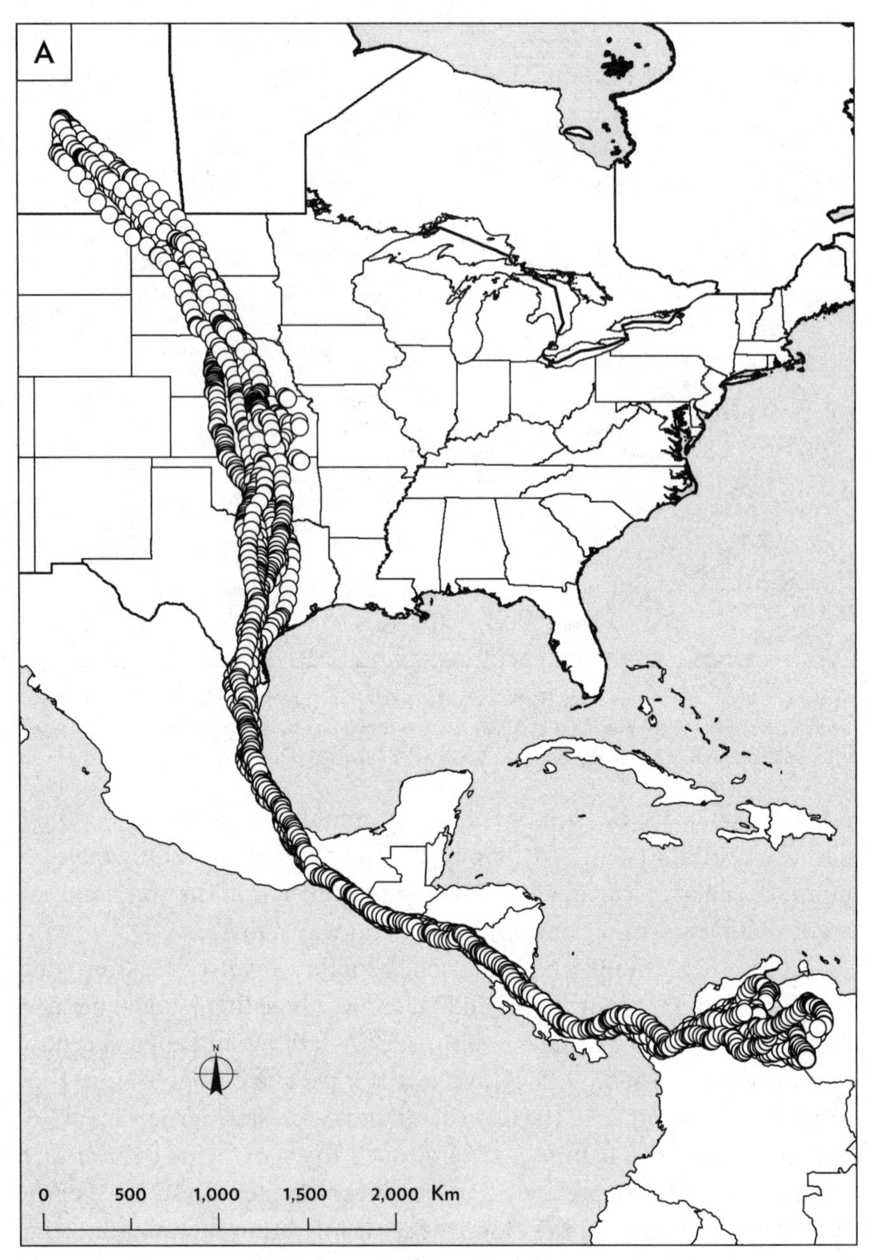

Map 7.4. Migratory movements of Leo, an adult female Turkey Vulture, between her nest site in central Saskatchewan and her wintering grounds in northwestern Venezuela. (A) From nest site to wintering grounds, 2007–2014 (B) From wintering grounds to nest site, 2008–2015

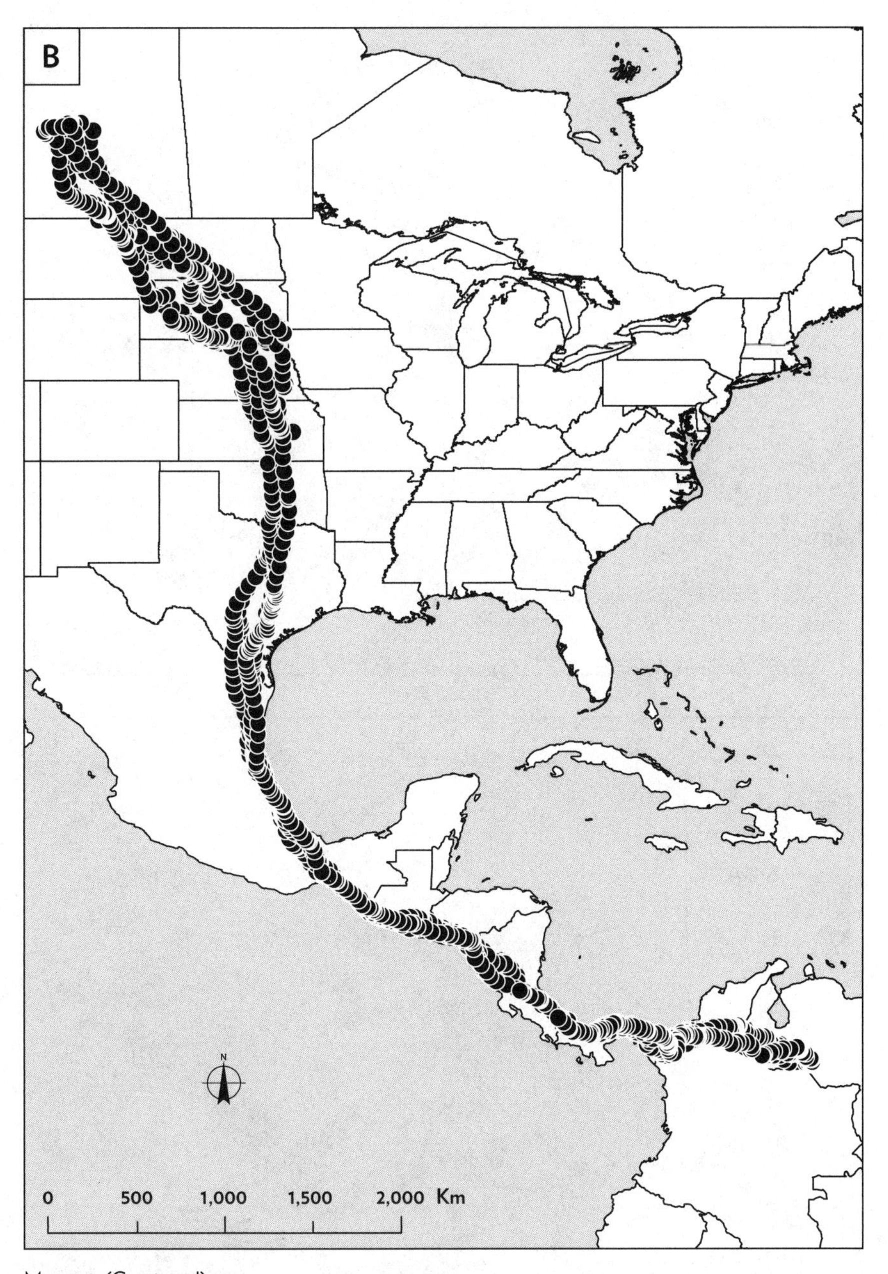

Map 7.4. (Continued)

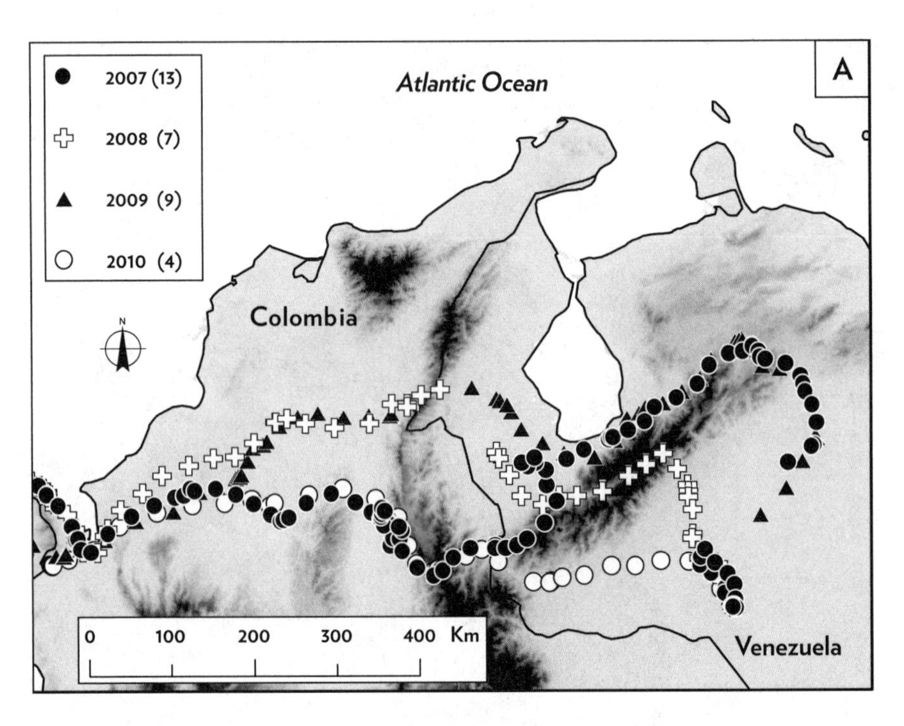

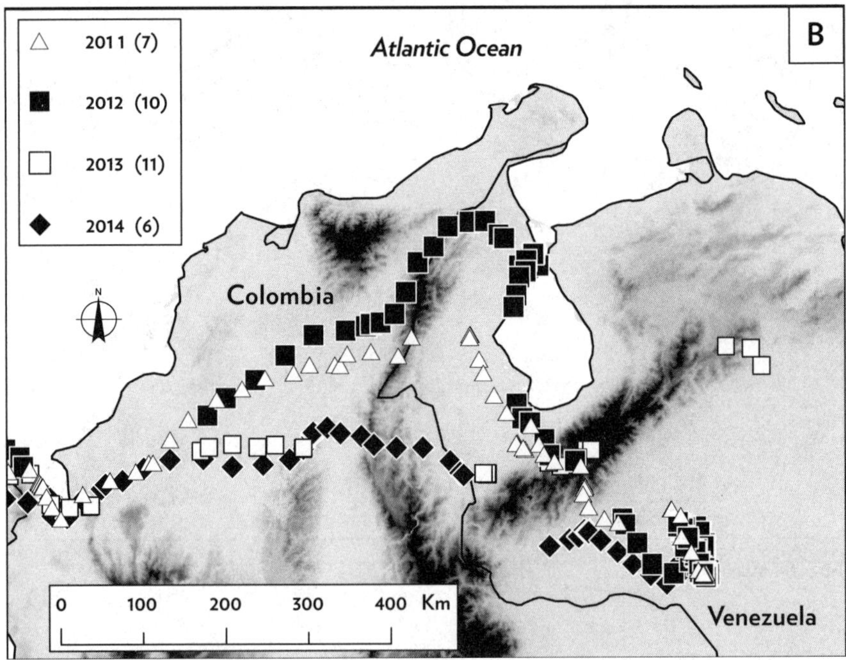

Map 7.5. Outbound migratory movements of Leo, an adult female Turkey Vulture, while the bird was navigating through and around the Andes Mountains. (A) 2007–2010 (B) 2011–2014

Although her flyway is measurably "wider" longitudinally north of the Texas-Mexico border, most likely because of strong winds in the Central Plains of the United States, it is even more varied among years in northern Colombia and Venezuela in autumn, but not in spring, when the bird approaches the Serranía de Perijá, a rugged northern branch of the Andes that forms the border between northern Colombia and northern Venezuela, and the even taller Cordillera de Merida, south of Lake Maracaibo, Venezuela (Map 7.5). In the eight years since 2007, her tortuous journeys through this mountainous region have ranged from 958 km (595 mi) in 2010 to 1746 km (1085 mi) in 2007, with travel times ranging from 4 days in 2010 to 13 days in 2007 (Map 7.5). Almost certainly, the differences reflect the high-mountain fog in the region, with Leo detouring around the higher elevations in her attempt to reach her preferred overwintering site in the lowland Llanos of western Venezuela.

I could go on with additional examples of flexibility in the migratory movements of both tracked Turkey Vultures and those of other tracked raptors. Overall the picture of raptor migration that is being revealed as a result of satellite telemetry is one of an extremely robust tool that allows us to see how birds of prey shuttle between appropriate breeding areas and wintering sites on a routine basis in ways that were largely unknown and unappreciated until the advent of satellite tracking.

SYNTHESIS AND CONCLUSIONS

1. Because of behavioral flexibility and rapid evolution, shifts in migration tendencies can appear and disappear within decades.
2. The atmosphere, climate, and weather play major roles in determining the flight strategies of birds of prey.
3. Soaring flight saves raptors considerable energy during their migrations, and many long-distance migrants depend on soaring flight to complete their journeys.
4. Raptors soar over both land and water.
5. Many long-distance migrants flock when migrating. Migrants in large flocks soar more efficiently than those flying alone or in smaller flocks.
6. Most raptors migrate entirely by day. A few species also migrate by night.

7. Migratory raptors fuel their migrations by laying down fat prior to migration, by feeding en route, and by extracting energy from the atmosphere while soaring.
8. Raptors flock more on migration that at other times of the year.
9. In spring, the "urge" to reproduce plays a role in the timing of departure from the wintering grounds, but not in the speed of migration.
10. Sex- and age-related differences in migration are widespread in raptors.
11. Many migrants that feed on small vertebrates on their breeding grounds switch to feeding on swarming insects on their wintering quarters.
12. The wintering ranges of most species of migrants are decidedly smaller than their breeding ranges.
13. Most raptors do not undertake water crossings of more than 25 kilometers (16 miles).
14. Many migrants take different routes on outbound and return migration, resulting in elliptical, or loop, migration.
15. Satellite tracking is vastly improving our understanding of raptor migration.

8 RAPTORS AND PEOPLE

[The] hawk-hater wrote that . . .
For him the only good hawk was a dead hawk.
. . . Now what can you do with a fellow like that?
Maurice Broun,1949

THE RELATIONSHIPS BETWEEN BIRDS of prey and people have had their ups and downs over the years. That said, compared with many other groups of birds, raptors have suffered disproportionately from human actions. Maurice Broun's remarks above, with which he concluded *Hawks Aloft,* his book about Hawk Mountain Sanctuary, distill the essence of one of those downs: deliberate human persecution.

The human-raptor relationship was not always that way. Anthropologists suggest that more than a million years ago, sub-Saharan protohumans followed soaring vultures converging on large-mammal carcasses to gather food for their families, in much the same way as jackals and hyenas do today. Vultures were "friends" in those days, and in some places they still are. Some anthropologists have even suggested that this ancient ecological relationship helped foster the communication skills in humans that allowed individuals returning from distant, out-of-sight carcasses to describe accurately feeding opportunities to group members. If this is so, then scavenging raptors may have played a significant role in the development of humanity.

On the other hand, evidence in the form of fatal skull punctures suggests "that a large raptor was responsible for the death of a circa two-million-year-old Taung child, a holotype of the early hominin species *Australopthecus africanus.*" (The conclusion, which seems reasonable, is based on comparisons of the wounds on the juvenile hominin's skull with more recent ones in monkeys killed by Crowned Eagles.) In New Zealand, the now-extinct Haast's Eagle—a 15-kg (30-lb) predatory raptor that fed on flightless Moas weighing up to fifteen times its body mass—is mentioned in Maori legends as taking early human settlers. Perhaps not surprisingly, this potential "man-eater" became extinct in the 1400s, fewer than 200 years after New Zealand was settled by Polynesians.

Raptors appear as either gods or messengers of gods in many ancient religions, with a number being featured as "symbols of strength and courage." The ancient Egyptian god Horus took the form of a falcon, and Christianity's Old Testament is full of references to various birds of prey, including several omens associated with vultures. Raptors also figure largely in the aboriginal lore of Australia. Black Kites, for example, were believed to spread flames purposely to increase their hunting success, a belief most likely derived from the fact that the species often congregates along fire lines while searching for burned prey. In North America, "thunderbirds," presumably large eagles or California Condors, were thought to be sources of thunder, and in South America, the famed Peruvian poet, Pablo Naruda, once described Black Vultures as "God's spies."

Falconry, which dates back six millennia, has been one of the tightest relationships between raptors and humans. Defined as the "capture of quarry with trained birds of prey," this ecological relationship is believed to have begun in central Asia thousands of years ago. Popular well into the Middle Ages and still practiced today, falconry and its techniques, particularly those involved in breeding young birds in captivity and subsequently releasing them into the wild, has played a major role in the reintroduction of endangered species globally.

Most conservationists date the era of human-induced extinction from the demise of the flightless Dodo on the Indian Ocean island of Mauritius, 330 years or so years ago, and we have seen an unfortunate shift in the human-raptor relationship in the years thereafter. For birds of prey, direct human persecution, land-use change, and environmental contaminants, including pesticides and heavy metals, have been potent forces in reducing raptor populations to fractions of what they otherwise naturally would be. Surprisingly, only one bird of prey has become extinct during this period, the Guadalupe Caracara. An island-dwelling scavenger, the caracara had the misfortune of being shot and poisoned to its eventual demise at the beginning of the twentieth century by a lethal combination of local shepherds, who feared for their livestock, and a series of overzealous museum curators, who sought the rarity for their collections.

Although not a single species of migratory raptor is known to have been lost at the hands of humans, many populations of migratory and sedentary birds of prey have been and remain threatened by a variety of human actions. By mid-2015, fifty-five species of birds of prey were listed as Critically Endangered (ten species), Endangered (fifteen), or Vulnerable (thirty) globally (Table 8.1).

Table 8.1 The world's Critically Endangered, Endangered, and Vulnerable diurnal birds of prey

Species	Regional distribution[a]	Conservation status[b]	Threats
Cathartidae			
California Condor	NA	CR	Lead, habitat degradation, poisoning (historical)
Sagittaridae			
Secretarybird	AF	VU	Trade, habitat degradation
Accipitridae			
White-collared Kite	SA	CR	Deforestation
Black Honey Buzzard	AU	VU	Deforestation
Cuban Kite	NA	CR	Deforestation
Sanford's Fish Eagle	AU	VU	Deforestation
Pallas's Fish Eagle	EU	VU	Habitat degradation and loss
Steller's Sea Eagle	EU	VU	Habitat degradation
Madagascar Fish Eagle	AF	CR	Deforestation
Egyptian Vulture	EU, AF	EN	Poisoning, direction persecution
Hooded Vulture	AF	EN	Poisoning, direct persecution, bush meat
White-backed Vulture	AF	EN	Poisoning, habitat degradation, persecution
White-rumped Vulture	OR	CR	Veterinary drug
Indian Vulture	OR	CR	Veterinary drug
Slender-billed Vulture	OR	CR	Poisoning, habitat degradation, persecution
Rüppell's Vulture	AF	EN	Poisoning, habitat degradation, persecution
Cape Vulture	AF	VU	Poisoning, habitat degradation, persecution
Red-headed Vulture	OR	CR	Veterinary drug, land-use change
White-headed Vulture	AF	VU	Habitat degradation, poisoning
Lappet-faced Vulture	AF	VU	Poisoning
Malagasy Marsh Harrier	AF	EN	Persecution, land-use change
Black Harrier	AF	VU	Land-use change
Beadouin's Snake Eagle	AF	VU	Land-use change
Madagascar Serpent Eagle	AF	EN	Deforestation
Mountain Serpent Eagle	OR	VU	Deforestation
Nicobar Sparrowhawk	OR	VU	Habitat degradation
Slaty-mantled Goshawk	AU	VU	Deforestation
Imitator Goshawk	AU	VU	Deforestation

(Continued)

Table 8.1 (Continued)

Species	Regional distribution[a]	Conservation status[b]	Threats
New Britain Goshawk	AU	VU	Deforestation
Grey Sparrowhawk	AU	VU	Deforestation
Gundlach's Hawk	NA	EN	Deforestation, persecution
Plumbeous Hawk	SA	VU	Deforestation
White-necked Hawk	SA	VU	Habitat fragmentation
Grey-backed Hawk	SA	EN	Deforestation
Crowned Solitary Eagle	SA	EN	Land-use change
Ridgeway's Hawk	NA	CR	Persecution, at least historically
Galapagos Hawk	SA	VU	Persecution, at least historically
Socotra Buzzard	EU	VU	Trade (?)
Papuan Eagle	AU	VU	Feather trade
Spanish Imperial Eagle	EU	VU	Electrocution, poisoning
Eastern Imperial Eagle	EU	VU	Land-use change
Indian Spotted Eagle	OR	EN	Deforestation
Greater Spotted Eagle	EU	VU	Land-use change
Martial Eagle	AF	VU	Persecution, poisoning
Flores Hawk-Eagle	AU	CR	Land-use change
Javan Hawk-Eagle	OR	EN	Land-use change, persecution
Philippine Hawk-Eagle	OR	EN	Land-use change
Wallace's Hawk-Eagle	OR	VU	Deforestation
Black-and-chestnut Eagle	SA	EN	Deforestation
Falconidae			
Plumbeous Forest Falcon	SA	VU	Deforestation
Seychelles Kestrel	AF	VU	Land-use change, pesticides
Mauritius Kestrel	AF	EN	Land-use change
Grey Falcon	AU	VU	Land-use change
Saker Falcon	EU	EN	Land-use change, falconry
Taita Falcon	AF	VU	Land-use change

Source: Birdlife Red List as of August 2015 (www.birdlife.org/datazone).
[a] AF = Africa, AU = Australasian Region, EU = Europe, NA = North America, OR = Oriental Region, SA = South America.
[b] CR = Critically Endangered, EN = Endangered, VU = Vulnerable.

It is against this backdrop that I introduce a series of ecological principles of raptor endangerment, and provide several examples of both historical and current human threats to birds of prey. My emphasis throughout the chapter is that gathering and sharing of ecological knowledge is key to practical and effective raptor protection.

PRINCIPLES OF RAPTOR ENDANGERMENT

Although the threat of imminent extinction looms for only a handful of raptors, populations of many of the world's birds of prey, including those of many common and widespread species, are now in decline as a result of human actions. Although each population's story is unique, a number of general principles apply. Understanding these principles of endangerment allows one to anticipate future endangerment and, more importantly, prevent it.

Principle One: There are three general categories of human threats to raptors, and their effects can be additive. These threats include (1) land-use change and the loss of natural habitats, (2) the misuse of environmental contaminants, including pesticides, and (3) direct, full-frontal assault, including poisoning, shooting, and trapping and subsequently killing birds of prey. At the beginning of the twenty-first century, populations of more than half of all migratory birds of prey were threatened by land-use change, more than one in five were threatened by environmental contaminants, and nearly a third were threatened by direct human persecution assault. Overall, two-thirds were threatened by at least one of these factors, close to a third by at least two of these factors, and almost one in ten by all three factors.

Multiple threats to raptors typically work additively. For example, direct persecution often forces raptors into areas far from human populations and their actions where, subsequently, raptors are threatened by land-use change when these "more natural" landscapes undergo development. Similarly, land-use change in the form of agricultural development can threaten raptors through the loss of more natural landscapes, as well as by increased exposure to environmental toxins and contaminants.

Principle two: Land-use change and, in particular, tropical deforestation, currently remains the principal threat to raptors overall. This is especially so in tropical regions, where many natural landscapes are being modified by humans. Recent global analyses suggest that the Indomalayan tropics of southeastern Asia and the Pacific Islands, and the African tropics of sub-Saharan Africa, both of which in northern winter host many species of migrants from Eurasia, currently are two of the most disturbed areas on earth.

This is a problem because more than half of all threatened and near-threatened tropical raptors are forest-dependent birds of prey, many of which require enormous tracts of forest to survive. Because many tropical habitats are threatened by increasing levels of human disturbance and land-use change, their degradation is likely to remain a concern for some time. Open-habitat migrants, including those that depend on prey associated with seasonally fluctuating wetlands, or on swarming locusts and other seasonally available upland prey, also are vulnerable to land-use change. This is so because many of these species have nomadic and gregarious lifestyles and, as a result, many individuals can be affected simultaneously by single pesticide applications. And even if pesticide applications themselves do not poison raptors, the successful removal of agricultural pests, including locusts that act as nutritional resources for birds of prey, can threaten raptors via reduced prey availability.

Principle three: Rare raptors are at a higher risk of extinction than common raptors. Species with small populations—large raptors, and island-dwelling species and other endemics—have a greater risk of extinction, not only because small populations provide little buffer against rapid environmental change, but also because the smaller ranges associated with small populations tend to place most, if not all, members of the species in the same small ecological boat geographically. An important corollary to this is that migratory raptors tend to have larger populations overall than do sedentary raptors and, therefore, are less prone to extinction. This is not to say, however, that common raptors are immune from human threats.

Principle four: Migratory species face particular kinds of risks. Although it is difficult to collect survival data on birds of prey during their migrations, the information that does exist suggests that migratory movements, particularly long-distance movements, are one of the more difficult and dangerous times in a bird's life. Although many migratory raptors reduce their energy needs while migrating by soaring and gliding together within well-established updraft corridors, the concentrations of raptors that result can create problems. These include increased competition, the spread of contagious diseases, and the increased likelihood that a single, local problem will negatively affect a large portion of a species' regional or, possibly, world population.

Another threat unique to migrants arises from "double endemism." This occurs when migratory raptors, such as Levant Sparrowhawks, both breed and overwinter in relatively small and restricted ranges and, therefore, depend heavily on the stability of not only one, but two, small regions for species survival. If either area undergoes rapid ecological degradation, many individuals will face the problem simultaneously, increasing significantly the likelihood of specieswide endangerment. Many migrants have large breeding ranges but far smaller overwintering ranges. Although such species are at lower intrinsic risk than are double endemics, they too face a greater risk of extinction than do species that have both large breeding ranges and large wintering ranges.

Principle five: Scavenging birds of prey are at greater risk than are predatory raptors. Although vultures are one of the world's most effective vertebrate scavengers, they also are one of its most endangered groups of birds of prey. As of mid-2015, eleven of the world's twenty-two species of obligate scavenging birds of prey, or half, were threatened with extinction, compared with less than 17% of all raptors, including obligate scavengers. In most cases, threats are linked to toxins found in the carcasses the birds feed on. In addition to the veterinary drug diclofenac (see Environmental Contaminants: Veterinary drugs, below), some have suggested that antibiotics fed to livestock as growth enhancers may speed the evolution of superpathogens that could negatively affect the health of vultures consuming treated carcasses.

In Africa, recent declines in at least eight species of vultures have been linked to a number of causes, including direct human persecution, such as hunting them for food, using them in witchcraft, and incidental killings at carcasses laced with poisons to eliminate mammalian predators such as lions, hyenas, and jackals. In southern Africa, vultures are poisoned illegally because poachers know that game wardens use large gatherings of vultures to help locate poachers at elephant and rhinoceros kills. Complicating the issue is that many vultures are highly social foragers that routinely feed in large, mixed-species groups. This means that large groups of several species can be exposed to dietary toxins at one "common table," enhancing the likelihood of rapid, large-scale declines. Overall, although the circumstances may differ considerably in different parts of the world, dietary toxins rank as the number-one threat to scavenging raptors globally.

The consequence of removing avian scavengers from both natural and human-dominated landscapes remains little studied, but data from East Africa indicate that vultures consume more meat than do the region's other vertebrate predators, suggesting an important role for these scavengers, at least in that part of the world. In addition, a recent study in India estimated that feral dog populations increased by nearly one-third during the collapse of vulture populations there. With this came a substantial increase in the number of human deaths from rabies, carrying an economic cost estimated at more than $30 billion between 1992 and 2003.

A BRIEF HISTORY OF HUMAN IMPACTS ON BIRDS OF PREY

People have been negatively affecting raptor populations for a long time. Hunter-gatherers along the shorelines of San Francisco Bay killed and deposited the remains of nine species of migratory raptors, including Bald Eagles and Peregrine Falcons, in shell middens dating from 600 BCE. Much the same is likely to have happened elsewhere. Some of humanity's greatest prehistoric impacts occurred on Pacific Islands settled by Polynesians and other seafarers. Species of eagles and hawks that were major ecological features in Hawaii disappeared soon after people arrived on the islands. Similarly, top avian predators disappeared from New Zealand shortly after most of their lowland habitats were burned and hunted free of potential prey by human colonists.

In sixteenth-century Europe, officials encouraged landowners to seek out and destroy raptors to protect livestock and royal game. The extermination campaigns that followed targeted most species of birds of prey, and many common raptors were extirpated from substantial parts of their historic ranges. The Red Kite, for example, which earlier had been one of the most ordinary and widespread raptors in Great Britain—and once thrived on the streets of London in the early 1500s—was reduced shortly thereafter to a single isolated population in remote central Wales. Conservation efforts to bolster this remnant population, which had been under way for more than a century, eventually succeeded, and there are now self-sustaining populations of Red Kites in Scotland and England, and, more recently, Ireland. Unfortunately, band returns and direct observations indicate that even today, this species suffers from poisoning, shooting, and other illegal persecution in Great Britain.

The full impact of the direct human persecution of raptors in Europe will probably never be known, not only because most efforts were never officially documented, but also because most of the mistreatment happened well before natural-resource conservationists cared about the losses. What is known indicates that populations of many large raptors were substantially affected and that regional extirpations occurred in both urban and rural areas. We also know that the primary habitats of many raptors shifted from open, human-dominated landscapes to remote, forest-dominated, high-elevation areas, and that in spite of more than 50 years of substantially reduced persecution in the latter twentieth and early twenty-first centuries, many species have yet to return to pre-persecution levels of abundance.

An All Too Typical New World Example: Raptor Persecution in Pennsylvania

Some of the strongest evidence of human persecution of birds of prey comes from the Commonwealth of Pennsylvania. As was true in much of the United States, raptors were totally unprotected throughout the eighteenth and nineteenth centuries, a time when persecution was quite common. Such persecution increased substantially in the middle of the nineteenth century, when most rural residents considered the birds to be threats and when the recent invention of breech-loading guns allowed people to kill them more easily. By 1885, hatred for raptors reached a fevered pitch, and the commonwealth offered a 50-cent bounty on the heads of all birds of prey, including owls. The "Scalp Act" was embraced by most rural residents, and polls suggest the law was also supported by at least 90% of the general public. Over the next two years, 180,000 raptor "scalps" were sent to Harrisburg, the capital, even as populations of destructive rodents and insects increased, and numerous fraudulent claims drained the state treasury. State legislators finally repealed what by then many, including the state veterinarian and author of *Diseases and Enemies of Poultry* Leonard Pearson, called an unjust, uneconomic, foolish, and simply wrong-headed law.

Pennsylvania reinstated bounties in 1913.

In 1921, a year in which the state paid a total of $128,269 in predator bounties, its own employees killed at least 603 hawks and destroyed forty-one nests. Birds of prey remained unprotected statewide until 1937, when, except for the three bird-and-game-eating accipiters—the Sharp-shinned Hawk, Cooper's Hawk, and Northern Goshawk—all raptors received legal protection.

The new protective measures were unpopular among farmers and hunters, however, and limited enforcement and misidentification continued to plague populations of legally protected species well into the 1960s. A specific bounty on Northern Goshawks established in 1929 was removed in 1951, but it was not until 1969 that the state granted full legal protection to all three accipiters. The Great Horned Owl and Snowy Owl remained unprotected statewide until 1972, when the United States ratified an amendment to the Migratory Bird Treaty Act of 1918 that extended federal protection to all species of birds of prey throughout the country.

Why did it take Pennsylvania so long to protect its birds of prey? Awkwardly, part of the answer lies within the conservation community of early twentieth century North America. Most conservationists who had opposed the "fool hawk law" of 1885 took exception principally to its nonselective nature: that is, that the diminutive American Kestrel was targeted along with the fierce Northern Goshawk, for example, rather than to the notion that some hawks needed to be destroyed. And indeed, the latter half of the nineteenth century and the first half of the twentieth century was the age of "good hawks" (the rodent-eating) and "bad hawks" (the bird- and game-eaters) in conservation. The latter, including so-called chicken hawks and duck hawks, needed to be controlled, while raptors that preyed on rodents and other agricultural pests did not need to be controlled. Consider, for example, the following passage from a 1913 autobiography of the father of modern North American conservation, John Muir:

> When I went to the stable to feed the horses, I noticed a big white-breasted hawk [probably a Northern Goshawk or Red-tailed Hawk] on a tall oak tree in front of our chicken house, evidently waiting for a chicken breakfast . . . I ran to the house for a gun, and when I fired, he fell . . . then managed to stand erect. I fired again to put him out of pain. He flew off . . . but then died suddenly in the air, and dropped like a stone.

Although the event that Muir refers to took place when he was a young boy in 1850s Wisconsin, the founder of the Sierra Club expressed absolutely no remorse (other than in finishing off the culprit quickly) when recalling this exploit more than 50 years later.

John Muir was not alone in this thinking. William T. Hornaday, the savior of the American bison, director of the New York Zoological Park, and founder of the Permanent Wildlife Protection Fund, had very much the same view in 1913:

> "Chicken hawk or hen hawk" are usually applied to the [Red-shouldered Hawk] or the [Red-tailed Hawk] species. Neither of these is really very destructive to poultry, but both are very

> destructive to mice, rats and other pestiferous creatures. . . . Neither of them should be destroyed—not even though they do once in a great while, take a chicken or wild bird, however [t]here are several species of birds that may at once be put under the sentence of death for their destructiveness of useful birds, without any extenuating circumstances worth mentioning. Four of these are cooper's hawk, the sharp-shinned hawk, pigeon hawk [Merlin] and duck hawk [Peregrine Falcon].

Hornaday's distinctions were both moralistic and utilitarian: "The ethics of men and animals are thoroughly comparative. . . . Guilty animals, therefore, must be brought to justice."

By 1931, Hornaday had dropped the Merlin from his list of "bad hawks," presumably because of its increased rarity, but retained the others, along with the Great Horned Owl, Barred Owl, and, amazingly enough, the tiny Eastern Screech Owl.

Scientific and bird-watching communities of the time, too, carefully selected the raptors they were concerned with. Mabel Osgood Wright, founding president of the Connecticut Audubon Society, recommended in *Birdcraft* (1936) that one could help songbirds by "shooting some of their enemies," including several kinds of hawks. Boy Scouts, too, received instruction in what was considered raptor "conservation taxonomy." George E. Hix, a scoutmaster in Brooklyn, New York, writing in 1933 in *Birds of Prey for Boy Scouts,* noted, "the beneficial hawks are the larger, slower species, [and] the smaller swifter hawks are the ones which are destructive to wildlife . . . [and these include] the [Northern] goshawk, Cooper's, sharp-shinned, and pigeon hawks [Merlin]."

Unfortunately, because shooters tended to be unable or unwilling to distinguish the "bad hawks" from the "good hawks," birds of prey, both "good" and "bad," were at risk.

The Special Case of America's Bald Eagles

Unfortunately, the conservation history of America's national emblem resembles that of several other birds of prey. Mark Catesby, writing in the *Natural History of Carolina, Florida, and the Bahama Islands* (1731–1743), lauded the species' "great strength and spirit," but also noted its tendency to feed on pigs, something his ornithological successors, Alexander Wilson

(1808–1814) and John James Audubon (1840), repeated in the early 1800s. And Audubon went further, characterizing the Bald Eagle as "exhibit[ing] a great degree of cowardice." He also agreed with the colonial statesman Benjamin Franklin, who stated that, "For my part, I wish the bald eagle had not been chosen as the representative of our country; he is a bird of bad moral character . . . [and] a rank coward."

In light of such comments, it is not surprising that by the beginning of the twentieth century, Bald Eagle numbers were low enough for William T. Hornaday, writing in 1931 in *Thirty Years War for Wild Life,* to declare: "will someone tell me of a spot in the United States where the [bald] eagle is a 'pest'? Can any man this side of the Pacific Northwest Coast go out and find a pair of these birds in less than a week of diligent search?" In fact there was such a place, and Hornaday both pointed it out and recommended a solution: "The [bald] eagle is, in a few places in Alaska, too numerous; and there it should be thinned out." And thinned it was.

Between 1917 and 1952, the Territory of Alaska paid bounties of 50 cents to 2 dollars a head on 128,000 Bald Eagles. This translates to an average take of slightly more than ten birds per day for 35 years. And Alaska was not the only place that eagles were shot in large numbers during the early twentieth century. Charles Broley, who began banding nestling Bald Eagles in central Florida in the late 1930s and continued to do so into the late 1940s, noted that forty-eight of the 814 nestlings he had banded between 1939 and 1946 were later recovered as "shot." Shooting—sometimes from airplanes—was not the only way that Bald Eagles were persecuted. In the 1920s and 30s, poultry farmers in southern New Jersey routinely cut down potential nesting trees for eagles whenever they came upon them. By the middle of the twentieth century, direct human persecution of the Bald Eagle, coupled with the growing misuse of the pesticide DDT (see below), were having quite an impact.

Concerned that the species was on the verge of extirpation outside of Alaska, Congress enacted the Bald Eagle Protection Act of 1940. The law, which initially excluded the Alaska Territory from its provisions, was amended in 1959, when Alaska became the forty-ninth state. It was again amended in 1962 to extend protection to Golden Eagles because of their resemblance to young Bald Eagles. The Bald Eagle, which was first listed as federally endangered in 1967, remained on the list as threatened in the lower forty-eight states in the early twenty-first century, and midwinter surveys in the conterminous United States suggest that populations increased at an annual rate of close to 2% between 1986 and 2000, with most growth occurring in the northeastern United States. All indications are that the species has rebounded both from its persecution and its pesticide-era lows, as the sanctuary's recent record-setting high counts indicate. The Bald Eagle was removed from the United States Endangered Species List in August 2007.

MORE RECENT THREATS TO RAPTORS

Land-Use Change

The ecological footprint of human civilization has negatively affected migratory raptors in many ways. Historically, the advent of cultivated crops and, soon thereafter, large-scale row-crop farming, increased the rate at which people co-opted the primary productivity of natural habitats. At first, the growth of such landscapes reduced both prey populations and nesting habitat for raptors. Eventually, the development of modern agriculture produced the additional threat of pesticides.

Today, land-use change challenges more raptors in the Old World than in the New World. At the beginning of the twenty-first century, slightly more than half of all African migratory raptors were threatened by land-use change, along with 61% of all European migrants and 58% of Australasian and Pacific Island migrants. By comparison, only 41% of all migrants were threatened by land-use change in North America and the Neotropics. The geographical differences largely reflect the rates of land-use changes in the regions involved.

As African grasslands and savannas continue to shift from natural grazing to a combination of row-crop agriculture and wild-game ranches, populations of migratory birds of prey, particularly those that depended heavily on seasonally abundant insects and the "double endemics," are likely to suffer heavily, particularly if pesticides are used to control insect pests. Similarly, the accelerated loss of tropical forests in Africa promises to threaten forest-dependent species such as European Honey Buzzards, almost all of which overwinter there. In Australasia, extensive habitat loss in the Indomalayan and Australasian tropics threatens several migratory species, including the Crested Honey Buzzard. In Europe, agricultural practices, including shifts from low-impact pastoral farming to high-impact industrialized agriculture are affecting many species, including the region's vultures, which depend on the carcasses of large mammals, as well as harriers and small falcons, which depend on natural grasslands for nesting and feeding.

Threats to Neotropical forests have attracted the attention of conservationists for many years. But although these losses continue to threaten several of the region's sedentary endemics, land-use changes in Neotropical savannas and semi-open woodlands pose an even greater threat to many of the region's short- and long-distance migrants, including North American populations of Turkey Vultures and Swallow-tailed Kites.

Environmental Contaminants

Humanity continues to contaminate ecosystems with numerous natural and synthetic substances, many of which threaten populations of birds of prey. As predators and scavengers, raptors are particularly vulnerable to such threats because they frequently encounter contaminants that accumulate in the prey and carcasses they feed on.

In most instances, the negative effects of this environmental contamination are unintended and unanticipated, but the costs for cleaning up and reversing population declines are significant and substantial and are therefore often delayed, compounding the impact. Much of the threat stems from the systematic misuse of agricultural pesticides, which impacts raptors indirectly when they feed on poisoned pests and other organisms targeted by the pesticides, and directly when they consume pesticides themselves as nontarget species.

Organochlorine pesticides. Perhaps the most infamous pesticide of all is DDT (**d**ichloro-**d**iphenyl-**t**richloroethane), a synthetic organochlorine pesticide that came into widespread use across much of Canada, the United States, and Europe in the late 1940s and early 1950s. An inexpensive, broad-spectrum pesticide—that is, it could be used against many insects and other invertebrate pests—DDT was a long-lasting pesticide that was far less toxic to vertebrates than the inorganic pesticides it replaced. In fact, DDT earned Paul Müller, a Swiss entomologist who developed the chemical for use in agriculture, a Nobel Prize in 1948. Although this "wonder chemical" and its widespread use raised alarms among conservationists as early as the late 1940s, it and other organochlorine pesticides quickly became the insecticidal agents of choice in the agricultural world of the 1950s and 1960s.

Field workers, including Hawk Mountain's Maurice Broun, began to notice regional declines in the reproductive success of Bald Eagles as early as the late 1940s, but it was not until 10 years later that anyone linked the declines to organochlorine pesticides. In the early 1960s, Rachel Carson's bestseller, *Silent Spring*, which included mention of declining counts of Bald Eagles at Hawk Mountain Sanctuary, placed the pesticide problem in the hearts and minds of Americans, and in 1965 a group of concerned raptor scientists met in Madison, Wisconsin, to discuss the situation. By then, populations of Peregrine Falcons and other raptors had been in rapid decline across much of eastern North America and Western Europe for almost a quarter of a century.

By 1964, a survey of more than a hundred historic Peregrine Falcon breeding sites across the eastern United States failed to locate a single parental bird. In addition to the speed with which this extirpation had occurred, what was particularly striking about the decline was that the Peregrine Falcon was a widespread, near-cosmopolitan, generalist predator, with a long-standing reputation for population stability. If the Peregrine Falcon could disappear so quickly, what raptor could survive?

The scientists that met in Wisconsin in 1965 focused their attention on the growing misuse (principally overuse) of modern synthetic "biocides" in pest control, and the ability of these new chemicals to increase in concentration among organisms in food chains. High levels of several of the pesticides in Peregrine Falcons suggested a link. This, together with an increase in eggshell breakage, suggested a mechanism for reproductive failure.

With science on the case, things began to happen quickly. A falcon specialist in Great Britain published an analysis tying the timing of eggshell thinning in British Peregrine Falcons to the widespread use of DDT there. Two years later, controlled experimental studies involving captive American Kestrels demonstrated a "smoking gun" relationship between the two phenomena. Armed with this new information, conservationists successfully forced bans on the widespread use of DDT and other organochlorine pesticides across most of North America and Western Europe in the early 1970s.

In most cases, the bans quickly led to reductions in contaminant levels and a reversal in eggshell thinning. Soon thereafter, many raptor populations that had suffered through pesticide-era lows began to rebound. In several cases, natural increases were assisted by reintroduction efforts. By the mid-1980s, numerous recoveries were well under way.

One of the more instructive aspects of this saga involves the declines and recoveries of European populations of Peregrine Falcons in former West Germany and East Germany at the height of the Cold War. Although both banned DDT in 1972, the pesticide continued to be used in East Germany for an additional 17 years. Not surprisingly, West German populations of Peregrine Falcons turned the corner and began to rebound shortly after the official ban, but the once-ubiquitous tree-nesting population of East Germany continued to decline through the mid-1970s and was extirpated later that decade, almost certainly because of its protracted exposure to the contaminant.

Widespread use of DDT and many other organochlorine pesticides, including the far more toxic cyclodienes, aldrin and dieldrin, ceased in North America and Western Europe in the early 1970s. Even so, broad-spectrum, persistent pesticides continue to pose a real threat to migratory raptors. For example, in less-developed areas in the tropics and in the Southern Hemisphere, even local use of these pesticides can impact large numbers of wintering birds of prey that feed on agricultural pests.

Organophosphate pesticides. The chemicals that replaced organochlorine pesticides created their own set of problems for raptors. Organophosphate pesticides, such as malathion and parathion, came into use in the 1960s and 1970s largely in response to growing concerns regarding the persistent nature of organochlorine pesticides, some of which, including DDT, had "half-lives" of up to 15 years (that is, 15 years after application, half the original DDT remains active; 15 years later, one-quarter remains active, etc.).

The good news about these newer pesticides was that because they had half-lives of a few hours to a few days, they were far less persistent than the organochlorines they replaced. The bad news was that organophosphates killed insects by inhibiting the enzyme that turns off the principal neurotransmitter in the nervous systems of both insects *and* vertebrates. That meant that new pesticides were acutely toxic to vertebrates as well as to insects. Indeed, organophosphate pesticides are ten to one hundred times as toxic to vertebrates, including fishes, birds, and mammals, as are the organochlorine pesticides they replaced. A second popular class of organochlorine replacements, carbamates, or urethanes, including Sevin, aldicarb, carbofuran, and mirex, share many of the properties of organophosphates, including enhanced vertebrate toxicity.

Unlike their predecessors, whose greatest threat was environmental persistence and eventual accumulation in the fatty body tissues of birds of prey, the new pesticides posed their greatest threat acutely and immediately whenever they were used in ways that allowed raptors to contact them in their undegraded form. Moreover, because these pesticides were absorbed through the skin and lungs, as well as through the gastrointestinal tract, virtually any contact, either indirectly as a result of eating recently poisoned prey, or directly as a nontarget species, posed a considerable threat.

In England, where organophosphates and carbamates were used illegally both to control mammal predators and to protect racing pigeons, 87% of 136 raptor kills reported between 1985 and 1994 were attributed to the use of these toxins. And in the eastern Mediterranean, where the highly toxic carbamate pesticide Lannate has been used to protect grapes from vineyard moths, 150 Eleonora's Falcons were killed in 1999. That the die-off occurred in the species' geographic stronghold, where most of the world's 7000 pairs of Eleonora's Falcons are known to breed, highlights the potentially disastrous consequences of even local pesticide use.

The most striking example of how organophosphates can kill large numbers of raptors involves Swainson's Hawks in the New World. An estimated 6000 to 20,000 Swainson's were killed by the organophosphates monocrotophos and dimethoate on their Argentine wintering grounds in the winter of 1995–1996. Forensic analyses at the time indicated that monocrotophos alone poisoned more than 4000 hawks at six different sites. Firsthand accounts indicated that the hawks died immediately after being sprayed with the pesticides while hunting grasshoppers in alfalfa and sunflower fields, or within days after eating poisoned insects. Neither pesticide involved had been approved for use in controlling grasshoppers on any crop. Such off-label pesticide abuse, including excessive use on approved crops, as well as applications in nonapproved circumstances, remains problematic in many parts of the world. Subsequent gatherings between conservationists and the company producing monocrotophos used in Argentina resulted in the removal of existing stock from the region in which the birds had been killed. Although the problem appears to have been solved in Argentina, this pesticide continues to be used elsewhere.

Veterinary drugs. Unfortunately, pesticides are not the only agricultural chemicals that threaten raptors. In fact, the most recent large-scale raptor kill attributed to environmental contaminants, and almost certainly the largest ever, did not involve pesticides at all, but rather a seemingly benign veterinary drug. The story begins on the Indian subcontinent, where in the mid-1990s the Indian ornithologist Vibhu Prakash noted dramatic declines in populations of several species of common and widespread vultures in a former vulture stronghold, Keoladeo National Park, in Rajasthan, India. The birds in question included several species of vultures in the genus *Gyps*, five of which occur in India.

Gyps species are large, highly social vultures that nest in colonies and often co-mingle while feeding on the bodies of large, decaying mammals. In areas where both carcasses and nest sites are abundant, the density of *Gyps* vultures can greatly exceed that typically associated with birds of prey. In the 1970s and early 1980s, for example, the White-rumped Vulture, one of the species currently in decline at Keoladeo and a bird that regularly associates with humans, was nesting at a density of approximately nine pairs per square mile in the city of Delhi, and at four times that density in Keoladeo National Park. Given that the White-rumped Vulture historically ranged throughout most of the Indian subcontinent and much of southeastern Asia, experts at the time suggested that it quite possibly was the most abundant large raptor on earth with an estimated population easily in excess of five million birds. Within 15 years, however, this species, together with two close relatives, the Indian Vulture and the Slender-billed Vulture, ranked among the most threatened of all birds of prey.

The catastrophic decline of three populations of once-common raptors alarmed the conservation community for two reasons. First was the unprecedented speed with which the birds were disappearing. Overall, White-rumped, Indian, and Slender-billed Vultures declined in the 1990s and 2000s by more than 95%. By comparison, although several regional populations of the Peregrine Falcon, particularly those in North America and Europe, had undergone similar declines during the DDT era, the species' fate was never in doubt. This was not the case for the vultures, all three of whose populations were in free fall. Second, and perhaps more disturbing, was that no one could explain why the birds were disappearing. The usual suspects, land-use change, pesticides, and direct persecution, offered no answers.

Biologists had known for some time that the increased and largely uncontrolled hunting of wild ungulates in southeastern Asia had decimated the food base for vultures there, but carcasses remained readily available in much of India and Pakistan, and vulture populations had crashed there as well. Although some conservationists pointed to direct persecution—an argument intuitively suggested by the region's high human densities—vultures were declining throughout their ranges, including in India, where all three species were valued for their role as environmental scavengers and held in cultural and religious esteem.

At first, pesticide or heavy-metal poisoning was thought likely, but exhaustive toxicology of carcasses failed to detect clinical levels of environmental contaminants. Given the dead ends, the conservation community began to focus on the possibility that an unknown and particularly virulent pathogen was killing the birds. The argument made sense. After all, the birds looked sick, with neck drooping and lethargy typical prior to death. And not one affected colony ever recovered, once high mortality set in. Autopsied carcasses from India and Pakistan revealed renal malfunction and visceral gout, along with a high frequency of enteritis, conditions that often reflect infectious diseases. As recently as 2003, most conservation scientists believed an epidemic disease, probably viral, was their most promising lead.

At about the same time, Lindsay Oaks reached a different conclusion. An avian pathologist at Washington State University, Oaks, whose earlier analyses of vulture tissues had ruled out the widespread use of pesticides and heavy metals as the culprit, turned to analyzing fresh carcasses for signs of fragile viruses and bacteria. Finding none, Oaks concluded an epidemic disease unlikely. Frustrated, but determined to explain the ecological tragedy, Oaks began to focus on a series of less likely contaminants, including several veterinary drugs that recently had come into widespread use in livestock in the region. In early 2003, the new investigation paid off: tests revealed that every one of the twenty-five vultures that Oaks examined and that had died of renal failure had been exposed to diclofenac, a nonsteroidal anti-inflammatory painkiller that had come into widespread use in India and Pakistan in the 1990s. None of the thirteen vultures that Oaks examined that had succumbed from other causes exhibited any exposure to the drug. Significantly, four nonreleasable vultures that had been given single oral doses of diclofenac, either at the normal mammalian dosage (two birds) or at one-tenth of that dosage (two birds), died of renal failure within 2 days.

Additional analyses on carcasses from much larger areas of India and Nepal revealed high residue of the drug in dead vultures there, too.

Moreover, a model of vulture demographics indicated diclofenac exposure rates as low as 1-in-760 available livestock carcasses could have produced the massive die off, meaning that it wouldn't take much to destroy more than 90% of existing populations. The population declines, which continued throughout the region for several years, led conservationists to call for an immediate ban on the drug, and restrictions in areas where it was not used. In early 2005, the Indian government enacted a 6-month phase-out of diclofenac and farmers there have been asked to replace it with alternatives believed less toxic to the birds. Prohibitions in Pakistan and Nepal followed. Because the drug already is widely distributed throughout the region, it is likely to remain in use for some time despite the ban on new sales and use.

Population modeling tells us that the likelihood of extinction, or at least extirpation, is high throughout most of the ranges of the three affected species. Such findings dictated an adequate number of vultures be removed from the wild as soon as possible and that captive breeding begin, not only to ensure the long-term survival of these species, but also to serve as breeding stock for future release efforts once the diclofenac has been "ecologically" removed. The latter approach helped to restore the Peregrine Falcon in several portions of their ranges following bans on DDT and other organochlorines. And, in time, populations of southern Asia's vultures, too, may rebound. Although population declines in the three species have halted, the crisis demonstrates how easy it is for human actions to eliminate functional populations of common raptors in ways that remain frustratingly difficult to predict.

Heavy metals. Although agricultural pesticides continue to rank as the most significant and widespread environmental contaminants, heavy metals, particularly lead and mercury, continue to threaten many populations of birds of prey. Lead poisoning by ingesting contaminated prey and carcasses is particularly problematic in areas where lead shot and lead pellets are used in sport and subsistence hunting.

In North America, California Condors and both Black and Turkey vultures, along with Red-tailed Hawks, Roughlegs, Golden Eagles, Prairie Falcons, and Peregrine Falcons, all have been diagnosed with lead poisoning. In Spain, eight species, including the Griffon Vulture, Red Kite, Black Kite, and Western Marsh Harrier, have been similarly diagnosed.

Clearly, scavenging and other species most likely to feed on hunter-injured waterfowl and upland game are at higher risk. Although the extent to which lead poisoning affects entire populations of raptors remains relatively little studied, the unintended poisoning of California Condors by the ingestion of lead shot and fragments consumed with carcasses almost certainly contributed to the decline of this critically endangered species in mid–twentieth century North America. And the problem with lead continues for condors. Ongoing availability of lead-tainted carcasses in areas near where captive-bred individuals are being released as part of an ongoing reintroduction effort continues to compromise the success of the project. Lead-awareness campaigns for hunters, who could become conservation heroes, are under way. Many hunters are unaware of this problem and simply need to be told why burying entrails and other carcass remains and switching to more benign shot, including copper, tungsten, and bismuth, can help protect the scavenging birds that are currently being affected. The success of this campaign ultimately relies on the hunting community, an essential conservation partner for raptors when it comes to lead poisoning.

ADDITIONAL THREATS TO RAPTORS

Power Lines

Numerous human activities continue to threaten birds of prey, and although most are local or regional, several, including the generation and distribution of electrical energy, can affect an entire species. Despite numerous attempts to reduce raptor deaths along power-distribution routes, power-line collisions and electrocution remain problematic, particularly at and near raptor-migration bottlenecks and other points where migrants concentrate. Electrocution becomes a larger issue in treeless areas, where power-line poles often provide the only elevated hunting perches. Large raptors are more vulnerable to electrocution because their broad wingspans allow them to contact conducting and ground wires simultaneously while perched on poles.

In some instances, power-line electrocutions can seriously impact raptor populations locally. One example is a decline in the number of the vulnerable Spanish Imperial Eagle in Doñana National Park, southern Spain, where, in the 1980s, 69% of all known eagle deaths were attributed to electrocution. In the United States, electrocutions were responsible for 25% of all Golden Eagle deaths reported by the National Wildlife Health

Center in the early 1960s through the mid-1990s. Other raptors whose populations have declined locally as a result of power-line electrocutions include Cape Vultures, Griffon Vultures, Egyptian Vultures, and Bonelli's Eagles.

Although many companies have redesigned and retrofitted power lines to reduce the threat of electrocution, others have not. Even in developed countries, the overwhelming majority of transmission poles fail to meet established standards for raptor safety. Deregulation of the power industry, with its increased focus on competition and cost cutting, suggests that power-line electrocutions will continue to threaten birds of prey for a long time.

Wind Power

As the number of wind farms continues to increase, so do their interactions with raptors. Wind turbines can harm raptors two ways: through collisions with the turbines themselves, and by the habitat destruction brought about by the construction and maintenance of the structures. Not all wind turbines are equal. First-generation turbines, many of which remain in use today, were mounted on 60- to 80-ft open latticework towers, had 50- to 60-ft rotors, and turned at rates of sixty to eighty times per minute. Newer turbines are mounted on taller, 200- to 260-ft, tubular towers, have larger, 150- to 260-ft, rotors, and turn at slower rates of eleven to twenty-eight times per minute. The new turbines, which produce considerably more energy per unit, are spaced more widely than the older units. Bird strikes at arrays of wind turbines were first studied at an installation of more than 5000 old-style turbines at Altamont Pass, California. Raptors appeared to be at higher relative risk at the site than were other species of birds, at least in part because they often perched on and hunted from the turbines' latticework support structures.

Two studies based on work conducted in and around Tarifa, Spain, a major concentration point for raptors travelling between Western Europe and Africa across the Strait of Gibraltar, where more than 1000 old and new turbines are in use, provide insights into the extent of the threat. Both studies measured mortality at the installations and the factors responsible for it. The first found that resident raptors, particularly Griffon Vultures and Common Kestrels, were more likely to be killed by turbines than were migrating raptors, that the vultures were most vulnerable in autumn and winter, when a lack of strong thermals forced them to slope-soar closer to the ridgetop arrays of turbines, and that the kestrels were most vulnerable in summer, when they were most abundant and were attracted to preferred hunting habitats at one of the turbine arrays.

Based on the numbers of carcasses found, and correcting for the removal of carcasses by scavenging animals between inspections, the authors estimated annual mortality at approximately thirty vulture kills and thirty-six kestrel kills per year at two wind farms with a combined total of 250 turbines. Because most of the collisions occurred at a small number of turbines and during predictable wind conditions, the authors of the report recommended suspending operations of specific turbines during high-risk situations.

A second study reported that most raptors noticeably avoided approaching active turbines, and found evidence of only two collisions (a Griffon Vulture and a Short-toed Snake Eagle) during 14 months of fieldwork, or an annual mortality of approximately 0.03 birds per turbine. The study concluded that the sixty-six-turbine wind farm in question was low risk, most likely because migrants in the area were flying well above the height of the turbines. Studies of wind power–bird interactions continue to grow as the number of operational turbines increases. Two major themes are emerging. The first is that old-style turbines kill more raptors than new-style turbines. The second is that regardless of the style of turbine, improperly sited and inappropriately managed wind farms can and almost certainly will cause problems for raptors, including migratory species. Key factors in reducing the potential impact of wind turbines include situating turbines away from high-density raptor populations and known migration corridors and bottlenecks, avoiding sites that will displace existing populations from important nesting and feeding areas, and using on-off cycles to reduce the likelihood of collisions during periods of peak migration.

Nuclear Power Plants

Although nuclear power does not inherently threaten birds of prey on a large scale, the accidental release of radionuclides from malfunctioning generation sites has the potential to expose raptors to dangerous chemicals, both directly via airborne plumes, and indirectly via contaminated prey bases.

Global Climate Change

One factor associated with the generation of electrical energy likely to affect migratory raptors on a large scale is global climate change. If the changes that climatologists currently suggest do occur, raptor movement ecology is likely to change in several ways. Milder winters in the Temperate Zone should result in greater survivorship among resident species there, as well as an increased tendency for year-round residency among current populations of partial migrants. As more sedentary raptor populations thrive and increase, long-distance migrants overwintering in tropical areas will face increased competition on the breeding grounds, and those overwintering in temperate areas in the Southern Hemisphere will face increased competition on both the breeding and wintering grounds. The extent to which such species survive will depend on their ability to shorten their migratory journeys, which, in turn, may allow them to compete more successfully with nonmigrants by departing later in the autumn and returning earlier in the spring. Whether or not such species survive in the long haul, the outcome will be much the same: a substantial shift from migratory behavior, particularly long-distance migration, toward increased nonmigratory behavior in raptors throughout many temperate regions. In subtropical regions, predicted increases in droughts are likely to lead to increased nomadic behavior and short-distance movements in some raptors, as well as to desertification and population declines for others. In coastal areas, ongoing sea-level rise will eliminate lowland breeding areas for many raptors. Although most regional scenarios are speculative, the overall impression of a warmer planet is one in which migratory raptors are far less common than now. In principle this could have implications for the rate at which raptors speciate and diversify.

On islands, the loss of migration eventually could affect raptor species diversity there. The theory of island biogeography suggests that relatively large islands, as well as those that are relatively close to continents or to other islands, will host larger numbers of species than will islands that are smaller and more remote. The theory also suggests that true island-endemic species—as opposed to continental species that also occur on islands—are more likely to occur on islands that are large enough to permit viable populations of ancestral stock from continental areas to speciate, but are remote enough to forestall regular gene flow from mainland landmasses. These principles appear to hold for most species of island raptors. Madagascar, a large continental island close to mainland Africa and its associated smaller islands, for example, is inhabited by sixteen species of raptors, including four migrants and eleven island endemics. On the other hand, New Zealand, a group of two large and many smaller continental islands that is currently far from any other continental landmass, has only two species of raptors, one of which is endemic. Similarly, in the New World, Cuba, a large continental island in the Greater Antilles close to North America, has sixteen species of raptors, including nine migrants and one or, possibly, two endemics, whereas Puerto Rico, a smaller island in the same archipelago that is farther from the mainland, has only nine species, all of which possibly migrate, and no endemics.

The phenomenon of migration-dosing speciation also appears to apply to island raptors (see Chapter 4 for details). If migration dosing does occur routinely on islands, then the phenomenon of raptor migration itself is an engine of raptor speciation, and its reduction in the face of climate change could, over evolutionary time, lead to a reduction in raptor diversity globally, particularly among island-dwelling species.

A FUTURE FOR BIRDS OFPREY

The list of phenomena outlined above demonstrates a long history of human actions that have negatively impacted birds of prey. Simply put, the degree to which people continue to engage in such actions, or choose not to, will shape the degree to which populations of raptors flourish or perish in the future. Almost certainly, land-use change will continue to be the most significant threat. Cultivated landscapes—areas in which one-third or more of the land is in crops, aquaculture, or livestock production—now cover more than a quarter of earth's land surface. Humanity's demand for food is expected to double in the next half century. Exactly how human populations will meet this challenge remains unclear. What is clear is that an increased portion of the world's primary productivity will be to feed people, which will leave less for other species, including raptors.

Agricultural growth, acting through the loss of natural habitats, environmental contamination, and, in some instances, direct persecution, already poses the greatest threat to migratory raptors, many of which depend on the same seasonally productive landscapes that humans prize for farmlands. This is likely to continue. Intensified farming and modern forestry, coupled with the widespread use of new chemicals, both to grow farm and forestry products more efficiently and to protect them from nonhuman users, almost certainly will produce new and often unanticipated threats to birds of prey. Thus, it appears inevitable that many raptor populations will decline as a result.

Although direct human persecution is on the wane overall, environmental contaminants continue to present a growing threat to many populations of birds of prey.

It might seem easy for one to grow increasingly discouraged in light of all of this, but I am not pessimistic. If anything, I am increasingly optimistic, for several reasons. First, recent advances in our understanding of the environmental needs of raptors strengthen our ability to protect them. Knowledge is key to on-the-ground conservation, and we are adding to our knowledge faster than ever. Science is now truly "on the case" in much of raptor conservation and, increasingly, it is helping practitioners in the field. An excellent example of this is the subdiscipline of raptor movement ecology, where the advent of satellite tracking has increased our understanding of raptor biology significantly over the past several decades.

Second, new generations of raptor biologists are growing quickly, not only in North America and Europe, but also in continental Asia, the Pacific Islands, Australia, Africa, and Latin America, where many raptors remain threatened. More adequately prepared and far better connected than their predecessors, this new cadre of conservationists is well positioned to protect the world's birds of prey.

Third, the recent push to engage local human populations in raptor protection and to incorporate the needs of birds of prey into regional and continental management schemes for natural resources promises to help ensure the raptors' future. My own experiences in the Falkland Islands, where Hawk Mountain has been working closely with government officials and the local BirdLife partner, Falklands Conservation, to change the attitudes of local residents and landowners toward scavenging birds of prey, confirms the adage that "local support is key to lasting conservation success."

Finally, and perhaps most importantly, raptors themselves have proved to be far more resilient than many conservationists have thought them to be. Consider the Alaskan population of Bald Eagles, which was able to withstand more than one-third of a century of incessant bounty hunting, or how the Peregrine Falcon rebounded from pesticide-era lows of mid–twentieth century North America and Western Europe. These examples demonstrate that, as long as a threat is identified and eventually reversed, and as long as the problem is not so swift that it overwhelms the birds entirely, then birds of prey are capable of withstanding the onslaught of both intentional and unintentional human insults. Although it is impossible to overstate the irreversible nature of species extinction and the many threats that raptors currently face, it is equally impossible to overstate the resourceful nature of the birds themselves.

If my optimism is tempered at all, it is only by the concern that we will fail to learn from past mistakes. If we choose to build on the existing and growing body of knowledge and continue to study and to learn more about how birds of prey function in natural and human-dominated landscapes, and if we continue to recognize that the time to save a species is while it is still common, adult and juvenile birds of prey will be breeding and feeding in large numbers well into the future.

SYNTHESIS AND CONCLUSIONS

1. Only one raptor, the island-dwelling Guadalupe Caracara, has become extinct in the last 330 years.
2. As of early 2016, fifty-five species of raptors remain threatened globally, and populations of more than one hundred species remain threatened in many parts the world.
3. The principal human threats to raptors are land-use change; environmental contaminants, including pesticides and other agricultural chemicals; and direct assault, including persecution and trapping for captive use. Of these, land-use change remains the greatest threat overall, particularly in the tropics.
4. Human activities, including direct persecution, have affected raptors since prehistoric times.
5. Migratory raptors face special problems because many are double endemics, whose livelihoods depend on the maintenance of two geographically disjoint habitats.
6. The world's twenty-two species of obligate scavenging raptors are by far the most threatened birds of prey.
7. Historically, direct persecution increased significantly once raptors were viewed as threats to domestic and game animals.

8. The misuse of agricultural pesticides reduced many populations of raptors in the middle of the twentieth century, and such misuse continues sporadically even today.
9. The large-scale generation and distribution of electrical energy continues to impact raptors in many ways, including directly via electrocution and collisions, and indirectly via global climate change.
10. Learning from previous mistakes, creating a better understanding of the ecological needs of raptors, and understanding the resilience of the birds themselves are three essential aspects of successful raptor conservation.

APPENDIX

Scientific Names of Raptors, Owls, and Other Birds and Their Distributions, and the Scientific Names of Other Animals Cited in the Text

Common name	Scientific name	Distribution
RAPTORS		
African Cuckoo-Hawk	*Aviceda cuculoides*	Africa
African Fish Eagle	*Haliaeetus vocifer*	Africa
African Goshawk	*Accipiter tachiro*	Africa
African Harrier-Hawk	*Polyboroides typus*	Africa
African Hawk-Eagle	*Hieraaetus spilogaster*	Africa
African Hobby	*Falco cuvierii*	Africa
African Marsh Harrier	*Circus ranivorus*	Africa
American Kestrel	*Falco sparverius*	North, Middle, and South America
Amur Falcon	*Falco amurensis*	Eurasia
Andaman Serpent Eagle	*Spilornis elgini*	Oriental Region
Andean Condor	*Vultur gryphus*	South America
Aplomado Falcon	*Falco femoralis*	Middle and South America
Asian Imperial Eagle	*Aquila heliacal*	Asia, Eurasia
Augur Buzzard	*Buteo augur*	Africa
Australian Hobby	*Falco longipennis*	Australasia
Ayres's Hawk-Eagle	*Hieraaetus ayresii*	Africa
Bald Eagle	*Haliaeetus leucocephalus*	North America
Banded Kestrel	*Falco zoniventris*	Africa
Barbary Falcon	*Falco pelegrinoides*	Eurasia
Barred Forest Falcon	*Micrastur ruficollis*	Middle and South America
Barred Hawk	*Leucopternis princeps*	Middle and South America
Barred Honey Buzzard	*Pernis celebensis*	Oriental Region, Australasia
Bat Falcon	*Falco rufigularis*	Middle and South America
Bat Hawk	*Macheiramphus alcinus*	Africa, Oriental Region
Bateleur	*Terathopius ecaudatus*	Africa
Bearded Vulture	*Gypaetus barbatus*	Africa, Eurasia
Beaudouin's Snake Eagle	*Circaetus beaudouini*	Africa
Besra	*Accipiter virgatus*	Oriental Region
Bicolored Hawk	*Accipiter bicolor*	Middle and South America

Common name	Scientific name	Distribution
Black Baza	*Aviceda leuphotes*	Oriental Region
Black Caracara	*Daptrius ater*	South America
Black Eagle	*Ictinaetus malayensis*	Oriental Region
Black Falcon	*Falco subniger*	Australasia
Black Harrier	*Circus maurus*	Africa
Black Hawk-Eagle	*Spizaetus tyrannus*	Middle and South America
Black Honey Buzzard	*Henicopernis infuscatus*	Australasia
Black Kite	*Milvus migrans*	Africa, Eurasia, Oriental Region, Australasia
Black Sparrowhawk	*Accipiter melanoleucus*	Africa
Black Vulture	*Coragyps atratus*	North, Middle, and South America
Black-and-chestnut Eagle	*Oroaetus isidori*	South America
Black-and-white Hawk-Eagle	*Spizastur melanoleucus*	Middle and South America
Black-breasted Buzzard	*Hamirostra melanosternon*	Australasia
Black-chested Buzzard-Eagle	*Geranoaetus melanoleucus*	South America
Black-chested Snake Eagle	*Circaetus pectoralis*	Africa
Black-collared Hawk	*Busarellus nigricollis*	Middle and South America
Black-faced Hawk	*Leucopternis melanops*	South America
Black-mantled Goshawk	*Accipiter melanochlamys*	Australasia
Black-shouldered Kite	*Elanus axillaris*	Australasia
Black-thighed Falconet	*Microhierax fringillarius*	Oriental Region
Black-winged Kite	*Elanus caeruleus*	Africa, Eurasia, Oriental Region
Blyth's Hawk-Eagle	*Spizaetus alboniger*	Oriental Region
Bonelli's Eagle	*Hieraaetus fasciatus*	Eurasia, Oriental Region
Booted Eagle	*Hieraaetus pennatus*	Eurasia
Brahminy Kite	*Haliastur indus*	Oriental Region, Australasia
Broad-winged Hawk	*Buteo platypterus*	North America
Brown Falcon	*Falco berigora*	Australasia
Brown Goshawk	*Accipiter fasciatus*	Australasia
Brown Snake Eagle	*Circaetus cinereus*	Africa
Buckley's Forest Falcon	*Micrastur buckleyi*	South America
California Condor	*Gymnogyps californianus*	North America
Cape Vulture	*Gyps coprotheres*	Africa
Carunculated Caracara	*Phalcoboenus carunculatus*	South America
Cassin's Hawk-Eagle	*Spizaetus africanus*	Africa
Changeable Hawk-Eagle	*Spizaetus limnaeetus*	Oriental Region
Chestnut-flanked Sparrowhawk	*Accipiter castanilius*	Africa
Chestnut-shouldered Goshawk	*Erythrotriorchis buergersi*	Australasia
Chilean Hawk	*Accipiter chilensis*	South America
Chimango Caracara	*Milvago chimango*	South America
Chinese Sparrowhawk	*Accipiter soloensis*	Oriental Region
Cinereous Harrier	*Circus cinereus*	South America
Cinereous Vulture	*Aegypius monachus*	Eurasia, Oriental Region
Collared Falconet	*Microhierax caerulescens*	Oriental Region
Collared Forest Falcon	*Micrastur semitorquatus*	Middle and South America
Collared Sparrowhawk	*Accipiter cirrocephalus*	Australasia
Common Black Hawk	*Buteogallus anthracinus*	North, Middle, and South America
Common Buzzard	*Buteo buteo*	Africa, Eurasia, Oriental Region

Common name	Scientific name	Distribution
Common Kestrel	*Falco tinnunculus*	Eurasia, Africa
Congo Serpent Eagle	*Dryotriorchis spectabilis*	Africa
Cooper's Hawk	*Accipiter cooperii*	North and Middle America
Crane Hawk	*Geranospiza caerulescens*	Middle and South America
Crested Eagle	*Morphnus guianensis*	Middle and South America
Crested Goshawk	*Accipiter trivirgatus*	Oriental Region
Crested Hawk-Eagle	*Spizaetus cirrhatus*	Oriental Region
Crested Honey Buzzard	*Pernis ptilorhynchus*	Eurasia
Crested Serpent Eagle	*Spilornis cheela*	Oriental Region
Crowned Eagle	*Stephanoaetus coronatus*	Africa
Crowned Solitary Eagle	*Harpyhaliaetus coronatus*	South America
Cryptic Forest Falcon	*Micrastur mintoni*	South America
Cuban Kite	*Chondrohierax wilsoni*	North America
Dark Chanting Goshawk	*Melierax metabates*	Africa
Dickinson's Kestrel	*Falco dickinsoni*	Africa
Doria's Goshawk	*Megatriorchis doriae*	Australasia
Double-toothed Kite	*Harpagus bidentatus*	Middle and South America
Dwarf Sparrowhawk	*Accipiter nanus*	Australasia
Eastern Chanting Goshawk	*Melierax poliopterus*	Africa
Eastern March Harrier	*Circus spilonotus*	Oriental Region
Egyptian Vulture	*Neophron percnopterus*	Africa, Eurasia, Oriental Region
Eleonora's Falcon	*Falco eleonorae*	Africa, Eurasia
Eurasian Hobby	*Falco subbuteo*	Africa, Eurasia, Oriental Region
Eurasian Sparrowhawk	*Accipiter nisus*	Africa, Eurasia
European Honey Buzzard	*Pernis apivorus*	Eurasia
Ferruginous Hawk	*Buteo regalis*	North America
Fiji Goshawk	*Accipiter rufitorques*	Australasia
Flores Hawk-Eagle	*Spizaetus floris*	Australasia
Forest Buzzard	*Buteo trizonatus*	Africa
Forest Fish Eagle	*Haliaeetus sanfordi*	Australasia
Fox Kestrel	*Falco alopex*	Africa
Frances's Sparrowhawk	*Accipiter francesii*	Africa
Gabar Goshawk	*Micronisus gabar*	Africa
Galapagos Hawk	*Buteo galapagoensis*	South America
Golden Eagle	*Aquila chrysaetos*	North and Middle America, Eurasia
Grasshopper Buzzard	*Butastur rufipennis*	Africa
Great Black Hawk	*Buteogallus urubitinga*	Middle and South America
Greater Kestrel	*Falco rupicoloides*	Africa
Greater Spotted Eagle	*Aquila clanga*	Eurasia
Greater Yellow-headed Vulture	*Cathartes melambrotus*	South America
Grey Falcon	*Falco hypoleucos*	Australasia
Grey Goshawk	*Accipiter novaehollandiae*	Australasia
Grey Kestrel	*Falco ardosiaceus*	Africa
Grey Sparrowhawk	*Accipiter brachyurus*	Australasia
Grey-backed Hawk	*Leucopternis occidentalis*	South America
Grey-bellied Hawk	*Accipiter poliogaster*	South America
Grey-faced Buzzard	*Butastur indicus*	Eurasia
Grey-headed Fish Eagle	*Ichthyophaga ichthyaetus*	Oriental Region
Grey-headed Goshawk	*Accipiter poliocephalus*	Australasia
Grey-headed Kite	*Leptodon cayanensis*	Middle and South America

Common name	Scientific name	Distribution
Grey-lined Hawk	*Buteo nitida*	North, Middle, and South America
Griffon Vulture	*Gyps fulvus*	Africa, Eurasia
Gundlach's Hawk	*Accipiter gundlachi*	North America
Gurney's Eagle	*Aquila gurneyi*	Australasia
Gyrfalcon	*Falco rusticolus*	North America, Eurasia
Haast's Eagle	*Harpagornis moorei* (extinct)	North America, Eurasia
Harpy Eagle	*Harpia harpyja*	Middle and South America
Hawaiian Hawk	*Buteo solitarius*	Pacific Ocean
Henst's Goshawk	*Accipiter henstii*	Africa
Himalayan Vulture	*Gyps himalayensis*	Oriental Region
Hooded Vulture	*Necrosyrtes monachus*	Africa
Hook-billed Kite	*Chondrohierax uncinatus*	North, Middle, and South America
Imitator Goshawk	*Accipiter imitator*	Australasia
Indian Spotted Eagle	*Aquila hastate*	Oriental Region
Indian Vulture	*Gyps indicus*	Oriental Region
Jackal Buzzard	*Buteo rufofuscus*	Africa
Japanese Sparrowhawk	*Accipiter gularis*	Eurasia
Javan Hawk-Eagle	*Spizaetus bartelsi*	Oriental Region
Jerdon's Baza	*Aviceda jerdoni*	Oriental Region
King Vulture	*Sarcoramphus papa*	Middle and South America
Laggar Falcon	*Falco jugger*	Oriental Region
Lanner Falcon	*Falco biarmicus*	Eurasia, Africa
Lappet-faced Vulture	*Aegypius tracheliotus*	Africa
Laughing Falcon	*Herpetotheres cachinnans*	Middle and South America
Lesser Fish Eagle	*Ichthyophaga humilis*	Oriental Region
Lesser Kestrel	*Falco naumanni*	Eurasia
Lesser Spotted Eagle	*Aquila pomarina*	Eurasia
Lesser Yellow-headed Vulture	*Cathartes burrovianus*	Middle and South America
Letter-winged Kite	*Elanus scriptus*	Australasia
Levant Sparrowhawk	*Accipiter brevipes*	Eurasia
Lined Forest Falcon	*Micrastur gilvicollis*	South America
Little Eagle	*Hieraaetus morphnoides*	Australasia
Little Sparrowhawk	*Accipiter minullus*	Africa
Lizard Buzzard	*Kaupifalco monogrammicus*	Africa
Long-crested Eagle	*Lophaetus occipitalis*	Africa
Long-legged Buzzard	*Buteo rufinus*	Eurasia, Africa
Long-tailed Hawk	*Urotriorchis macrourus*	Africa
Long-tailed Honey Buzzard	*Henicopernis longicauda*	Australasia
Long-winged Harrier	*Circus buffoni*	South America
Madagascar Buzzard	*Buteo brachypterus*	Africa
Madagascar Cuckoo-Hawk	*Aviceda madagascariensis*	Africa
Madagascar Fish Eagle	*Haliaeetus vociferoides*	Africa
Madagascar Harrier-Hawk	*Polyboroides radiatus*	Africa
Madagascar Serpent Eagle	*Eutriorchis astur*	Africa
Madagascar Sparrowhawk	*Accipiter madagascariensis*	Africa
Malagasy Kestrel	*Falco newtoni*	Africa
Malagasy Marsh Harrier	*Circus maillardi*	Africa
Mantled Hawk	*Leucopternis polionotus*	South America
Martial Eagle	*Polemaetus bellicosus*	Africa

Common name	Scientific name	Distribution
Mauritius Kestrel	*Falco punctatus*	Africa
Merlin	*Falco columbarius*	North America, Eurasia
Meyer's Goshawk	*Accipiter meyerianus*	Australasia
Mississippi Kite	*Ictinia mississippiensis*	North America
Moluccan Goshawk	*Accipiter henicogrammus*	Australasia
Montagu's Harrier	*Circus pygargus*	Eurasia
Montane Solitary Eagle	*Harpyhaliaetus solitarius*	Middle and South America
Mountain Buzzard	*Buteo oreophilus*	Africa
Mountain Caracara	*Phalcoboenus megalopterus*	South America
Mountain Hawk-Eagle	*Spizaetus nipalensis*	Oriental Region
Mountain Serpent Eagle	*Spilornis kinabaluensis*	Oriental Region
Nankeen Kestrel	*Falco cenchroides*	Australasia
New Britain Goshawk	*Accipiter princeps*	Australasia
New Zealand Eagle (Extinct)	*Harpagornis moorei*	New Zealand
New Zealand Falcon	*Falco novaeseelandiae*	Australasia
Nicobar Serpent Eagle	*Spilornis minimus*	Oriental Region
Nicobar Sparrowhawk	*Accipiter butleri*	Oriental Region
Northern Crested Caracara	*Caracara cheriway*	North, Middle, and South America
Northern Goshawk	*Accipiter gentilis*	North and Middle America, Eurasia
Northern Harrier	*Circus cyaneus*	North America, Eurasia
Orange-breasted Falcon	*Falco deiroleucus*	Middle and South America
Oriental Hobby	*Falco severus*	Oriental Region, Australasia
Ornate Hawk-Eagle	*Spizaetus ornatus*	Middle and South America
Osprey	*Pandion haliaetus*	Worldwide
Ovambo Sparrowhawk	*Accipiter ovampensis*	Africa
Pacific Baza	*Aviceda subcristata*	Australasia
Pale Chanting Goshawk	*Melierax canorus*	Africa
Pallas's Fish Eagle	*Haliaeetus leucoryphus*	Eurasia
Pallid Harrier	*Circus macrourus*	Eurasia
Palm-nut Vulture	*Gypohierax angolensis*	Africa
Papuan Eagle	*Harpyopsis novaeguineae*	Australasia
Pearl Kite	*Gampsonyx swainsonii*	South America
Peregrine Falcon	*Falco peregrinus*	Worldwide
Philippine Eagle	*Pithecophaga jefferyi*	Oriental Region
Philippine Falconet	*Microhierax erythrogenys*	Oriental Region
Philippine Hawk-Eagle	*Spizaetus philippensis*	Oriental Region
Philippine Serpent Eagle	*Spilornis holospilus*	Oriental Region
Pied Falconet	*Microhierax melanoleucos*	Oriental Region
Pied Goshawk	*Accipiter albogularis*	Australasia
Pied Harrier	*Circus melanoleucos*	Eurasia
Plain-breasted Hawk	*Accipiter ventralis*	South America
Plumbeous Forest Falcon	*Micrastur plumbeus*	South America
Plumbeous Hawk	*Leucopternis plumbeus*	South America
Plumbeous Kite	*Ictinia plumbea*	Middle and South America
Prairie Falcon	*Falco mexicanus*	North and Middle America
Pygmy Falcon	*Polihierax semitorquatus*	Africa
Red Goshawk	*Erythrotriorchis radiatus*	Australasia
Red Kite	*Milvus milvus*	Eurasia
Red-chested Goshawk	*Accipiter toussenelii*	Africa

Common name	Scientific name	Distribution
Red-footed Falcon	*Falco vespertinus*	Eurasia
Red-headed Vulture	*Sarcogyps calvus*	Oriental Region
Red-necked Buzzard	*Buteo auguralis*	Africa
Red-necked Falcon	*Falco chicquera*	Africa, Oriental Region
Red-shouldered Hawk	*Buteo lineatus*	North America
Red-tailed Hawk	*Buteo jamaicensis*	North and Middle America
Red-thighed Sparrowhawk	*Accipiter erythropus*	Africa
Red-throated Caracara	*Ibycter americanus*	Middle and South America
Ridgway's Hawk	*Buteo ridgwayi*	North America
Roadside Hawk	*Buteo magnirostris*	Middle and South America
Roughleg	*Buteo lagopus*	North America, Eurasia
Rufous Crab Hawk	*Buteogallus aequinoctialis*	South America
Rufous-bellied Eagle	*Hieraaetus kienerii*	Oriental Region
Rufous-breasted Sparrowhawk	*Accipiter rufiventris*	Africa
Rufous-necked Sparrowhawk	*Accipiter erythrauchen*	Australasia
Rufous-tailed Hawk	*Buteo ventralis*	South America
Rufous-thighed Hawk	*Accipiter erythronemius*	South America
Rufous-thighed Kite	*Harpagus diodon*	South America
Rufous-winged Buzzard	*Butastur liventer*	Oriental Region
Rüppell's Vulture	*Gyps rueppellii*	Africa
Saker Falcon	*Falco cherrug*	Eurasia
Savanna Hawk	*Buteogallus meridionalis*	Middle and South America
Scissor-tailed Kite	*Chelictinia riocourii*	Africa
Secretarybird	*Sagittarius serpentarius*	Africa
Semicollared Hawk	*Accipiter collaris*	South America
Semiplumbeous Hawk	*Leucopternis semiplumbeus*	Middle and South America
Seychelles Kestrel	*Falco araeus*	Africa
Sharp-shinned Hawk	*Accipiter striatus*	North and Middle America
Shikra	*Accipiter badius*	Africa, Oriental Region
Short-tailed Hawk	*Buteo brachyurus*	Middle and South America
Short-toed Snake Eagle	*Circaetus gallicus*	Eurasia, Oriental Region
Slate-colored Hawk	*Leucopternis schistaceus*	South America
Slaty-backed Forest Falcon	*Micrastur mirandollei*	Middle and South America
Slaty-mantled Goshawk	*Accipiter luteoschistaceus*	Australasia
Slender-billed Kite	*Rostrhamus hamatus*	Middle and South America
Slender-billed Vulture	*Gyps tenuirostris*	Oriental Region
Snail Kite	*Rostrhamus sociabilis*	North, Middle, and South America
Sooty Falcon	*Falco concolor*	Africa, Eurasia
Southern Crested Caracara	*Caracara plancus*	South America
Southern Banded Snake Eagle	*Circaetus fasciolatus*	Africa
Spanish Imperial Eagle	*Aquila adalberti*	Eurasia
Spot-tailed Sparrowhawk	*Accipiter trinotatus*	Australasia
Spotted Harrier	*Circus assimilis*	Australasia
Spotted Kestrel	*Falco moluccensis*	Australasia
Spot-winged Falconet	*Spiziapteryx circumcincta*	South America
Square-tailed Kite	*Lophoictinia isura*	Australasia
Steller's Sea Eagle	*Haliaeetus pelagicus*	Eurasia
Steppe Eagle	*Aquila nipalensis*	Eurasia
Striated Caracara	*Phalcoboenus australis*	South America
Sulawesi Goshawk	*Accipiter griseiceps*	Australasia
Sulawesi Hawk-Eagle	*Spizaetus lanceolatus*	Australasia

Common name	Scientific name	Distribution
Sulawesi Serpent Eagle	*Spilornis rufipectus*	Oriental Region
Swainson's Hawk	*Buteo swainsoni*	North and Middle America
Swallow-tailed Kite	*Elanoides forficatus*	North, Middle, and South America
Swamp Harrier	*Circus approximans*	Australasia
Taita Falcon	*Falco fasciinucha*	Africa
Tawny Eagle	*Aquila rapax*	Africa
Tiny Hawk	*Accipiter superciliosus*	Middle and South America
Turkey Vulture	*Cathartes aura*	North, Middle, and South America
Upland Buzzard	*Buteo hemilasius*	Eurasia
Variable Hawk	*Buteo polyosoma*	South America
Verreaux's Eagle	*Aquila verreauxii*	Africa
Vinous-breasted Sparrowhawk	*Accipiter rhodogaster*	Australasia
Wahlberg's Eagle	*Aquila wahlbergi*	Africa
Wallace's Hawk-Eagle	*Spizaetus nanus*	Oriental Region
Wedge-tailed Eagle	*Aquila audax*	Australasia
Western Banded Snake Eagle	*Circaetus cinerascens*	Africa
Western Marsh Harrier	*Circus aeruginosus*	Eurasia
Whistling Kite	*Haliastur sphenurus*	Australasia
White Hawk	*Leucopternis albicollis*	Middle and South America
White-backed Vulture	*Gyps africanus*	Africa
White-bellied Goshawk	*Accipiter haplochrous*	Australasia
White-bellied Sea Eagle	*Haliaeetus leucogaster*	Oriental Region, Australasia
White-breasted Hawk	*Accipiter chionogaster*	Middle America
White-browed Hawk	*Leucopternis kuhli*	South America
White-collared Kite	*Leptodon forbesi*	South America
White-eyed Buzzard	*Butastur teesa*	Oriental Region
White-fronted Falconet	*Microhierax latifrons*	Oriental Region
White-headed Vulture	*Aegypius occipitalis*	Africa
White-necked Hawk	*Leucopternis lacernulatus*	South America
White-rumped Falcon	*Polihierax insignis*	Oriental Region
White-rumped Hawk	*Buteo leucorrhous*	South America
White-rumped Vulture	*Gyps bengalensis*	Oriental Region
White-tailed Eagle	*Haliaeetus albicilla*	Eurasia
White-tailed Hawk	*Buteo albicaudatus*	North, Middle, and South America
White-tailed Kite	*Elanus leucurus*	North, Middle, and South America
White-throated Caracara	*Phalcoboenus albogularis*	South America
White-throated Hawk	*Buteo albigula*	South America
Yellow-headed Caracara	*Milvago chimachima*	Middle and South America
Zone-tailed Hawk	*Buteo albonotatus*	North, Middle, and South America

OWLS

Common name	Scientific name	Distribution
Abyssinian Owl	*Asio abyssinicus*	Africa
African Barred Owlet	*Glaucidium capense*	Africa
African Scops Owl	*Otus senegalensis*	Africa
African Wood Owl	*Strix woodfordii*	Africa
Akun Eagle-Owl	*Bubo leucostictus*	Africa
Amazonian Pygmy Owl	*Glaucidium hardyi*	South America
Andaman Hawk-Owl	*Ninox affinis*	Oriental Region

Common name	Scientific name	Distribution
Andaman Scops Owl	*Otus balli*	Oriental Region
Andean Pygmy Owl	*Glaucidium jardinii*	South America
Ashy-faced Owl	*Tyto glaucops*	North America
Asian Barred Owlet	*Glaucidium cuculoides*	Oriental Region
Austral Pygmy Owl	*Glaucidium nana*	South America
Australian Masked Owl	*Tyto novaehollandiae*	Australasia
Balsas Screech Owl	*Megascops seductus*	Middle America
Band-bellied Owl	*Pulsatrix melanota*	South America
Bare-legged Owl	*Gymnoglaux lawrencii*	North America
Bare-shanked Screech Owl	*Megascops clarkii*	Middle and South America
Barking Boobook	*Ninox connivens*	Australasia
Barn Owl	*Tyto alba*	Worldwide
Barred Boobook	*Ninox variegata*	Australasia
Barred Eagle-Owl	*Bubo sumatranus*	Oriental Region
Barred Owl	*Strix varia*	North and Middle America
Bearded Screech Owl	*Megascops barbarus*	Middle America
Biak Scops Owl	*Otus beccarii*	Australasia
Black-and-white Owl	*Strix nigrolineata*	Middle and South America
Black-banded Owl	*Strix huhula*	South America
Black-capped Screech Owl	*Megascops atricapilla*	South America
Blakiston's Fish Owl	*Bubo blakistoni*	Eurasia
Boreal Owl	*Aegolius funereus*	North America, Eurasia
Brown Fish Owl	*Ketupa zeylonensis*	Oriental Region, Eurasia
Brown Hawk-Owl	*Ninox scutulata*	Oriental Region, Eurasia
Brown Wood Owl	*Strix leptogrammica*	Oriental Region
Buff-fronted Owl	*Aegolius harrisii*	South America
Buffy Fish Owl	*Ketupa ketupu*	Oriental Region
Burrowing Owl	*Athene cunicularia*	North, Middle, and South America
Cape Eagle-Owl	*Bubo capensis*	Africa
Cape Pygmy Owl	*Glaucidium hoskinsii*	Middle America
Central American Pygmy Owl	*Glaucidium griseiceps*	Middle America
Chaco Owl	*Strix chacoensis*	South America
Chestnut-backed Owlet	*Glaucidium castanonotum*	Oriental Region
Choco Screech Owl	*Megascops centralis*	South America
Christmas Boobook	*Ninox natalis*	Indian Ocean
Cinnabar Boobook	*Ninox ios*	Australasia
Cinnamon Screech Owl	*Megascops petersoni*	South America
Cloud-forest Pygmy Owl	*Glaucidium nubicola*	South America
Cloud-forest Screech Owl	*Megascops marshalli*	South America
Colima Pygmy Owl	*Glaucidium palmarum*	Middle America
Collared Owlet	*Glaucidium brodiei*	Oriental Region
Collared Scops Owl	*Otus bakkamoena*	Oriental Region
Congo Bay Owl	*Phodilus prigoginei*	Africa
Costa Rican Pygmy Owl	*Glaucidium costaricanum*	Middle America
Crested Owl	*Lophostrix cristata*	Middle and South America
Cuban Pygmy Owl	*Glaucidium siju*	North America
Dusky Eagle-Owl	*Bubo coromandus*	Oriental Region
East Brazilian Pygmy Owl	*Glaucidium minutissimum*	South America
Eastern Screech Owl	*Megascops asio*	North and Middle America
Elf Owl	*Micrathene whitneyi*	North and Middle America

Common name	Scientific name	Distribution
Eurasian Eagle-Owl	*Bubo bubo*	Eurasia
Eurasian Pygmy Owl	*Glaucidium passerinum*	Eurasia
Eurasian Scops Owl	*Otus scops*	Eurasia
Fearful Owl	*Nesasio solomonensis*	Australasia
Ferruginous Pygmy Owl	*Glaucidium brasilianum*	Middle and South America
Flammulated Owl	*Megascops flammeolus*	North and Middle America
Flores Scops Owl	*Otus alfredi*	Australasia
Foothill Screech Owl	*Megascops roraimae*	South America
Forest Owlet	*Heteroglaux blewitti*	Oriental Region
Fraser's Eagle-Owl	*Bubo poensis*	Africa
Fulvous Owl	*Strix fulvescens*	Middle America
Giant Scops Owl	*Mimizuku gurneyi*	Oriental Region
Golden Masked Owl	*Tyto aurantia*	Australasia
Grass Owl	*Tyto capensis*	Africa, Oriental Region, Australasia
Great Grey Owl	*Strix nebulosa*	North America, Eurasia
Great Horned Owl	*Bubo virginianus*	North, Middle, and South America
Greater Sooty Owl	*Tyto tenebricosa*	Australasia
Greyish Eagle-Owl	*Bubo cinerascens*	Africa
Hume's Owl	*Strix butleri*	Eurasia
Jamaican Owl	*Pseudoscops grammicus*	North America
Javan Owlet	*Glaucidium castanopterum*	Oriental Region
Javan Scops Owl	*Otus angelinae*	Oriental Region
Jungle Boobook	*Ninox theomacha*	Australasia
Jungle Owlet	*Glaucidium radiatum*	Oriental Region
Karthala Scops Owl	*Otus pauliani*	Africa
Koepcke's Screech Owl	*Megascops koepckeae*	South America
Lesser Horned Owl	*Bubo magellanicus*	South America
Lesser Sooty Owl	*Tyto multipunctata*	Australasia
Little Owl	*Athene noctua*	Eurasia, Africa
Little Sumba Hawk-Owl	*Ninox sumbensis*	Australasia
Long-eared Owl	*Asio otus*	North and Middle America, Eurasia
Long-tufted Screech Owl	*Megascops sanctaecatarinae*	South America
Long-whiskered Owlet	*Xenoglaux loweryi*	South America
Luzon Scops Owl	*Otus longicornis*	Oriental Region
Madagascar Owl	*Asio madagascariensis*	Africa
Maned Owl	*Jubula lettii*	Africa
Mantanani Scops Owl	*Otus mantananensis*	Oriental Region
Manus Boobook	*Ninox meeki*	Australasia
Manus Masked Owl	*Tyto manusi*	Australasia
Marsh Owl	*Asio capensis*	Africa
Mayotte Scops Owl	*Otus mayottensis*	Africa
Mentawai Scops Owl	*Otus mentawi*	Oriental Region
Minahassa Masked Owl	*Tyto inexspectata*	Australasia
Mindanao Scops Owl	*Otus mirus*	Oriental Region
Mindoro Scops Owl	*Otus mindorensis*	Oriental Region
Moheli Scops Owl	*Otus moheliensis*	Africa
Moluccan Masked Owl	*Tyto sororcula*	Australasia
Moluccan Boobook	*Ninox squamipila*	Australasia

Common name	Scientific name	Distribution
Moluccan Scops Owl	*Otus magicus*	Australasia
Morepork	*Ninox novaeseelandiae*	Australasia
Mottled Owl	*Strix virgata*	Middle and South America
Mottled Wood Owl	*Strix ocellata*	Oriental Region
Mountain Pygmy Owl	*Glaucidium gnoma*	North and Middle America
Mountain Scops Owl	*Otus spilocephalus*	Oriental Region
Nicobar Scops Owl	*Otus alius*	Oriental Region
Northern Hawk-Owl	*Surnia ulula*	North America, Eurasia
Northern Pygmy Owl	*Glaucidium californicum*	North America
Northern Saw-whet Owl	*Aegolius acadicus*	North America
Northern White-faced Owl	*Ptilopsis leucotis*	Africa
Ochre-bellied Boobook	*Ninox ochracea*	Australasia
Oriental Bay Owl	*Phodilus badius*	Oriental Region
Oriental Scops Owl	*Otus sunia*	Oriental Region
Pacific Pygmy Owl	*Glaucidium peruanum*	South America
Pacific Screech Owl	*Megascops cooperi*	Middle America
Palau Owl	*Pyrroglaux podarginus*	Australasia
Palawan Scops Owl	*Otus fuliginosus*	Oriental Region
Pallid Scops Owl	*Otus brucei*	Eurasia
Papuan Hawk-Owl	*Uroglaux dimorpha*	Australasia
Pearl-spotted Owlet	*Glaucidium perlatum*	Africa
Pel's Fishing Owl	*Scotopelia peli*	Africa
Pemba Scops Owl	*Otus pembaensis*	Africa
Pere David's Owl	*Strix davidi*	Eurasia
Pernambuco Pygmy Owl	*Glaucidium mooreorum*	South America
Pharaoh Eagle-Owl	*Bubo ascalaphus*	Africa
Philippine Eagle-Owl	*Bubo philippensis*	Oriental Region
Philippine Hawk-Owl	*Ninox philippensis*	Oriental Region
Philippine Scops Owl	*Otus megalotis*	Oriental Region
Powerful Boobook	*Ninox strenua*	Australasia
Puerto Rican Screech Owl	*Megascops nudipes*	North America
Rainforest Scops Owl	*Otus rutilus*	Africa
Rajah Scops Owl	*Otus brookii*	Oriental Region
Red Owl	*Tyto soumagnei*	Africa
Red-chested Owlet	*Glaucidium tephronotum*	Africa
Reddish Scops Owl	*Otus rufescens*	Oriental Region
Rufescent Screech Owl	*Megascops ingens*	South America
Rufous Boobook	*Ninox rufa*	Australasia
Rufous Fishing Owl	*Scotopelia ussheri*	Africa
Rufous-banded Owl	*Strix albitarsis*	South America
Rufous-legged Owl	*Strix rufipes*	South America
Rusty-barred Owl	*Strix hylophila*	South America
Ryukyu Scops Owl	*Otus elegans*	Oriental Region
Sandy Scops Owl	*Otus icterorhynchus*	Africa
Sangihe Scops Owl	*Otus collari*	Australasia
Sao Tome Scops Owl	*Otus hartlaubi*	Africa
Serendib Scops Owl	*Otus thilohoffmanni*	Oriental Region
Seychelles Scops Owl	*Otus insularis*	Indian Ocean
Shelley's Eagle-Owl	*Bubo shelleyi*	Africa
Short-eared Owl	*Asio flammeus*	Worldwide
Siau Scops Owl	*Otus siaoensis*	Australasia

Common name	Scientific name	Distribution
Simeulue Scops Owl	*Otus umbra*	Oriental Region
Sjostedt's Barred Owlet	*Glaucidium sjostedti*	Africa
Snowy Owl	*Bubo scandiaca*	North America, Eurasia
Sokoke Scops Owl	*Otus ireneae*	Africa
Solomons Boobook	*Ninox jacquinoti*	Australasia
Southern Boobook	*Ninox boobook*	Australasia
Southern White-faced Owl	*Ptilopsis granti*	Africa
Spangled Boobook	*Ninox odiosa*	Australasia
Speckled Boobook	*Ninox punctulata*	Australasia
Spectacled Owl	*Pulsatrix perspicillata*	Middle and South America
Spot-bellied Eagle-Owl	*Bubo nipalensis*	Oriental Region
Spotted Eagle-Owl	*Bubo africanus*	Africa
Spotted Owl	*Strix occidentalis*	North and Middle America
Spotted Owlet	*Athene brama*	Oriental Region, Eurasia
Spotted Wood Owl	*Strix seloputo*	Oriental Region
Striped Owl	*Pseudoscops clamator*	Middle and South America
Stygian Owl	*Asio stygius*	Middle and South America
Subtropical Pygmy Owl	*Glaucidium parkeri*	South America
Sula Scops Owl	*Otus sulaensis*	Australasia
Sulawesi Masked Owl	*Tyto rosenbergii*	Australasia
Sulawesi Scops Owl	*Otus manadensis*	Australasia
Sumba Boobook	*Ninox rudolfi*	Australasia
Taliabu Masked Owl	*Tyto nigrobrunnea*	Australasia
Tamaulipas Pygmy Owl	*Glaucidium sanchezi*	Middle America
Tawny Fish Owl	*Ketupa flavipes*	Oriental Region
Tawny Owl	*Strix aluco*	Eurasia
Tawny-bellied Screech Owl	*Megascops watsonii*	South America
Tawny-browed Owl	*Pulsatrix koeniswaldiana*	South America
Togian Boobook	*Ninox burhani*	Australasia
Torotoroka Scops Owl	*Otus madagascariensis*	Africa
Tropical Screech Owl	*Megascops choliba*	Middle and South America
Unspotted Saw-whet Owl	*Aegolius ridgwayi*	Middle America
Ural Owl	*Strix uralensis*	Eurasia
Usambara Eagle-Owl	*Bubo vosseleri*	Africa
Vermiculated Fishing Owl	*Scotopelia bouvieri*	Africa
Vermiculated Screech Owl	*Megascops guatemalae*	Middle and South America
Verreaux's Eagle-Owl	*Bubo lacteus*	Africa
Wallace's Scops Owl	*Otus silvicola*	Australasia
West Peruvian Screech Owl	*Megascops roboratus*	South America
Western Screech Owl	*Megascops kennicottii*	North and Middle America
Whiskered Screech Owl	*Megascops trichopsis*	North and Middle America
White-browed Hawk-Owl	*Ninox superciliaris*	Africa
White-fronted Scops Owl	*Otus sagittatus*	Oriental Region
White-throated Screech Owl	*Megascops albogularis*	South America
Yungas Pygmy Owl	*Glaucidium bolivianum*	South America
Yungas Screech Owl	*Megascops hoyi*	South America
OTHER BIRDS		
(African) Grey Parrot	*Psittacus erithacus*	Africa
Albatrosses	Diomedeidae	Oceans globally
American Cliff Swallow	*Petrichelidon pyrrhonota*	North and Middle America
American Coot	*Fulica americana*	North and Middle America

Common name	Scientific name	Distribution
American Crow	*Corvus brachyrhynchos*	North America
American Golden Plover	*Pluvialis dominica*	North America
American Robin	*Turdus migratorius*	North and Middle America
American Wigeon	*Anas americana*	North America
American Yellow Warbler	*Setophaga petechia*	North and Middle America
Anhinga	*Anhinga anhinga*	North, Middle, and South America
Arctic Tern	*Sterna paradisaea*	North America, Eurasia
Arctic Warbler	*Phylloscopus borealis*	North America, Eurasia
Australian Raven	*Corvus coronoides*	Australia
Auklets	Alcidae	North America, Eurasia
Azure-winged Magpie	*Cyanopica cyanus*	Eurasia
Band-tailed Pigeon	*Patagioenas fasciata*	North and Latin America
Barn Swallow	*Hirundo rustica*	Worldwide
Bar-tailed Godwit	*Limosa lapponica*	North America, Eurasia
Black-capped Chickadee	*Poecile atricapillus*	North America
Black-crowned Night Heron	*Nycticorax nycticorax*	Worldwide
Black-throated Loon	*Gavia arctica*	Eurasia
Blue Jay	*Cyanocitta cristata*	North America
Blue-and-white Swallow	*Notiochelidon cyanoleuca*	South America
Bluethroat	*Luscinia svecica*	North America, Eurasia
Bohemian Waxwing	*Bombycilla garrulous*	North America, Eurasia
Boobies	*Sula* spp.	South America, Pacific and Tropical oceans
Brambling	*Fringilla montifringilla*	Eurasia
Brown-headed Cowbird	*Molothrus ater*	North and Middle America
Buff-bellied Pipit	*Anthus rubescens*	North America, Eurasia
Canada Goose	*Branta canadensis*	North America
Canada Warbler	*Wilsonia canadensis*	North America
Cattle Egret	*Bubulcus ibis*	Worldwide
Cockatoos	*Cacatua* spp.	Australasia
Common Chaffinch	*Fringilla coelebs*	Eurasia
Common Crane	*Grus grus*	Eurasia
Common Ground Dove	*Columbina passerina*	North and Latin America
Common Pigeon	*Columba livia*	Worldwide
Common Snipe	*Gallinago gallinago*	North America, Eurasia
Common Starling	*Sturnus vulgaris*	North America, Eurasia
Common Swift	*Apus apus*	Eurasia
Corellas	*Cacatua* spp.	Australasia
Cormorants	*Phalacrocorax* spp.	Worldwide
Cuckoos	Cuculidae	Worldwide
Ducks	Anatidae	Worldwide
Eastern Meadowlark	*Sturnella magna*	North and Latin America
Eastern Yellow Wagtail	*Motacilla tschutschensis*	North America, Eurasia
Emperor Penguin	*Aptenodytes forsteri*	Antarctica
Eurasian Magpie	*Pica pica*	Eurasia
Fieldfare	*Turdus pilaris*	Eurasia
Finches	Emberizidae, Fringillidae	Worldwide
Flamingoes	Phoenicopteridae	Eurasia, Africa, Latin America
Fox Sparrow	*Passerella iliaca*	North America
Fulmars	*Fulmarus* spp.	Northern and southern oceans

Common name	Scientific name	Distribution
Great Blue Heron	*Ardea herodias*	North and Middle America
Great Grey Shrike	*Lanius excubitor*	North America, Eurasia
Great-tailed Grackle	*Quiscalus mexicanus*	North and Latin America
Greater Prairie Chicken	*Tympanuchus cupido*	North America
Greater Scaup	*Aythya marila*	North America, Eurasia
Green Heron	*Butorides virescens*	North and Middle America
Green-winged Teal	*Anas carolinensis*	North America
Grey Catbird	*Dumetella carolinensis*	North America
Grey-cheeked Thrush	*Catharus minimus*	North America
Grey Jay	*Perisoreus candensis*	North America
Gulls	Laridae	Worldwide
Herring Gull	*Larus argentatus*	North and Middle America, Eurasia
Hooded Crow	*Corvus cornix*	Eurasia
Hooded Pitui	*Pitohui dichrous*	Australasia
Horned Lark	*Eremophila alpestris*	North and Middle America, Eurasia
House Finch	*Carpodacus mexicanus*	North and Middle America
House Sparrow	*Passer domesticus*	Eurasia, Oriental Region, Africa
House Wren	*Troglodytes aedon*	North and Latin America
Kea	*Nestor notablis*	Australasia
King Rail	*Rallus elegans*	North and Middle America
Kingfishers	Alcendinidae	Worldwide
Kiwis	Apterygidae	Australasia
Lapland Longspur	*Calcarius lapponicus*	North America
Laughing Gull	*Larus atricilla*	North and Latin America
Least Tern	*Sterna antillarum*	North and Middle America
Lesser Scaup	*Aythya affinis*	North America
Lesser Yellowlegs	*Tringa flavipes*	North America
Long-billed Dowitcher	*Limnodromus scolopaceus*	North America, Eurasia
Long-tailed Jaeger	*Stercorarius longicaudus*	North America, Eurasia
Macaws	Psittacidae	Middle and South America
Magnificent Frigatebird	*Fregata magnificens*	Atlantic and Pacific Oceans
Mallard	*Anas platyrhynchos*	North America, Eurasia
Moas	Dinornithiformes (extinct)	Pacific Ocean
Monk Parakeet	*Myiopsitta monachus*	South America
Mourning Dove	*Zenaida macroura*	North and Middle America
Mute Swan	*Cygnus olor*	North America, Eurasia
Nightjars	Caprimulgidae	Worldwide
Northern Cardinal	*Cardinalis cardinalis*	North America
Northern Bobwhite	*Colinus virginianus*	North and Middle America
Northern Flicker	*Colaptes auratus*	North and Middle America
Northern Fulmar	*Fulmaris glacialis*	Northern oceans
Northern Pintail	*Anas acuta*	North America, Eurasia
Northern Raven	*Corvus corax*	North and Middle America, Eurasia
Northern Shoveler	*Anas clypeata*	North America, Eurasia
Ostriches	*Struthio* spp.	Africa
Parasitic Jaeger	*Stercorarius parasiticus*	North America, Eurasia
Parrots	*Psittaciformes*	North America

Common name	Scientific name	Distribution
Passenger Pigeon	*Ectopistes migratorius* (extinct)	North America
Pectoral Sandpiper	*Caldris melanotos*	North America, Eurasia
Petrels	Procellariidae	Oceans globally
Phalaropes	Scolopacidae	North America, Eurasia
Pomarine Skua	*Stercorarius pomarinus*	North America, Eurasia
Ptarmigan	*Lagopus* spp.	North America, Eurasia
Pyrrhocorax	Corvidae	Eurasia
Rails	Raillidae	Worldwide
Ravens	*Corvus* spp.	Australia, Africa, Eurasia, North and Middle America
Redhead	*Aythya Americana*	North America
Redpolls	*Carduelis* spp.	North America, Eurasia
Red-billed Chough	*Pyrrhocorax pyrrhocorax*	Eurasia
Red-breasted Goose	*Branta ruficollis*	Eurasia
Red-breasted Merganser	*Mergus serrator*	North America, Eurasia
Red-necked Phalarope	*Phalaropus lobatus*	North America, Eurasia
Red-winged Blackbird	*Agelaius phoeniceus*	North and Middle America
Royal Tern	*Sterna maxima*	North and Latin America, Africa
Ruby-crowned Kinglet	*Regulus calendula*	North America
Ruby-throated Hummingbird	*Archilochus colubris*	North America
Rusty Blackbird	*Euphagus carolinus*	North America
Sabine's Gull	*Xema sabini*	North America, Eurasia
Say's Phoebe	*Sayornis saya*	North and Middle America
Scoters	*Melanitta* spp.	North America, Eurasia
Semipalmated plover	*Charadrius semipalmatus*	North America
Semipalmated sandpiper	*Calidris pusilla*	North America
Shearwaters	Procellariidae	Worldwide
Shrikes	Laniidae	Africa, Eurasia, Asia, North and Middle America
Skuas	Stercorariidae	South and North America, Eurasia, Antarctica, southern ocean
Snowy Egret	*Egretta thula*	North and Latin America
Songbirds	Passeriformes	Worldwide
Sparrows	Emberizidae, Passeridae	Worldwide
Spotted Sandpiper	*Actitis macularius*	North America
Storks	Ciconiidae	Worldwide
Subantarctic Skua	*Stercorarius antarcticus*	Southern ocean
Swifts	Apodidae	Worldwide
Tree Swallow	*Tachycineta bicolor*	North America
Tundra Swan	*Cynus columbianus*	North America, Eurasia
Upland Goose	*Chloephaga picta*	South America
Vermilion Flycatcher	*Pyrocephalus rubinus*	North and Latin America
Wandering Albatross	*Doimedea exulans*	Pacific Ocean
Western Jackdaw	*Corvus monedula*	Eurasia
White-collared Swift	*Streptoprocne zonaris*	Latin America
White-winged Dove	*Zenaida asiatica*	North and Middle America
Wild Turkey	*Meleagris gallopavo*	North and Middle America
Whip-poor-will	*Caprimulgus vociferous*	North America
Willet	*Catoptrophorus semipalmatus*	North America

Common name	Scientific name	Distribution
Wood Warbler	*Phylloscopus sibilatrix*	Eurasia
Yellow-billed Cuckoo	*Coccyzus americanus*	North and Middle America
Yellow-legged Gull	*Larus cachinnans*	Eurasia
OTHER ANIMALS		
Apple snails	*Pomacea* spp.	
Arctic fox	*Alopex lagopus*	
Arctic ground squirrel	*Spermophilus parryii*	
Armadillos	Dasypodidae	
Bats	Chiroptera	
Black-tailed jackrabbit	*Lepus californicus*	
Chimpanzee	*Pan troglodytes*	
Chipmunks	Rodentia	
Crabs	Crustacea	
Crappies	*Pomoxis* spp.	
Common vole	*Microtus arvalis*	
Crayfishes	Crustacea	
Desert cottontail	*Sylvilagus auduboni*	
Dolphins	Cetacea	
Domestic pig	*Sus scrofa*	
Grasshoppers	Orthoptera	
Earthworms	Oligochaeta	
Elephants	Elephantidae	
Hares	Leporidae	
Hyenas	Hyaenidae	
Jackals	*Canis* spp.	
Lion	*Panthera leo*	
Locusts	Orthoptera	
Meadow vole	*Microtus pennsylvanicus*	
Mole Rat	Batyergidae spp.	
Mollusks	Mollusca	
Mullets	Mugilidae	
Rabbits	Leporidae	
Raccoons	Procyonidae	
Red Fox	*Vulpes vulpes*	
Sloths	Bradypodidae	
Snails	Gastropoda	
Short-tailed shrew	*Blarina brevicauda*	
Snowshoe hare	*Lepus americanus*	
Southern cotton rat	*Sigmodon hirsutus*	
Striped skunk	*Mephitis mephitis*	
Squirrels	Rodentia	
Susliks	Rodentia	
Termites	Isoptera	
Tundra vole	*Microtus oeconomus*	
Whales	Cetacea	
Wildebeests	Bovidae	

GLOSSARY

accipiters. Members of the most common genus of raptors, *Accipiter*, a group of approximately fifty largely forest-dwelling species of raptors, most of which have short, rounded wings and long tails.

adaptive wind drift. Voluntary wind drift that enables migrants to complete their journeys at lower costs than if they attempted to compensate completely for wind drift en route. Adaptive wind drift is more likely early in migration than later on.

aerodynamic lift. An upward force on the wing created as the wing deflects and forces air downward.

aggressive mimicry. Mimicry in which a predator mimics a nonpredatory model in order to deceive and more easily approach its prey.

air sacs. Thin walled out pockets of the lungs that store and moisten incoming air prior to its passage into the lungs. Air sacs enable full ventilation of the lungs during each inspiration and expiration, permitting more efficient gas exchange than is found in mammals, which lack this anatomical feature.

albedo. A measure of surface reflectivity. The fraction of electromagnetic radiation reflected by a surface expressed as a percentage.

albino. An individual that lacks all pigments.

allopreen. Two birds preening each other's feathers. Usually involves the feathers of the head and neck.

altitudinal migration. Occurs when migrants shuttle between high-altitude breeding areas and low-altitude wintering areas.

altricial. Young birds, including raptors, that hatch blind, cannot walk or thermoregulate, and require intense parental care.

antioxidants. A molecule capable of slowing or preventing oxidation. Antioxidants can counteract the effect of *free radicals.*

aposematic. Warning coloration in distasteful species and their mimics.

archaebacteria. Single-celled organisms lacking a cell nucleus and other organelles. One of three main branches of evolutionary descent together with bacteria and eukaryotes.

aspect ratio. A measure of the wing defined as wingspan squared divided by wing area. Raptors with long, narrow wings have *high-aspect ratios.*

asynchronous. Not synchronous or simultaneous. In birds, asynchronous hatching of a clutch of eggs means that individual eggs hatch on different days, creating a situation in which earlier hatchlings have a "head start" on later hatchlings.

austral summer. Summertime south of the equator, beginning in December and ending in March.

barbs. The branches extending from either side of the shaft of a feather that make up the vane.

barbules. The branchlets extending from either side of *barbs* that hold the vane together.

Batesian mimicry. The close resemblance of a palatable or harmless species (the mimic) to a distasteful or harmful species (the model) to deceive a predator.

bifoveate. Having two foveae of cone-packed depressions on the retina. Raptors have two foveae, a central fovea for monocular vision and a temporal fovea for binocular vision.

biogeography. The study of the geographical distributions of species and the historical and biological factors that led to and continue to shape them.

biological accumulation. The accumulation of persistent environmental toxins, including heavy metals and some pesticides in the body tissues of organisms that consume them. See *biological magnification.*

biological indicator. A species whose numerical presence in an ecosystem is used to indicate the conservation status of an ecosystem. Biological indicators are more easily measured and monitored "surrogates" for the ecosystem processes they seek to indicate. Many conservation biologists consider raptors to be indicators of ecosystem health.

biological magnification (also known as biological amplification). The concentration of persistent environmental toxins, including heavy metals and some pesticides, in organisms in food chains that results in increased levels of a toxin in each successive trophic level. *Persistent* and *fat-soluble* pesticides, in particular, are prone to high rates of biological magnification. See *biological accumulation*.

boreal summer. Summertime north of the equator, beginning in June and ending in September.

Breeding Bird Surveys (BBSs). In North America, Breeding Bird Surveys represent a continent-wide breeding-bird population monitoring system developed and administered by the USGS Patuxent Wildlife Research Center, in Laurel, Maryland. The North American Breeding Bird Survey began in 1966, when approximately 600 surveys were conducted in the United States and Canada east of the Mississippi River. As of 2014, there were about 3700 active BBS routes in the continental United States and Canada, of which nearly 2900 are surveyed annually. Each survey route is 24.5 miles (39.2 km) long with fifty stops at 0.5-mile intervals. A 3-minute point count is conducted at each stop, during which the observer records all birds heard or seen within 0.25 mile of the stop. Routes are randomly located in order to sample habitats that are representative of the entire region. Other requirements such as consistent methodology and observer expertise, visiting the same stops each year, and conducting surveys under suitable weather conditions are necessary to produce comparable data over time. A large sample size (number of routes) is needed to average local variations and reduce the effects of sampling error. The density of BBS routes varies considerably across the continent, reflecting regional densities of skilled birders.

broad-spectrum. Among pesticides, those that kill many kinds of pests. The organochlorine pesticide DDT, for example, kills many kinds of insects, as well as other organisms.

brooding. Sitting on and covering hatched young to keep them warm and to protect them from the sun, rain, and potential predators.

buteos. Members of the genus *Buteo*, which includes twenty-eight species of largely open-habitat raptors with long, broad wings and short tails. Buteos are called hawks in North America and buzzards in Europe and Asia.

caecum. A blind pouch or cul-de-sac at the junction of the small and large intestine in small mammals that use fermentation to digest plant matter.

calendar migrant. A species whose migrations are triggered by and largely driven by day of year, rather than by weather, prey availability, or body condition; an obligate migrant. Calendar migrants typically flock when migrating.

carotenoids. A group of more than 600 organic pigments that are created by and occur in the chromoplasts of plants, algae, fungi, and some bacteria, and that absorb light energy for photosynthesis. In humans, they include beta-carotene, a precursor to vitamin A. Raptors and other birds use them to color their feathers and featherless areas to signal general health and fitness.

carrion. Dead, putrefying flesh that serves as the food of scavenging birds of prey, including both New and Old World vultures.

cere. A fleshy and sometimes colorful swelling of featherless skin that surrounds the nares (or nostrils) of raptors, parrots, skuas, and pigeons, and protects the nostrils.

chain migration. Occurs when migratory populations that breed at high latitudes migrate approximately the same distance as those that breed at lower latitudes, thereby maintaining their latitudinal relationship between seasons. Compare with *leapfrog migration*.

charismatic megafauna. Sometimes called flagship species, charismatic megafauna are symbolic species whose conservation status is used by conservation biologists to capture public attention. The protection of charismatic species can help protect the ecosystems on which they depend. Large raptors are thought by many to be charismatic megafauna.

Christmas Bird Counts (CBCs). A bird-monitoring program in North America, administered by the National Audubon Society. Each CBC consists of a "count circle," 24 kilometers (15 miles) in diameter, within which teams of observers count birds on a single day between December 14th and January 5th each year. The CBC began on Christmas Day 1900, when the ornithologist Frank Chapman proposed the Christmas Bird Count as a new holiday tradition.

circle soaring. Soaring in circles. Raptors often circle-soar to remain within thermals.

cloaca. The common opening at the lower end of a bird's digestive tract for the digestive, excretory, and reproductive systems.

cold front. A large-scale, synoptic weather event in which cold, dense air passes through an area. In eastern North America, such movements are typically from northwest to southeast and are accompanied by northwesterly winds. In autumn in North America, some of the best hawk flights occur following the passage of a cold front.

commensal relationship. A symbiotic relationship in which one species benefits from a common resource while the other species is not adversely affected.

complete migrants. A species or population in which at least 90% of all individuals migrate regularly. A species or population, and not an individual, characteristic. See also *irruptive* and *partial* migrants.

contour feathers. The feathers that make up a bird's exterior surface, including its wings and tail.

convergent evolution. The independent evolution of similar behavioral, physiological, or anatomical traits in distantly related species. Raptors and owls share many traits derived by convergent evolution because of the similar niches they occupy.

Coriolis force (also known as the Coriolis effect). A deflecting force resulting from the earth's daily rotation, and that causes winds, oceanic currents, and migrating birds to be "deflected" to the right in the Northern Hemisphere and to the left in the Southern Hemisphere. The "effect" of the Coriolis force is greater poleward.

corvids. A family of typically gregarious small to large black, black-and-white, or brightly colored birds, that includes crows, ravens, jays, and magpies.

cosmetic. Topical products that enhance the appearance of animals, including both humans and some birds, including at least one raptor.

cosmopolitan. Occurring worldwide on all habitable landmasses, including at least some islands. For raptors, habitable landmasses include North and South America, Europe, Africa, Asia, Australia, and many islands, but not Antarctica. The Peregrine Falcon and the Osprey are considered to be cosmopolitan or near-cosmopolitan species.

crepuscular. Active during the twilight hours of dawn and dusk.

Critically Endangered. A species facing an extremely high risk of extinction in the wild in the immediate future. See also *Endangered* and *Vulnerable.*

crosswinds. Winds that intersect the preferred direction of travel at perpendicular or near-perpendicular angles, and that often alter the direction of travel via a process called wind drift. See also *head winds* and *tailwinds.*

DDT (dichlorodiphenyltrichloroethane). A synthetic organochlorine insecticide once widely used in agriculture and human-disease control. A low-cost, persistent, broad-spectrum compound often termed a "miracle" pesticide, DDT came into widespread use in the late 1940s. DDT was banned in many countries, including Canada, the United States, and much of Western Europe, in the early 1970s, when it was found to be a contaminant in human body tissue and in many ecosystems globally. DDT negatively affects the reproductive success of birds, including raptors, by disrupting a female's ability to produce sufficient eggshell material for her eggs.

DDT era. The time between the late 1940s and the early 1970s when DDT was in widespread use in North America and Western Europe and when populations of many birds of prey in these areas, including Peregrine Falcons and Bald Eagles, declined precipitously.

deflection updrafts. Pockets of rising air created when horizontal winds encounter and deflect up and over mountains, ridges, escarpments, buildings, and even tall vegetation.

delayed maturation. Breeding for the first time, not in the first year of life, but in later years. May differ between sexes within species.

delayed-return migration. Occurs when juvenile raptors remain on their wintering grounds during their entire second and, sometimes, third year before returning to the breeding grounds. Delayed return migration often occurs in species with delayed maturation. The phenomenon is believed to save subadults the expense of migrating and to eliminate competition with adults during the breeding season.

diclofenac. A nonsteroidal anti-inflammatory drug approved by the US Food and Drug Administration (FDA) for use in humans, which, when used in veterinary medicine, kills many species of vulture that consume carcasses of livestock treated with the drug.

dihedral. In ornithology, the upward angle of a bird's wings when soaring.

dispersal. Purposeful movement away from population centers that acts to separate members of populations. Often undertaken by recently fledged individuals, dispersal acts to increase population ranges, while reducing population densities overall.

distensible crop. An expandable and, when filled, externally visible sack on the "throat side" or ventral surface of the esophagus, capable of holding large amounts of recently consumed food. In many scavenging raptors, fully distended crops are thought to signal recent feeding by an individual bird.

diurnal. Active during daylight hours. Compare with *nocturnal.* Hawks, eagles, falcons, vultures, and their allies, are diurnal birds of prey.

diversion line. A geographic feature, including mountain ranges, bodies of water, deserts, and large unbroken forested or open areas, which raptors avoid traveling across and that deflect migrants from their preferred directions of travel, serving to concentrate them.

double endemism. Occurs in migratory species that are native to and restricted to relatively small breeding and wintering areas that are geographically separated from each other.

down feathers. Fluffy, soft feathers that insulate raptors and other birds, particularly recently hatched young.

drag. A force that resists the motion of a body through a gas or liquid. Birds encounter three types of drag when flying: "induced drag" created by trailing vortexes, "parasite drag" created by the outline of a bird's body, and "profile drag" created by the bird's flapping wings. Induced drag decreases as speed increases, parasite drag starts at zero and increases as a cube of speed, and profile drag is constant across speeds.

dynamic soaring. Soaring that takes advantage of the vertical wind gradient that occurs close to a flat surface, when friction slows the layer of air in contact with the surface. Birds engaged in dynamic soaring alternately soar upward into increasing head winds to gain altitude and then turn and glide downwind to gain airspeed, before turning again into the head wind and using the kinetic energy gained on the downwind leg to gain altitude on the upwind leg. Albatrosses and other pelagic seabirds are believed to use dynamic soaring while at sea.

ecological equivalent. A species that can replace another species "ecologically" when the other species is absent or inactive. In birds of prey, owls are said to be ecological equivalents of raptors and vice versa, because owls are active principally at night, whereas raptors are active principally during the day. In North America, Great Horned Owls are said to be ecological equivalents of Red-tailed Hawks.

ecological pyramid (also known as food pyramid). A graphical representation of the food relationship of an ecological community, expressed quantitatively as numbers, mass, or energy at each trophic level, with producers at the bottom, and primary, secondary, and higher-level consumers ascending, in order, to the top of the pyramid.

ectoparasites. Parasites such as fleas, flies, lice, mites, and ticks that live on the skin and feathers of birds.

egg tooth. A short, pointed calcareous structure at the tip of the upper beak of a developing chick that it uses to pound the inner surface of the eggshell during hatching. The egg tooth is usually sloughed off or reabsorbed within a few days of hatching.

Endangered. A species facing a very high risk of extinction in the near future. See also *Critically Endangered* and *Vulnerable.*

endemics. Species that are native to, and are restricted to, a relatively small geographic region. *Island endemics* are species limited to a single island or archipelago. Endemics are often called "limited-range" species.

engorged crop. A crop that is filled to excess. In birds the crop is a dilation of the lower esophagus that stores food.

euryphagic. Dietary generalists that feed on a wide variety of prey. For comparison, see *stenophagic.*

extirpation. Total elimination of a species from part, but not all, of its range. Not to be confused with extinction.

extra-pair copulation. Copulation with a bird other than one's mate.

eye ring. Distinctively colored bare skin or feathering surrounding the eye.

eyrie. The nest of a bird of prey, especially large eagles and cliff-nesting falcons.

facial disk. The dish-shaped and somewhat round forward-facing part of the head of harriers and most owls that funnels sound into a bird's ears.

facultative scavenger. A raptor that feeds on both living prey and carrion, depending on the relative availability of the two resources, but does not depend on either food resource entirely.

falcons. Members of the genus *Falco*, a group of thirty-seven raptors with long, pointed wings and long tails.

fat soluble. The ability to dissolve in fat. Pesticides that are fat soluble, but water insoluble, accumulate in the fatty tissues of animals, and can result in *biological magnification* of pesticides in animals higher in the food chain, including raptors.

femoral-tract feathers. The feathers that cover the femurs (upper legs) of birds.

fledge. When a nestling first flies from the nest.

flex gliding. High-speed gliding during which a bird reduces its wingspan, wing slotting, and overall flight surface by flexing its wings inward and toward its body. Flex gliding increases aerodynamic performance during high-speed gliding flight.

flight feathers. The long, stiff feathers of the wings and tail. On the wings, these feathers are called remiges and include the outermost *primaries* and innermost *secondaries.* On the tail, they are called rectrices.

floaters. In population biology, subordinate individuals of breeding age that do not breed because of their inability to establish a territory and attract a mate.

flocking. Joining together in groups as a result of social attractions. Flocks differ from "aggregations" in that the latter result when birds are attracted to locations by physical or ecological factors alone.

food chain. A sequence of organisms, including primary producers (mainly plants), herbivores (plant eaters), and carnivores (meat eaters), through which energy moves in ecosystems.

free radicals. Atoms, molecules, or ions with an unpaired electron in their outer shell, which causes them to be particularly chemically reactive. In organisms, free radicals have been associated with oxidative damage and aging.

generation time. The average age of breeding females within a population.

genus. A group of one or more closely related species.

gizzard. The lower muscular and grinding part of a raptor's two-part stomach.

gonads. An organ in an animal that produces sperm or unfertilized eggs. The ovaries and testicles.

gular fluttering. Heat-dissipation mechanism in which an individual vibrates its wetted gular (or throat) surface with its mouth open. Particularly common in owls and nestling raptors.

habitat. A species habitat is the ecological location where a species can be found, or its "address" in nature. Compare with *niche.*

half-life. The time it takes for half the amount of a substance, including a synthetic pesticide, to degrade in the environment into its by-products and, therefore, reduce its initial effect.

head winds. Winds aligned against the preferred direction of travel that strike birds in the head and hinder them by "pushing" against them and slowing down their ground speed, or rate of travel. See also *crosswinds* and *tailwinds.*

hibernate. To spend the winter in an inactive or resting state at a lower body temperature, with slower breathing, and a lower metabolic rate, to conserve energy.

high-aspect ratio. High-aspect-ratio wings are long, narrow, and pointed. They occur principally in birds that live in open habitats and that often fly at high speeds.

homeotherm. Warm-blooded. An animal that is able to keep its internal body temperature constant as ambient temperatures vary. Raptors are homeotherms. Compare with *poikilotherm.*

human commensal. An organism, including a raptor, that benefits from the presence of humans without negatively affecting them.

human subsidies. An ecological condition in which human activities (e.g., herding, farming, disposal of organic trash, etc.) provide food for otherwise nutritionally stressed wildlife populations, including birds of prey.

imprinting. A form of learning that occurs only during a short time early in a young animal's life. Imprinting is often irreversible.

information center. The information center hypothesis suggests that communal roosts and breeding sites function as information centers for resource availability with unsuccessful hunters at such sites watching for successful hunters to return and then departing in the direction they have returned from.

instinctive. Highly complex, stereotypic behavior with a large genetic component.

insular. An island form. A species that is found on islands, but not on the mainland.

island endemics. See *endemics.*

island hopping. Migrating along archipelagos, presumably to reduce the risk of being lost at sea.

irruptive migrant. Species or regional populations in which the extent of migratory movement varies annually, typically due to among-year shifts in prey abundance, and whose migrations are less regular than those of partial and complete migrants. A species or population, not an individual, characteristic. Compare with *complete migrant* and *partial migrant.*

keratin. A protein that forms the scales, claws, and most of mature feathers in birds.

kettle. Among bird-watchers, a flock of raptors, especially when the birds are circling in a thermal. Raptors in kettles are sometimes said to be "kettling." Both terms owe their origins to a boulder field called "The Kettle" at Hawk Mountain Sanctuary in eastern Pennsylvania, over which migrating Broad-winged Hawks and other raptors often circle upwards while soaring in thermals created by the dry, unvegetated surface of the field.

key innovation. An evolutionary change in an organism that significantly increases access to ecological resources and that subsequently causes range expansion, significant and, sometimes, spectacular population increase, and, in many instances, enhanced adaptive radiation. Lungs, which developed in aquatic vertebrates, enabled the rapid diversification of land-based vertebrate life. Feathers, which developed in reptiles, enabled the rapid radiation of flighted birds. Both are considered key innovations.

keystone species. Species that have a major influence on the community structure of an ecosystem.

leading line. A geographic feature, including mountain ranges and river systems, which serve to attract, channel, and concentrate birds during their migrations.

leapfrog migration. Occurs when migratory populations that breed at high latitudes migrate substantially farther and "leap over" migratory and nonmigratory populations breeding at lower latitudes, thereby reversing their latitudinal relationship between seasons. Compare with *chain migration.*

lee waves. Oscillating airstreams of various amplitudes that form on the lee, or downwind sides, of mountain ranges under specific wind and temperature conditions. Lee waves form most readily when horizontal winds are deflected over steep, high barriers. Higher barriers create lee waves with greater amplitudes, and amplitudes are sometimes enhanced when winds are deflected over a series of parallel ridges.

leucistic. An individual with reduced pigmentation, resulting in a "washed-out" plumage and pale skin.

lifetime reproductive success. The number of offspring that survive to sexual maturity that are produced by one individual over the course of its lifetime.

linear soaring. Long-distance, straight-line soaring in *thermal streets* and mountain ridges are more typical of tropical than temperate regions. Called slope soaring along mountain ridges.

lores. The area between the eye and upper beak. Featherless in some raptors.

melanin. A class of polymers derived from the amino acid tyrosine that function as blackish pigments in birds and other animals.

melanistic. An individual showing high levels of melanin in its feathering, resulting in a dark color morph.

metabolism. All the synthetic (anabolic) and degradative (catabolic) chemical processes of living organisms necessary to sustain life.

midday lull (a.k.a. noon-day lull). A hawk-watching term that describes the phenomenon in which the magnitude of a migratory flight increases from early morning to late morning, decreases near noon, and increases again in early afternoon before again decreasing later in the day. The midday lull is usually ascribed to the presence of stronger thermals at midday, which can carry birds to higher flight altitudes, making them more difficult to see from the ground.

migration bottleneck. A site where migration corridors converge and subsequently diverge, and at which large numbers of migrants concentrate. Mountain passes, isthmuses, and narrow coastal plains and water crossings, including the Isthmus of Panama, the Strait of Gibraltar, and the Bosporus, are examples of migration bottlenecks. Major migration bottlenecks create "funnel shaped" migration flyways.

migration connectivity. In migration geography, a phenomenon in which migratory individuals that breed close to one another on the breeding grounds also overwinter close together on their wintering grounds.

migration corridors (or flyways). Routes used by massive numbers of migrating raptors that are created when leading lines and diversion lines act to concentrate migrants.

migration-dosing speciation. A series of events in which "doses" of diverted long-distance migrants arrive in areas tangential to or beyond their major migration flyways, remain there the following breeding season rather than returning to traditional breeding areas, breed, and eventually diverge, genetically, from parental stock in geographic isolation.

migratory restlessness (or *Zugunruhe*). The increased rate of to-and-fro hopping and flitting movements and wing fluttering in caged birds during the species' periods of migration.

mimicry. The close resemblance of one species (the mimic) to a second species (the model) to deceive one or more other species (the operator[s]).

mobbing. A collective or individual approach or attack by birds on a potential or perceived predator, including raptors, that often involves vocalizations and may or may not include actual contact by the mobbers.

molecular phylogenetics. Analyses of hereditary molecular differences, principally of DNA sequences, to learn about an organism's ancestral relationships.

monogamy. A social system in which a single male and a single female form a breeding relationship. The individuals involved are said to be monogamous.

monophyletic group. A group of species descended from a single ancestral species. Usually used in the restricted sense to mean the ancestral species and all of its descendants.

mountain (or **deflection**) **updrafts.** Pockets of rising air created when horizontal winds encounter and deflect up and over mountains, ridges, escarpments, buildings, and even tall vegetation.

narrow-front migration. Migration in which dispersed migrants deviate from their initial directions, either because they are attracted to or are avoiding certain geographic features such as mountain ranges, river systems, and coastlines, or because they seek out and join conspecifics en route. Narrow-front migration results in concentrated movements of migrants.

navigation. Movement toward a goal that often requires reorientation en route and includes knowledge of the distance between present location and the goal.

neoplasms. Benign, premalignant, and malignant tumors. Neoplasms have been found in many captive raptors. Their distribution and abundance in free-ranging raptors remains largely unstudied.

neurotransmitter. Internally produced chemicals that relay and often amplify signals among neurons (nerve cells) and between neurons and other cells. Acetylcholine, for example, is the principal neurotransmitter that communicates signals between neurons and muscle cells that induce muscle contraction. Acetylcholinesterase is a chemical that breaks down acetylcholine, thereby shutting off the signal and allowing the muscle to relax.

Cholinesterase inhibitors, including many organophosphate pesticides, kill by inhibiting the action of acetylcholinesterase, thereby creating a form of tetanus.

New World. North America, Central America, South America, and associated islands, including Greenland. Also known as the Western Hemisphere. Compare with *Old World.*

niche. The ecological role of a species in a community. The ecological "profession" of the species. Some niches are more specialized than others. Compare with *habitat.*

nictitating membrane. The third and translucent eyelid in birds.

nocturnal. Active during the hours of darkness. Compare with *diurnal.* Owls are nocturnal birds of prey.

nomadic. Wandering, sometimes seasonally, over large areas, without specific or obvious directional components.

obligate scavenging bird of prey. A raptor that depends on carrion to survive. Vultures and condors are obligate scavenging birds of prey.

obligate soaring migrant. A migratory raptor that depends on the use of energy-efficient soaring flight to complete its migratory journeys.

oceanic islands. Islands formed by volcanic activity from the floor of the ocean that have never been attached to a continent.

Old World. Europe, Asia, Africa, Australia, and associated islands. Also known as the Eastern Hemisphere. Compare with *New World.*

open-program species. A term used to describe a species that begins and continues life relatively open to learning by experience and whose behavioral patterns reflect such learning.

organochlorine pesticides. Also known as chlorinated hydrocarbons, organochlorines are a class of synthetic, *broad-spectrum*, contact pesticides that include *DDT* and dieldrin, and that first came into widespread use in the 1940s and 1950s. Usually used against insect pests, organochlorines resist biodegradation, persist for long periods in ecosystems, and often accumulate in nontarget organisms, including birds of prey.

orientation. Direct movement, typically along a single compass direction.

outbound migration. Migratory movements occurring immediately after the breeding season. Compare with *return migration.*

palaeospecies. A species known only from fossils.

partial migrants. A species or a population in which fewer than 90% of all individuals regularly migrate. A species or population, and not an individual, characteristic. Compare with *irruptive* and *complete* migrants.

patagium. A fold of vascularized skin on the leading edge of the inner wing of a bird that connects the shoulder to the wrist.

peninsular effect. The tendency for many migrants to concentrate at the tip of a peninsula before initiating a water crossing or to make landfall at the tip of a peninsula, presumably because doing so shortens overwater journeys.

peristaltic egestion. Alternating radially symmetrical contraction and relaxation of esophageal muscles that propagate a series of waves from the stomach to the throat, thereby permitting raptors to regurgitate boluses of indigestible wastes from their stomachs.

persistent. Among pesticides, those having long a relative long *half-life.* Synthetic organochlorine pesticides, in particular are persistent pesticides. *DDT*, for example, has a half-life of 15 years.

photosynthesis. The biochemical process that uses the radiant energy in sunlight to synthesize carbohydrates from carbon dioxide and water in the presence of chlorophyll. Green plants engage in photosynthesis.

phylogenetic extinction. The termination of a phyletic lineage (species) without the formation of any descendent lineage. Sometimes called *terminal extinction*. Compare with *pseudoextinction.*

phylogeny. The evolutionary history of a group. Its ancestral relationships.

plumage polymorphism. The situation in which two or more distinctive plumages exist within the same population.

poikilotherm. Cold-blooded. An animal that is largely or totally unable to keep its internal body temperature constant as ambient temperatures vary. Compare with *homeotherm.*

pole trap. A trap placed on the top of a fence post or other pole-like structure in an area with limited above-ground perches used to catch raptors by snaring them when they land on the top of the pole and trigger the snare.

polyandrous. A *polygamous* situation known as polyandry in which one female simultaneously has two or more male mates.

polygamous. A mating situation known as polygamy, in which one individual simultaneously has two or more mates.

polygynous. A *polygamous* situation called polygyny in which one male simultaneously has two or more female mates.

polyphyletic. A group of species that is derived from two or more distinct lineages, rather than from a common ancestor.

powder down. Down feathers that never molt, grow continually, and disintegrate at their tips to produce a powder that may help waterproof feathers.

precocial. Young birds that hatch relatively well developed, with eyes open, thick down, and the capacity to walk.

primaries. The outermost flight feathers on the wing. All raptors have ten primaries.

primary cavity-nester. A bird that excavates its own nesting cavity (hole). Compare with *secondary cavity-nester.*

principal axis of migration. A regional straight compass heading (loxodrome) or rhumb line between a bird's breeding and wintering area. A single principal axis of migration is the shortest route between breeding and winter areas only when those areas fall along a direct north-south axis. In all other cases, a great-circle route is the shortest path. Principal axes of migration often change with latitude.

propagules. The minimum number of individuals of a species needed for colonization.

proventriculus. The upper part of the bird's two-part stomach that secretes mucus, hydrochloric acid, and enzymes to help chemically digest prey.

pseudo- (or **phyletic**) **extinction.** The "extinction," or loss, of an ancestral species as the result of substantial changes in a continuing lineage, such that modern individuals have diverged from their ancestors to the point that they could not breed with one another. See *phylogenetic extinction.*

recrudescence. To reactivate activity. Usually applied to the gonads.

return migration. Migratory movements occurring immediately before the breeding season. Compare with *outbound migration.*

reversed size dimorphism (RSD). The situation in which the females of a species are larger than their male counterparts. RSD occurs in many predatory birds including skuas, frigatebirds, boobies, and most birds of prey (including owls). The function of RSD is not clear.

satellite tracking. Following the movements of animals, including raptors, to which a UHF (ultra-high-frequency) transponder has been attached. Used together with an array of satellites these tracking devices can monitor the movements of raptors across the surface of the globe.

scramble competition. Direct competition for a limiting resource, in which competitors scramble for access to the resource and some, but not all, competitors gain access to it.

sea thermals. Updrafts of warm rising air that form above tropical and subtropical oceans and seas 5° to 30° north and south of the equator dominated by easterly *trade winds.* Within these latitudinal belts, predominant northeasterly winds in the Northern Hemisphere and southeasterly winds in the

Southern Hemisphere blow relatively cool, subtropical surface air toward the equator. As they do, the increasingly warmer surface waters of the Equatorial Zone heat the cooler air, and in doing so produce sea thermals. Because the temperature differential between the cool winds and the warmer sea exists during both day and night, sea thermals are generated 24 hours a day.

secondaries. The innermost flight feathers on the wing. In raptors, the number of secondaries varies among species.

secondary cavity-nester. Cavity-nesting birds that nest in cavities (holes) created by decaying wood or erosion, or that were excavated by other species known as *primary cavity-nesters.*

selfish-herd effect. Proposed by the evolutionary biologist W. D. Hamilton in 1971. A selfish herd is said to occur when individuals in an area attempt to reduce their predation risk by situating themselves such that other conspecifics are between them and potential predators. When numbers of conspecifics do so, such behavior inevitably results in aggregations that include flocks and herds.

shell midden. An old, human-constructed dump, usually for bivalve shells.

short stopping. A phenomenon first described in migratory waterfowl that occurs when migrants shorten the lengths of their outbound movements to take advantage of newly available wintering areas that are closer to their breeding grounds than traditional sites.

siblicide. The killing of an organism by one of its siblings. In raptors, siblicide is said to occur when a nestling kills one of its nest mates.

slope soaring. Soaring in updrafts that are created when horizontal winds strike and are deflected over isolated mountains and mountain ranges.

slotted wingtips. Also called emarginated wingtips, these are wingtips in which the outermost primaries, or flight feathers, when fully spread, create gaps in the margin of the wingtip, giving the tip a fingered appearance. Birds with slotted wingtips have deeply notched outermost primaries that are wide at the base but that narrow abruptly about halfway to the tip and, therefore, are much narrower at the tip. Although their function remains debated, many scientists believe slotting reduces drag near the wingtips by creating a series of mini-airfoils.

soaring. Level or ascending nonflapping flight on outstretched wings in which upward air movements are equal to or greater than the bird's rate of descent in a glide.

species. A group of interbreeding organisms that is reproductively isolated from other groups.

stenophagic. Dietary specialists that feed on one or only a few types of prey. For comparison, see *euryphagic feeders.*

stopover site. A place where a bird stops and interrupts its migration for one or more days.

successional speciation. Evolution within a phylogenetic lineage of one species from another, in which the new species replaces the old.

sunning. Exposing plumage to the sun, usually by spreading the wings and tail, erecting back feathers, and remaining motionless.

supercilliary. Of or pertaining to the eyebrow.

superflocking species. Long-distance, typically transequatorial migrants that travel in large flocks of hundreds to tens of thousands of birds, particularly in the tropics and subtropics.

symbiotic cellulolytic intestinal bacteria. Beneficial, cellulase-producing bacteria that live in the digestive tracts of animals, including voles, and allow the animals to gain metabolic energy from cellulose-rich foods, including grasses.

syringeal. Having to do with the avian syrinx, or song box, the sound-producing organ of birds.

tailwinds. Winds aligned with the preferred direction of travel. Tailwinds help raptors by increasing their ground speed, or rate of travel by "pushing" them forward in the preferred direction of travel. See also *crosswinds* and *head winds.*

talons. The deeply curved, grasping claws of a bird of prey.

Temperate Zone. Latitudinally, the temperate regions of the earth between the tropics (23.5° north and south of the equator) and the Polar Zones (greater than 66.5° degrees north and south). Climates in the Temperate Zone are characterized as moderate and are not usually excessive or extreme.

teratorns. A group of very large extinct vultures that lived in the Americas until the Pleistocene. Members of the avian subfamily Teratornithinae.

terminal extinction. The termination of a phylogenetic lineage (the total loss of all members of a species) without any descendent lineage. Compare with *pseudoextinction.*

theory of island biogeography. A biogeographical theory suggesting that the numbers of species inhabiting an island is a function of the size of the island and its distance from the mainland, and is maintained by the relationship of the rates of immigration onto the island and extinction on the island.

thermal. A pocket of warm, rising air created by the differential heating of the earth's surface.

thermal soaring. Soaring flight in thermals, or pockets or warm, rising air. Raptors often circle when soaring in thermals to remain within them.

thermal streets. Linear arrays of thermals that are often aligned with prevailing winds. When thermal streets are oriented along the *principal axis of migration,* they enable migrants to soar linearly within them in much the same way as migrants *slope soar.*

thermoregulation. The ability of an animal to keep its body temperature within certain limits. Most raptors thermoregulate their internal body temperature at between about 39.5 °C in large species to about 41 °C in small species.

tomial tooth. A notch in the upper mandible or beak of falcons and falconets just behind the tip that is thought to aid in breaking the neck of prey.

tool use. The use of physical objects other than the animal's own body or appendages as a means to extend the physical influence realized by the animal. Raptors are known to use stones as tools in their feeding behavior.

top predator. Sometimes called apex predators, these are predators that as adults are rarely preyed on by other predators.

torpor. Metabolic state in which the core body temperature of a homeotherm drops several degrees and metabolic processes slow.

trade winds. Predominant northeasterly and southeasterly subtropical and tropical oceanic winds that encircle the globe 5° to 30° north and south of the equator, respectively. Thermal convection produced by these winds creates *sea thermals.*

trade-wind thermals. Sea thermals created when predominant northeasterly winds in the Northern Hemisphere Trade-Wind Zone (5° to 30° north) and predominant southeasterly winds in the in the Southern Hemisphere Trade-Wind Zone, pass over increasingly warmer subtropical and tropical waters. Raptors sometimes take advantage of such overwater thermals when migrating in the Trade-Wind Zones north and south of the equator.

Trade-Wind Zone. See *trade-wind thermals.*

trophic cascade. Ecological effects that occur across several tropic levels, many of which affect primary as well as secondary productivity.

tropics. The geographical zone between the Tropic of Cancer (23.5° north) and the Tropic of Capricorn (23.5° south). The tropics include all of the areas on earth where the sun is directly overhead at least once a year. They tend to have hot and humid climates.

umbrella species. Species whose protection are believed to lead to the protection of many other species.

under-tail coverts. Relatively short feathers that cover the underside base of a raptor's tail feathers.

urohydrosis. Urinating on one's legs.

vagrant. In birds, a species that does not regularly breed, overwinter, or migrate through an area, but occurs there infrequently, presumably when lost.

velocity of minimum power. The flapping flight speed at which a raptor expends the minimum amount of energy per unit of time. This is the speed at which a raptor can remain in the air the longest in flapping flight before running out of metabolic energy.

visual acuity. The sharpness of vision or resolution of images. The ability of an eye to resolve details.

voles. Mouselike small rodents with stout bodies, smallish eyes and ears, and a short tail. Voles are frequently preyed on by raptors in the northern Temperate Zone.

Vulnerable. A species facing a high risk of extinction in the wild in the medium-term future. See *Critically Endangered* and *Endangered.*

vulture restaurants. Supplementary feeding sites for vultures at which livestock carcasses, offal, etc., are made available to scavenging birds of prey on a regular basis.

wind drift. When migrants encountering *crosswinds* are pushed off their intended course even while maintaining the same heading.

wing chord. The distance between the carpal joint and the wingtip.

wing loading. A bird's mass divided by the combined area of it outstretched wings. Species with low wing loadings have relatively large wings for their body mass. Those with high wing loading have relatively small wings. Small birds tend to have lighter wing loadings than larger birds. For their body masses, raptors tend to be more lightly wing loaded than other birds.

wing slotting. Occurs when the distal ends of the outermost primaries on a bird's wing are separated horizontally and vertically in flight, thereby creating separate aerodynamic surfaces. Wing slotting is enhanced in some species by the presence of emarginate, or notched, outermost primaries. Wing theory suggests that wing slotting increases the "aerodynamic wingspan" of a bird, and that it reduces wingtip drag at low speeds.

wingspan. The distance from wingtip to wingtip on fully extended wings.

wingtip. The distal end of a bird's wing.

Zugunruhe (or **migratory restlessness**). The increased rate of to-and-fro hopping and flitting movements and wing fluttering in caged birds during the species' periods of migration.

REFERENCES AND RECOMMENDED READINGS BY CHAPTER

1. Introducing Raptors

Amadon, D., and J. Bull. 1988. Hawks and owls of the world: A distributional and taxonomic list. Proc. West. Found. Vert. Zool. 3:294–357.
Banks, R.C., R.T. Chesser, C. Cicero, J.L. Dunn, A.W. Kratter, I.J. Lovette, P.C. Rasmussen, J.V. Remsen, Jr., J.D. Rising, and D.F. Stotz. 2007. Forty-eighth supplement to the American Ornithologists' Union check-list of North American birds. Auk 124:1109–1115.
Borror, D.J. 1960. Dictionary of word roots and combining forms. Palo Alto: Mayfield Publishing.
Brown, L. 1976. Birds of prey: Their biology and ecology. Middlesex: Hamlyn.
Brown, L., and D. Amadon. 1968. Eagles, hawks and falcons of the world. New York: McGraw-Hill.
Burton, J.A. 1973. Owls of the world. New York: E.P. Dutton.
Choate, E.A. 1973. The dictionary of American bird names. Boston: Gambit.
Collar, N.J., M.J. Crosby, and A.J. Sattersfield. 1994. Birds to watch 2: The world list of threatened birds. Cambridge: Birdlife International.
del Hoyo, J., A. Elliott, and J. Sargatal. 1994. Handbook of the birds of the world. Vol. 2. Barcelona: Lynx Edicions.
Dunning, J.B., Jr. 1993. CRC handbook of avian body masses. Boca Raton: CRC Press.
Ericson, P.G.P., C.L. Anderson, T. Britton, A. Kallersjo, J.I. Ohlson, T.J. Parsons, D. Zuccon, and G. Mayr. 2006. Diversification of Neoaves: Integration of molecular sequence data and fossils. Biol. Letters. doi:10.1098/rsbl.2006.0523.
Feduccia, A. 1996. The origin and evolution of birds. New Haven: Yale University Press.
Gill, F., and M. Wright. 2006. Birds of the world: Recommended English names. Princeton: Princeton University Press.
Greenway, J.C. 1958. Extinct and vanishing birds of the world. New York: American Committee for International Wild Life Protection.
Griffiths, C.S. 1994. Monophyly of the Falconiformes based on syringeal morphology. Auk 111:787–805.

Griffiths, C.S., G.F. Barrowclough, J.G. Groth, and L.A. Mertz. 2007. Phylogeny, diversity, and the classification of the Accipitridae based on DNA sequences of the RAG-1 exon. J. Avian Biol. 38:587–602.
Holmgren, V.C. 1972. Bird walk through the Bible. New York: Seabury Press.
Hudson, G. E. 1948. Studies on the pelvic appendages in birds II: The heterogeneous order Falconiformes. Am. Midl. Nat. 39:102–127.
Jobling, J.A. 1991. A dictionary of scientific bird names. Oxford: Oxford University Press.
Ligon, J.D. 1967. Relationships of the Cathartid vultures. Occas. Pap. Mus. Zool. Univ. Mich. 651:1–26.
Martin, P.S. 1973. The discovery of America. Science 179:969–974.
Olsen, P. 1995. Australian birds of prey. Baltimore: Johns Hopkins Press.
Rea, A.M. 1983. Cathartid affinities: A brief overview. *In* Vulture biology, ed. S.R. Wilbur and J.A. Jackson. Berkeley: University of California Press.
Sibley, C.G., and J.E. Ahlquist. 1990. Phylogeny and classification of birds. New Haven: Yale University Press.
Sibley, C.G., and B.L. Monroe, Jr. 1990. Distribution and taxonomy of birds of the world. New Haven: Yale University Press.
Stattersfield, A.J. and D.R. Capper. 2000. Threatened birds of the world. Barcelona: Lynx Edicions.
Weidensaul, S. 1996. Raptors: The birds of prey. New York: Lyons & Burford.

2. Form and Function

Amlaner, C., Jr., and N.J. Ball. A synthesis of sleep in wild birds. Behaviour 87:85–119.
Anderson, M.D., A.W.A. Maritz, and E. Oosthuysen. 1999. Raptors drowning in farm reservoirs in South Africa. Ostrich 70:139–144.
Barja, G., S. Cadenas, C. Rojas, R. Pérez-Campo, and M. López-Torres. 1994. Low mitochondrial free radical production per unit O2 consumption can explain the simultaneous presence of high longevity and high aerobic metabolic rate in birds. Free Radical Res. 21:317–327.
Bildstein, K.L. 1992. Causes and consequences of reversed size dimorphism in raptors: The head-start hypothesis. J. Raptor Res. 26:115–123.
Bildstein, K.L. 1995. Redtail 0877–17127. Hawk Mountain News 83:22–23.
Boerner, M., and O. Krüger. 2009. Aggression and fitness differences between plumage morphs in the common buzzard (*Buteo buteo*). Behav. Ecol. 20:180–185.
Bortolotti, G.R., J.L. Tella, M.G. Forero, R.D. Dawson, and J.J. Negro. 2000. Genetics, local environment, and health as factors influencing plasma carotenoids in wild American Kestrels (*Falco sparverius*). Proc. R. Soc. Lond. B. Biol. 267:1433–1438.
Brodkorb, P. 1955. Numbers of feathers and weights of various systems in a Bald Eagle. Wilson Bull. 67:142.
Broley, C.L. 1947. Migration and nesting of Florida Bald Eagles. Wilson Bull. 59:3–20.
Brown, L. 1976. Eagles of the world. Newton Abbot: David & Charles.
Brown, L., and D. Amadon. 1968. Eagles, hawks and falcons of the world. New York: McGraw Hill.
Burton, J.A. 1973. Owls of the world. New York: E.P. Dutton.
Cade, T.J. 1982. The falcons of the world. Ithaca: Comstock Press.
Campbell, B., and E. Lack. 1985. A dictionary of birds. Vermillion: Buteo Books.

Carpenter, J.W., O.H. Pattee, S.H. Fritts, B.A. Rattner, S.N. Weimeyer, J.A. Royle, and M.R. Smith. 2003. Experimental lead poisoning in Turkey Vultures (*Cathartes aura*). J. Wildl. Dis. 39:96–104.

Casagrande, S., D. Csermely, E. Pini, V. Bertacche, and J. Tagliavini. 2006. Skin carotenoid correlates with male hunting skill and territory quality in the kestrel *Falco tiunnunculus*. J. Avian Biol. 37:190–196.

Chakarov, N., M. Boerner, and O. Krüger. 2008. Fitness in common buzzards at the cross-point of opposite melanin-parasite interactions. Funct. Ecol. 22:1062–1069.

Chaplin, S.B., D.B. Diesel, and J.A. Kasparie. 1984. Body temperature regulation in Red-tailed Hawks and Great Horned Owls: Responses to air temperature and food deprivation. Condor 86:175–181.

Clark, R.J. 1975. A field study of the Short-eared Owl, *Asio flammeus* (Pontoppidan) in North America. Wildl. Monogr. 47:1–67.

Clark, R.G., and R.D. Ohmart. 1985. Spread-winged posture of Turkey Vultures: single or multiple function? Condor 87:350–355.

Clark, W.S. 1998. First North American record of a melanistic Osprey. Wilson Bull. 110:289–290.

Clark, W.S. 2004. Is the Zone-tailed hawk a mimic? 2004. Birding 36:494–498.

Clark, W.S., and B.K. Wheeler. 1987. A field guide to hawks North America. Boston: Houghton Mifflin.

Clark, W.S., and B.K. Wheeler. 2001. A field guide to hawks of North America. 2nd ed. Boston: Houghton Mifflin.

Cooper, J.E. 1973. Post-mortem findings in East African birds of prey. J. Wildl. Dis. 9:368–375.

Craig, T.H., and L.R. Powers. 1976. Raptor mortality due to drowning in a livestock watering tank. Condor 78:412.

Delhey, K., A. Peters, and B. Kempenaers. 2007. Cosmetic coloration in birds: Occurrence, function, and evolution. Am. Nat. 169 (Supplement):S145–158.

Duff, D.G. 2006. Has the plumage of juvenile Honey-buzzards evolved to mimic that of Common Buzzard? Br. Birds 99:118–128.

Dunne, P., D.A. Sibley, and C. Sutton. 2000. Identification review: Zone-tailed Hawk. Birding 32:234–241.

Dunning, J.B., Jr., 1993. CRC handbook of avian body masses. Boca Raton: CRC Press.

Edelstam, C. 2001. Raptor plumages and external structure. *In* Raptors: Birds of prey of the world, ed. J. Ferguson-Lees and D.A. Christie. Boston: Houghton Mifflin.

Edwards, Jr., T.C., and M.W. Collopy. 1983. Obligate and facultative brood reduction in eagles: An examination of factors that influence fratricide. Auk 100:630–635.

Elowson-Haley, A. M. 1982. Osprey spreads wings after fishing. Wilson Bull. 94:366–368.

Feduccia, A. 1980. The age of birds. Cambridge: Harvard University Press.

Ferguson-Lees, J., and D. Christie. 2005. Raptors of the world: A field guide. London: Christopher Helm.

Forbes, N.A., J.E. Cooper, and R.J. Higgins. 2000. Neoplasms of birds of prey. *In* Raptor biomedicine III, ed. J.T. Lumeij, J.D. Remple, P.T. Redig, M. Lierz, and J.E. Cooper. Lake Worth: Zoological Education Network.

Fox, N. 1995. Understanding the bird of prey. Surrey: Hancock House.

Garcia, E. 2006. Birds in Gibraltar 2005. Gibraltar Bird Rep. 2005 5:9–45.

Garcia, E.F., and K.J. Bensusan. 2006. Northbound migrant raptors in June and July at the Strait of Gibraltar. Br. Birds 99:569–575.

Gautier-Clerc, M., A. Tamisier, and F. Cézilly. 2000. Sleep-vigilance trade-off in Gadwall during the winter period. Condor 102:307–313.

Greenwood, A. 1977. The role of disease in the ecology of British raptors. Bird Study 24:259–265.

Hall, J.S., H.S. Ip, J.C. Franson, C. Meteyer, S. Nashold, J.L. Teslaa, J. French, P. Redig, and C. Brand. 2009. Experimental infection of a North American raptor, American Kestrel (*Falco sparverius*), with highly pathogenic avian influenza virus (H5N1). PLoS ONE 4:e7555.

Hamerstrom, F., and K. Janick. 1973. Diurnal sleep rhythm of a young Barred Owl. Auk 90:899–900.

Hatch, D.E. 1970. Energy conserving and heat dissipating mechanisms of the Turkey Vulture. Auk 87:111–124.

Heidenreich, M. 1995. Birds of prey: medicine and management. Oxford: Blackwell Science.

Henny, C.J., and H.M. Wight. 1972. Population ecology and environmental pollution: Red-tailed and Cooper's hawks. Population ecology of migratory birds. Report 2. Washington: U.S. Fish & Wildlife Service, U.S. Department of Interior.

Herkenrath, P. 2007. Recovery of Peregrine Falcon after hitting the sea. Br. Birds 100:304–305.

Hill, G.E., and K.J. McGraw. 2006. Bird Coloration. Volume 1. Cambridge: Harvard University Press.

Hindman, L.J., W.F. Harvey IV, G.R. Costanzo, K.A. Converse, and G. Stein, Jr. 1997. Avian cholera in Ospreys: First occurrence and possible mode of transmission. J. Field Ornithol. 68:503–662.

Houston, D. C. 1980. A possible function of sunning behavior by Griffon Vultures, *Gyps* spp., and other large soaring birds. Ibis 122:366–369.

Kalmbach, E.R. 1939. American vultures and the toxin of *Clostridium botulinum*. J. Am. Vet. Med. Assoc. 94:187–191.

Kerlinger, P. 1989. Flight strategies of migrating hawks. Chicago: Chicago University Press.

Kirk, D.A., and M.J. Mossman. 1998. Turkey Vulture (*Cathartes aura*). *In* The birds of North America, No. 339, ed. A. Poole and F. Gill. Philadelphia: Birds of North America.

Kocan, A.A., J. Snelling, and E.C. Greiner. 1977. Some infectious and parasitic diseases in Oklahoma raptors. J. Wildl. Dis. 13:304–306.

Komi, P. 2008. More dark morph Montagu's Harriers in Spain. Birding World 21:117–120.

Lincoln, F.C. 1936. Recoveries of banded birds of prey. Bird-Banding 7:38–45.

Lindstedt, S.L., and W.A. Calder. 1976. Body size and longevity in birds. Condor 78: 91–145.

Luttich, S.N., L.B. Keith, and J.D. Stephenson. 1971. Population dynamics of the Red-tailed Hawk (*Buteo jamaicensis*) at Rochester, Alberta. Auk 88:75–87.

Lyons, D.M., and J.A. Mosher. 1987. Morphological growth, behavioral development, and parental care of Broad-winged Hawks. *J. Field Ornithol.* 58:334–344.

MacNabb, B.K. 2012. Extreme measures: The ecological energetics of birds and mammals. Chicago: University of Chicago Press.

MacWhirter, R.B., and K.L. Bildstein. 1996. Northern Harrier (*Circus cyaneus*). *In* The birds of North America, No. 210, ed. A. Poole and F. Gill. Philadelphia: The Academy of Natural Sciences, Washington: The American Ornithologists' Union.

Margalida, A., J. Bertran, J. Boudet, and R. Heredia. 2004. Hatching asynchrony, sibling aggression and cannibalism in Bearded Vultures *Gypaetus barbatus*. Ibis 146:386–393.

Mascdetti, G.G., and G. Vallortigara. 2001. Why do birds sleep with one eye open? Light exposure of the chick embryo as a determinant of monocular sleep. Curr. Biol. 11:971–974.

McGraw, K.J. 2006. Bird coloration: Mechanisms and measurements. Vol. 1. Cambridge: Harvard University Press.

Medica, D.L., R. Clauser, and K.L. Bildstein. 2007. Prevalence of West Nile antibodies in a breeding population of American Kestrels (*Falco sparverius*) in Pennsylvania. J. Wildl. Dis. 43:538–541.

Morrison, J.L. 1996. Crested Caracara (*Caracar plancus*). *In* The birds of North America, No. 249, ed. A. Poole and F. Gill. Philadelphia: Birds of North America.

Mueller, H. 1972. Zone-tailed Hawk or Turkey Vulture: Mimicry or aerodynamics? Condor 74:221–222.

Mueller, H.C., D.D. Berger, N.S. Mueller, W. Robichaud, and J.L. Kaspar. 2002. Age and sex differences in wing loading and other aerodynamic characteristics of Merlins. Wilson Bull. 114:272–275.

Mueller, H.C., and K. Meyer. 1985. The evolution of reversed sexual dimorphism in size; a comparative analysis of Falconiforms of the Western Palearctic. Curr. Ornithol. 2:65–101.

Mundy, P., D. Butchart, J. Ledger, and S. Piper. 1992. The vultures of Africa. London: Academic Press.

Negro, J.J. 2008. Two aberrant serpent-eagles may be visual mimics of bird-eating raptors. Ibis 150:307–314.

Negro, J.J., G.R. Bortolotti, J.L. Tella, K.J. Fernie, and D.M. Bird. 1998. Regulation of integumentary colour and plasma carotenoids in American Kestrels consistent with sexual selection theory. Funct. Ecol. 12:307–312.

Negro, J.J., J.M. Grande, J.L. Tella, J. Garrido, D. Hornero, J.A. Donázar, J.A. Sanchez-Zapata, J.R. Benítez, and M. Barcell. 2002. An unusual source of essential carotenoids. Nature 416:807–808.

Negro, J.J., A. Margalida, F. Hiraldo, and R. Heredia. 1958. The function of the cosmetic coloration of bearded vultures: When art imitates life. Anim. Behav. 58:F14-F17.

Negro, J.J., J.H. Sarasola, F. Fariñas, and I. Zorrilla. 2006. Function and occurrence of facial flushing in birds. Comp. Biochem. Physiol., Pt. A 143:78–84.

Nemeth, N., D. Gould, R. Bowen, and N. Komar. 2006. Natural and experimental West Nile virus infection in five species of raptors. J. Wildl. Dis. 42:1–13.

Newton, I. 1979. Population ecology of raptors. Vermillion: Buteo Books.

Pennycuick, C.J. 1972. Animal flight. London: Edward Arnold.

Poole, A.F. 1989. Ospreys: a natural and unnatural history. Cambridge: Cambridge University Press.

Prinzinger, R., B. Nagel, O. Bahat, R. Bogel, E. Karl, D. Weihs, and C. Walzer. 2002. Energy metabolism and body temperature in the Griffon Vulture (*Gyps fulvus*) with comparative data on Hooded Vultures (*Necrosyrtes monachus*) and the White-backed Vulture (*Gyps africanus*). J. Ornithol. 143:456–467.

Rattenborg, N.C. 2006. Do birds sleep in flight? Naturwissenschaften 93:413–425.

Rattner, B.A., M.A. Whitehead, G. Gaspar, C.U. Meteyer, W.A. Link, M.A. Taggart, A.A. Meharg, O.H. Pattee, and D.J. Pain. 2008. Apparent tolerance of Turkey Vultures (*Cathartes aura*) to the non-steriodal anti-inflammatory drug diclofenac. Environ. Toxicol. Chem. 27:2341–2345.

Roth II, T.C., J.A. Lesku, C.J. Amlaner, and S.L. Lima. 2006. A phylogenic analysis of the correlates of sleep in birds. J. Sleep Res. 15:395–402.
Simmons, K.E.L. 1986. The sunning behavior of birds: a guide for ornithologists. Exeter: Bristol Ornithological Club.
Simmons, R.E., T. Hardaker, and W.S. Clark. 2008. Partial albino African Marsh Harrier at Langebaan, South Africa. Gabar 19:60–65.
Snyder, H. 1988. Mimicry (Zone-tailed Hawk). *In* Handbook of North American birds, vol. 5. ed. R.S. Palmer. New Haven: Yale University Press.
Snyder, N.F.R., and J.W. Wiley. 1976. Sexual size dimorphism in hawks and owls of North America. Ornithological Monographs No. 20. Washington: American Ornithologists' Union.
Sonter, C. 1987. Dusting and sunning behavior in Black and Whistling Kites *Milvus migrans* and *Haliastur shenurus*. Aust. Bird Watcher 12:96–97.
Swann, D., T. Mahoney, and R.D. Applegate. 1993. Swimming behavior by immature Peregrine Falcons. Maine Nat. 1:36–37.
Tennekes, H. 1997. The simple science of flight. Cambridge: MIT Press.
Thomsett, S. 2007. A record of a first year dark plumage Augur Buzzard moulting into normal plumage. Gabar 18:2528.
Van Tyne, J., and A.J. Berger. 1976. Fundamentals of ornithology, 2nd ed. New York: J. Wiley & Sons.
Walther, H. 1979. Eleonora's Falcon. Chicago: University of Chicago Press.
Ward, F.P., and D.G. Fairchild. 1972. Air sac parasites of the genus *Serratospiculum* in falcons. J. Wildl. Dis. 8:165–168.
Wetmore, A. 1936. The number of contour feathers in passerine and related birds. Auk 53:159.
Wheeler, B.K. 2003. Raptors of eastern North America. Princeton: Princeton University Press.
White, T.H. 1951. The goshawk. London: Lyons & Burford.
Willis, E.O. 1963. Is the Zone-tailed Hawk a mimic of the Turkey Vulture? Condor 65:313–317.
Zu-Aretz, S. and Y. Leshem. 1980. The sea—a trap for gliding birds. Torgos 1:14–19.

3. Senses and Intelligence

Audubon, J.J. 1827. Account of the habits of the turkey buzzard *Vultur aura*, particularly with the view of exploding the opinion generally entertained of its extraordinary power of smelling. Edinb. New Philos. J. 2:172–184.
Baker-Gabb, D. 1993. Auditory location of prey by three Australian raptors. *In* Australian Raptor Studies, ed. P. Olsen. Melbourne: Australian Raptor Association.
Bang, B.G., and S. Cobb. 1968. The size of the olfactory bulb in 108 species of birds. Auk 85:55–61.
Birkhead, T. 2012. Bird sense: What it's like to be a bird. New York: Walker & Company.
Clark, R.J., and B.L. Stanley. 1976. Facial feathering of the Marsh Hawk (*Circus cyaneus hudsonius*), Long-eared (*Asio otus*) and Short-eared Owl (*Asio flammeus*) compared. Proc. Pennsylvania Acad. Sci. 50:86–88.
Clarke, R. 1996. Montagu's Harrier. Chelmsford: Arlequin Press.
DeCandido, R., and D. Allen. 2006. Nocturnal hunting by Peregrine Falcons at the Empire State Building, New York City. Wilson Bull. 118:53–58.

Diamond, J., and A. B. Bond. 1999. Kea: Bird of paradox. Berkeley: University of California Press.
Dumbacher, J. P., B. M. Beehler, T. F. Spande, H. M. Garraffo, and J. W. Dalt. 1992. Homobatrachotoxin in the Genus *Pitohui*: chemical defense in birds? Science 258:799–801.
Ellis, D. H., and S. Brunson. 1993. "Tool" use by the Red-tailed Hawk (*Buteo jamaicensis*). J. Raptor Res. 27:128.
Emery, N. J. 2005. Cognitive ornithology: The evolution of avian intelligence. Philos. Trans. R. Soc. B Biol. 361:23–43.
Fox, N. 1995. Understanding the bird of prey. Surrey: Hancock House Publishers.
Houston, D.C. 1984. Does the King Vulture *Sarcoramphus papa* use a sense of smell to locate food? Ibis 126:67–69.
Jones, M. P., K. E. Pierce Jr., and D. Ward. 2007. Avian vision: A review of form and function with special consideration to birds of prey. J. Exotic Pet Med. 16:69–87.
Koivula, M., and J. Viitala. 1999. Rough-legged Buzzards use vole scent marks to assess hunting areas. J. Avian Biol. 30:329–332.
Malakoff, D. 1999. Following the scent of avian olfaction. Science 286:704–705.
Martin, G. R., S. J. Portugal, and C. P. Murn. 2012. Visual fields, foraging and collision vulnerability in Gyps vultures. Ibis 154:626–631.
Mayr, E. 1974. Behavior programs and evolution strategies. Am. Sci. 62:650–659.
Mundy, P., D. Butchart, J. Ledger, and S. Piper. 1993. The vultures of Africa. London: Academic Press.
Olsen, P. 1995. Australian birds of prey. Baltimore: Johns Hopkins University Press.
Payne, R. S. 1971. Acoustic location of prey by Barn Owls (*Tyto alba*). J. Exp. Biol. 54:535–573.
Penteriani, V., C. Alonso-Alvarez, M. Delgado, F. Sergio, and M. Ferrer. 2006. Brightness variability in the white badge of the eagle owl *Bubo bubo*. J. Avian Biol. 37:110–116.
Podulka, S., R. W. Rohrbaugh, Jr., and R. Bonney, eds. 2004. Handbook of bird biology. Princeton: Princeton University Press.
Rice, W. R. 1982. Acoustical location of prey by the Marsh Hawk: Adaptation to concealed prey. Auk 99:403–413.
Stager, K. E. 1964. The role of olfaction in food location by the Turkey Vulture (*Cathartes aura*). Los Ang. C. Mus. Contrib. Sci. 81:1–63.
Van Lawick-Goodall, J., and H. van Lawick-Goodall. 1966. Use of tools by Egyptian Vultures *Neophron percnopterus*. Nature 212:1468–1469.
Viitala, J., E. Korpamäki, P. Palokangas, and M. Kolvula. 1995. Attraction of kestrels to vole scent marks visible in ultraviolet light. Nature 373:425–427.
Wood, S. R., K. J. Sanderson, and C. S. Evans. 2000. Perception of terrestrial and aerial alarm calls by honeyeaters and falcons. Aust. J. Zool. 48:127.134.
Yang, S-Y., B. A Walther, and G-J. Weng. 2015. Stop and smell the pollen: The role of olfaction and vision of the Oriental Honey Buzzard in identifying food. PLOS ONE DOI:10.1371/journal.pone.0130191.

4. Distribution and Abundance

Bildstein, K. L. 2006. Migrating raptors of the world: Their ecology and conservation. Ithaca: Cornell University Press.
Bildstein, K. L., W. Schelsky, J. Zalles, and S. Ellis. 1998. Conservation status of tropical raptors. J. Raptor Res. 32:3–18.

BirdLife International. 2014. Species factsheets downloaded from http://www.birdlife.org in October 2014.
Carrascal, L. M., and J. Seone. 2009. Factors affecting large-scale distribution of the Bonelli's eagle *Aquila fasciata* in Spain. Ecol. Res. 24:565–573.
Colinvaux, P. 1978. Why big fierce animals are rare. Princeton: Princeton University Press.
Collar, N. J., M. J. Crosby, and A. J. Sattersfield. 1994. Birds to watch 2: The world list of threatened birds. Cambridge: BirdLife International.
Cook, R. E. 1977. Raymond Lindeman and the trophic-dynamic concept in ecology. Science 198:22–26.
del Hoyo, J., A. Elliot, and J. Sargatal. 1994. Handbook of birds of the world. Vol. 2. New World vultures to guineafowl. Barcelona: Lynx Edicions.
Denes, F. V., L. F. Silvera, S. Seipke, R. Thorstrom, W. S. Clark, and J.-M. Thiollay. 2011. The White-collared Kite (*Leptodon Forbesi* Swann, 1922) and a review of the taxonomy of the Grey-headed Kite (*Leptodon cayanensis* Latham, 1790). Wilson J. Ornithol. 123:323–331.
Ferguson-Lees, J., and D. A. Christie. 2001. Raptors of the world. Boston: Houghton Mifflin.
Gaston, K. J. 1994. Rarity. London: Chapman & Hall.
Gaston, K. J., R. G. Davies, C. E. Gascoigne, and M. Williamson. 2005. The structure of global species-range size distributions: raptors & owls. Global Ecol. Biogeogr. 14:67–75.
Gill, F., and M. Wright. 2006. Birds of the world: Recommended English names. Princeton: Princeton University Press.
Kerlinger, P. 1989. Flight strategies of migrating hawks. Chicago: Chicago University Press.
Newton, I. 2003. Speciation and biogeography in birds. Amsterdam: Academic Press.
Rabinowitz, D., S. Cairns, and Theresa Dillon. 1986. Seven forms of rarity and their frequency in the flora of the British Isles. *In* Conservation biology: The science of scarcity and diversity, ed. M. E. Soulé. Sunderland: Sinauer.
Stauffer, R. C. 1975. Charles Darwin's natural selection: Being the second part of his species book written from 1856 to 1858. Cambridge: Cambridge University Press.
Watts, B. D., M. A. Byrd, and M. U. Watts. 2004. Status and distribution of breeding Ospreys in the Chesapeake Bay: 1995–1996. J. Raptor Res. 38:47–54.

5. Breeding Ecology

Beske, A. E. 1982. Local and migratory movements of radio-tagged juvenile harriers. Raptor Res.16:39–53.
Bildstein, K. L. 1992. Causes and consequences of reversed sexual size dimorphism in raptors: The head-start hypothesis. J. Raptor Res. 26:115–123.
Bird, D., D. Varland, and J. Negro. 1996. Raptors in human landscapes: Adaptations to built and cultivated landscapes. London: Academic Press.
Blanco, G., and J. L. Tella. 1997. Protective association and breeding advantages of choughs nesting in lesser kestrel colonies. Anim. Behav. 54:335–342.
Brown, L. 1980. The African Fish Eagle. Folkstone: Baily Bros. and Swiffin.
Burnham, K. K., W. A. Burnham, and I. Newton. 2009. Gyrfalcon *Falco rusticolus* post-glacial colonization and extreme long-term use of nest-sites in Greenland. Ibis 151:514–522.
Bustamante, J. 1993. Post-fledging dependence period and development of flight and hunting behavior in the Red Kite *Milvus milvus*. Bird Study 40:181–188.
Cade, T. J. 1960. Ecology of the Peregrine and Gyrfalcon populations in Alaska. Berkeley: University of California Press.

Campobello, D., M. Sara, and J. F. Hare. 2012. Under my wing: Lesser kestrels and jackdaws derive reciprocal benefits in mixed-species colonies. Behav. Ecol. 23:425–433.

Dennis, R. 2008. A life of Ospreys. Caithness: Whittles Publishing.

Dixon, A., B. Munkhjargal, D. Shijirmaa, A. Saruul, and G. Purev-Ochir. 2010. Artificial nests for Saker Falcons II: Progress and plans. Falco 35:6–8.

Dunne, P. 1995. The wind masters: The lives of North American birds of prey. Boston: Houghton Mifflin.

Faaborg, J. 1986. Reproductive success and survivorship of the Galapagos Hawk *Buteo galapagoensis*: Potential costs and benefits of cooperative polyandry. Ibis 128:337–347.

Fernández, C., and J.A. Donázar. 1991. Griffon Vultures *Gyps fulvus* occupying eyries of other cliff-nesting raptors. Bird Study 38:42–44.

Gangoso, L., and C.-J. Palacios. 2005. Ground nesting by Egyptian Vultures (*Neophron percnopterus*) in the Canary Islands. J. Raptor Res. 39:186–187.

Gargett, V. 1990. The Black Eagle: Verreaux's Eagle in Southern Africa. Academic Press: London.

Gerrard, J. M., and G. R. Bortolotti. 1988. The Bald Eagle: Haunts and habits of a wilderness monarch. Washington: Smithsonian Institution Press.

Hamerstrom, F. 1986. Harrier, hawk of the marshes: The hawk that is ruled by a mouse. Washington: Smithsonian Institution Press.

Heinrich, B. 2013. Why does a hawk build with green nesting material? Northeast. Nat. 20:209–218.

Henny, C. J., M. M. Smith, and V. D. Stotts. 1974. The 1973 distribution and abundance of breeding Ospreys in the Chesapeake Bay. Chesapeake Sci. 15:125–133.

Katzner, T., S. Robertson, B. Robertson, J. Klucsarits, K. McCarty, and K. L. Bildstein. 2005. Results from a long-term nest-box program for American Kestrels: Implications for improved population monitoring and conservation. J. Field Ornithol. 76:217–226.

Kenward, R. 2006. The Goshawk. London: T & A D Poyser.

Mee, A., B. A. Rideout, J. A. Hamber, J. N. Todd, G. Austin, M. Clark, and M. P. Wallace. 2007. Junk ingestion and nestling mortality in a reintroduced population of California Condors *Gymnogyps californianus.* Bird Conserv. Int. 17:119–130.

Mougeot, F. 2000. Territorial intrusions and copulatory patterns in red kites, *Milvus milvus,* in relation to breeding density. Anim. Behav. 59:633–642.

Mougeot, F. 2001. Decoy presentations as a means to manipulate the risk of extrapair copulations: An experimental study in a semicolonial raptor, the Montagu's Harrier (*Circus pygargus*) Behav. Ecol. 12:1–7.

Nagy, A. 1963. Population density of Sparrow Hawks in eastern Pennsylvania. Wilson Bull. 75:93.

Newton, I. 1979. Population ecology of raptors. Vermillion: Buteo Books.

Newton, I. 1986. The Sparrowhawk. Calton: T & A D Poyser.

Olsen, J. 2014. Australian high country raptors. Collinwood: CSIRO Publishing.

Ontiveros, D., J. Caro, and J. M. Pleguezuelos. 2008. Possible functions of alternative nests in raptors: the case of the Bonelli's Eagle. J. Ornithol. 149:253–259.

Poole, A. F. 1989. Ospreys: A natural and unnatural history. Cambridge: Cambridge University Press.

Quinn, J. L., J. Prop, Y. Kokorev, and J. M. Black. 2003. Predator protection or similar habitat selection in red-breasted goose nesting associations: Extremes along a continuum. Anim. Behav. 65:297–307.

Richardson, C. T., and C. K. Miller. 1997. Recommendations for protecting raptors from human disturbance. Wildl. Soc. Bull. 25:634–638.

Simmons, R.E. 2000. Harriers of the world: Their behavior and ecology. Oxford: Oxford University Press.

Simmons, R.E., and J.M. Mendelsohn. 1993. A critical review of cartwheeling flights of raptors. Ostrich 64:13–24.

Stone, W. 1937. Bird studies at Old Cape May. Philadelphia: Delaware Valley Ornithological Club.

Villarroel, M., D.M. Bird, U. Kuhnlein. 1998. Copulatory behavior and paternity in the American Kestrel: The adaptive significance of frequent copulations. Anim. Behav. 56:289–299.

Watson, D. 1977. The Hen Harrier. Berkhamsted: T & A D Poyser.

Watson, J. 1977. The Golden Eagle. London: T & A D Poyser.

Whitacre, D.F., ed. 2012. Neotropical birds of prey: Biology and ecology of a forest raptor community. Ithaca: Cornell University Press.

White, C.M., T.J. Cade, and J.H. Enderson. 2013. Peregrine Falcons of the world. Barcelona: Lynx Ediciones.

Wiens, J.D., B.R. Noon, and R.T. Reynolds. 2006. Post-fledging survival of Northern Goshawks: The importance of prey abundance, weather, and dispersal. Ecol. Appl. 16:406–418.

6. Feeding Behavior

Alerstam, T. 1987. Radar observation of the Peregrine Falcon *Falco peregrinus* and the Goshawk *Accipter gentilis*. Ibis 129:267–273.

Bildstein, K.L. 1978. Behavioral ecology of Red-tailed Hawks (*Buteo jamaicensis*), Rough-legged Hawks (*B. lagopus*), Northern Harriers (*Circus cyaneus*), and American Kestrels (*Falco sparverius*) in south central Ohio. Ohio Biol. Surv. Biol. Notes 18:1–53.

Bildstein, K.L. 1980. Corn cob manipulation in Northern Harriers. Wilson Bull. 92:128–130.

Bildstein, K.L., and M.W. Collopy. 1987. Hunting behavior of Eurasian (*Falco tinnunculus*) and American Kestrels (*F. sparverius*): A review. Raptor Res. Rep. 6:1–178.

Black, H.L., G. Howard, and R. Stjernstedt. 1979. Observations of the feeding behavior of the Bat Hawk. Biotropica 11:18–21.

Collopy, M.W., and K.L. Bildstein. 1987. Foraging behavior of northern Harriers wintering in southeastern salt and freshwater marshes. Auk 104:11–16.

Cresswell, W., and J.L. Quinn. 2013. Contrasting risks from different predators change the overall non-lethal effects of predation risk. Behav. Ecol. 24:871–876.

Daan, S., and S. Slopsema. 1978. Short-term rhythms in foraging behavior of the common vole, *Microtus arvalis*. J. Comp. Physiol. 127:215–227,

DeCandido, R., and D. Allen. 2006. Nocturnal hunting by Peregrine Falcons at the Empire State Building, New York City. Wilson J. Ornithol. 118:53–58.

del Hoyo, J., A. Elliott, and J. Sargatal. 1994. Handbook of the birds of the world. Volume 2. Barcelona: Lynx Edicions.

Einoder, L., and A. Richardson. 2006. An eco-morphological study of the raptorial digital locking mechanism. Ibis 148:515–525.

Engelbrecht, D. 2010. Object play of an immature Martial Eagle *Polemaetus bellicosus*. Ornithol. Observ. 1:52–54.

Fox, N. 1995. Understanding the bird of prey. Surrey: Hancock House.

Grubb, T.C. 1977. Weather-dependent foraging in Ospreys. Auk 94:146–149.

Grubb, T.C. 1977. Why Ospreys hover. Wilson Bull. 89:149–150.
Hamerstrom, F. 1957. The influence of a hawk's appetite on mobbing. Condor 59:192–194.
Hamerstrom, F. 1986. Harrier, hawk of the marshes: The hawk that is ruled by a mouse. Washington: Smithsonian Institution Press.
Hewitt, S. 2013. Avian drop-catch play: A review. Brit. Birds 106:206–216.
Hilgert, N. 1988. Aspects of breeding and feeding behavior of Peregrine Falcons in Guayllabamba, Ecuador. *In* Peregrine Falcon populations: their management and recovery, ed. T.J. Cade, J.H. Enderson, C.G. Thelander, and C.M. White. Boise: The Peregrine Fund.
Hope D.D., D.B. Lank, B.D. Smith, and R.C. Ydenberg. 2011. Migration of two calidrid sandpiper species on the predator landscape: How stopover time and hence migration speed vary with geographical proximity to danger. J. Avian Biol. 42:52–529.
Houston, D.C., A. Mee, and M. McGrady. 2007. Why do condors and vultures eat junk? The implications for conservation. J. Raptor Res. 41:235–238.
Hunt, G.W., R.R. Rogers, and D.J. Slowe. 1975. Migratory and foraging behavior of Peregrine Falcons on the Texas coast. Can. Field-Nat. 89:111–123.
Lotem, A., E. Schectman, and G. Katzir. 1991. Capture of submerged prey by little egrets *Egretta garzetta garzetta*: strike depth, strike angle and the problem of light reflection. Anim. Behav. 42:341–346.
Mundy, P., D. Buthchart, J. Ledger, and S. Piper. 1992. The vultures of Africa. London: Academic Press.
Negro, J.J., J. Bustamante, C. Melguizo, J.L. Ruiz, and J.M. Grande. 2000. Nocturnal activity of Lesser Kestrels under artificial lighting conditions in Seville, Spain. J. Raptor Res. 34:327–329.
Newton, I. 1979. Population ecology of raptors. Vermillion: Buteo Books.
Newton, I. 1986. The Sparrowhawk. Calton: T & A D Poyser.
Pierson, J.E., and P. Donahue. Peregrine Falcon feeding on bats in Suriname, South America. 1983. Am. Birds 37:257–259.
Ponitz, B., A. Schmitz, D. Fischer, H. Bleckmann, and C. Brücker. 2014. Diving-flight aerodynamics of a Peregrine Falcon (*Falco peregrinus*). PLOS ONE e86506.
Rexer-Huber, K., and K.L. Bildstein. 2013. Winter diet of Striated Caracara *Phalcoboenus australis* at a farm settlement on the Falkland Islands. Polar Biol. 36:437–443.
Rijnsdorp, A., S. Daan, and C. Dijkstra. 1981. Hunting in the Kestrel, *Falco tunnunculus*, and the adaptive significance of daily habits. Oecologia 50:391–406.
Ruppell, G. 1981. Analysis of prey-catching in the Osprey (*Pandion haliaetus*). J. Ornithol. 122:S. 285–305.
Siverio, F., M. Siverio, and J.J. Hernandez. 2009. Slime and algae ingestion by Ospreys. Br. Birds 102:36.
Strange I.J. 1996. The Striated Caracara *Phalcoboenus australis* in the Falkland Islands. Falkland Islands: Author's Edition.
White, C.M., and T.J. Cade. 1971. Cliff-nesting raptors and ravens along the Colville River in arctic Alaska. Living Bird 10:107–150.

7. Migration

Alerstam, T. 1990. Bird migration. Cambridge: Cambridge University Press.
Berthold, P. 2001. Bird migration: A general study. 2nd. Ed. Oxford: Oxford University Press.

Bildstein, K.L. 1987. Behavioral ecology of Red-tailed Hawks (*Buteo jamaicensis*), Rough-legged Hawks (*Buteo lagopus*), Northern Harriers (*Circus cyaneus*), and American Kestrels (*Falco sparverius*) in south central Ohio. Ohio Biol. Surv. Biol. Notes 18:1–53.

Bildstein, K.L. 1999. Racing with the sun: The forced migration of the Broad-winged Hawk. *In* Gatherings of angels: migrating birds and their ecology, ed., K.P. Able. Ithaca: Cornell University Press.

Bildstein, K.L. 2006. Migrating raptors of the world: Their ecology and conservation. Ithaca: Cornell University Press.

Bildstein, K.L., M.J. Bechard, C. Farmer, and L. Newcomb. 2009. Narrow sea crossings present a significant barrier to migrating Griffon Vultures *Gyps fulvus*. Ibis 151:382–391.

Blem, C.R. 1980. The energetics of migration. *In* Animal migration, orientation, and navigation, ed., S.A. Gauthreaux, Jr. New York: Academic Press.

Broun, M. 1949. Hawks aloft: The story of Hawk Mountain. Cornwall: Cornwall Press.

Careau, V., J.-F. Therrien, P. Porras, D. Thomas, and K. Bildstein. 2006. The use of soaring and gliding versus flapping flight in migrating Broad-winged Hawks: Behavior in the Nearctic and Neotropics compared. Wilson J. Ornithol. 118:471–477.

Cox, G.W. 1968. The role of competition in the evolution of migration. Evolution 22:180–192.

Cox, G.W. 1985. The evolution of avian migration systems between temperate and tropical regions of the New World. Am. Nat. 126:451–474.

Drost, R. 1938. Über den Einfluss von Verfrachtungen zur Herbstzugzeit auf den Sperber Accipiter nisus (L.). Zugleich ein Beitrag zur Frage nach der Orientierung der Vögel auf dem Zuge ins Winterquartier. Proc. Int. Ornithol. Congr. 9:503–521.

Duncan, C.D. 1996. Changes in the winter abundance of Sharp-shinned Hawks, *Accipiter striatus*, in New England. J. Field Ornithol. 67:254–262.

Emlen, S.T. 1967. Migratory orientation in the Indigo Bunting, *Passerina cyanea*. Auk 81:309–342, 463–489.

Ferguson-Lees, J., and D.A. Christie. 2001. Raptors of the world. Boston: Houghton Mifflin.

Geyr von Schweppenburg, H.F. 1963. Zut Terminologie und Theorie der Leitlinie. J. Ornithol. 104:191–204.

Hedenstrom, A. 1993. Migration by soaring or flapping flight in birds: the relative importance of energy cost and speed. Philos. Trans. R. Soc. Lond. B Biol. 342:353–361.

Kerlinger, P. 1984. Flight behaviour of Sharp-shinned Hawks during migration. 2. Over water. Anim. Behav. 32:1029–1034.

Kerlinger, P. 1985. Water-crossing behavior of raptors during migration. Wilson Bull. 97:109–113.

Kerlinger, P. 1989. Flight strategies of migrating hawks. Chicago: University of Chicago Press.

Kerlinger, P., and S.A. Gauthreaux. 1985. Flight behavior of raptors during spring migration in south Texas studied with radar and visual observations. J. Field Ornithol. 56:394–402.

Kjellén, N. 1992. Differential timing of autumn migration between sex and age groups in raptors at Falsterbo, Sweden. Ornis Scand. 23:420–434.

Kjellén, N. 1994. Differences in age and sex ratio among migrating and wintering raptors in southern Sweden. Auk 111:274–284.

Leshem, Y., and Y. Yom-Tov. 1996. The magnitude and timing of migration by soaring raptors, pelicans and storks over Israel. Ibis 138:188–203.

Maransky, B.P., and K.L. Bildstein. 2001. Follow your elders: age-related differences in the migration behavior of Broad-winged Hawks at Hawk Mountain Sanctuary, Pennsylvania. Wilson Bull. 113:350–353.

Maransky, B., L. Goodrich, and K. Bildstein. 1997. Seasonal shifts in the effects of weather on the visible migration of Red-tailed Hawks at Hawk Mountain, Pennsylvania, 1992–1994. Wilson Bull. 109:246–252.

Meyburg, B.-U., X. Eichaker, C. Meyburg, and P. Paillat. 1995. Migrations of an adult Spotted Eagle tracked by satellite. Br. Birds 88:357–361.

Newton, I. 1979. Population ecology of raptors. Vermillion: Buteo Books.

Newton, I. 1986. The Sparrowhawk. Calton: T&AD Poyser.

Smith, N.G. 1980. Hawk and vulture migrations in the Neotropics. *In* Migrant birds in the Neotropics, ed., A. Keast and E. Morton. Washington, D.C.: Smithsonian Institution Press.

Smith, N.G. 1985. Thermals, cloud streets, trade winds, and tropical storms: How migrating raptors make the most of atmospheric energy in Central America. *In* Proceedings of Hawk Migration Conference IV. ed. M. Harwood. Lynchburg: Hawk Migration Association of North America.

Sodhi, N.S., L.W. Oliphant, P.C. James, and I.G. Warkentin. 1993. Merlin. *In* The birds of North America, No. 44, ed. A. Poole and F. Gill. Philadelphia: Birds of North America.

Spaar, R. 1997. Flight strategies of migrating raptors: a comparative study of interspecific variation in flight characteristics. Ibis 139:523–535.

Spaar, R., and B. Bruderer. 1996. Soaring migration of Steppe Eagles *Aquila nipalensis* in southern Israel: Flight behaviour under various wind and thermal conditions. J. Avian Biol. 27:289–301.

Spaar, R., and B. Bruderer. 1997. Optimal flight behavior of soaring migrants: a case study of migrating Steppe Buzzards, *Buteo buteo vulpinus*. Behav. Ecol. 8:288–297.

Spaar, R., H. Stark, and F. Liechti. 1998. Migratory strategies of Levant Sparrowhawks: Time or energy minimization? Anim. Behav. 56:1185–1197.

Trowbridge, C.C. 1902. The relation of wind to bird migration. Am. Nat. 36:735–753.

Viverette, C.B., S. Struve, L.J. Goodrich, and K.L. Bildstein. 1996. Decreases in migrating Sharp-shinned Hawks (*Accipiter striatus*) at traditional raptor-migration watch sites in eastern North America. Auk 113:32–40.

Yosef, R. 1996. Raptors feeding on migration at Eilat, Israel: opportunistic behavior or migratory strategy? J. Raptor Res. 30:242–245.

Zalles, J.I., and K.L. Bildstein. 2000. Raptor watch: A global directory of raptor migration sites. Cambridge: BirdLife International; Kempton: Hawk Mountain Sanctuary.

8. Raptors and People

Berger, L.R., and W.S. McGraw. 2007. Further evidence for eagle predation of, and feeding damage on, the Taung child. South Afr. J. Sci.103:496–498.

Bickerton, D., and E. Szathmáry. 2011. Confrontational scavenging as a possible source for language and cooperation. Evol. Biol. 11:261.

Bijleveld, Maarten. 1974. Birds of prey in Europe. London: Macmillan.

Bildstein, K.L. 2006. Migrating raptors of the world: Their ecology and conservation. Ithaca: Cornell University Press.

Bildstein, K.L., W. Shelsky, J. Zalles, and S. Ellis. 1998. Conservation status of tropical raptors. J. Raptor Res. 32:3–18.

BirdLife International. 2015. Species factsheets. http//www.birdlife.org searched in August 2015.

Broun, M. 1949. Hawks aloft: The story of Hawk Mountain. Cornwall: Cornwall Press.

Burnham, W.A. 1990. Raptors and People. *In* Birds of Prey, ed. by I. Newton. New York: Facts on File.

Campbell, M.O. 2015. Vultures: Their evolution, ecology and conservation. Boca Raton: CRC Press.

D'Elia, J., and S.M. Haig. 2013. California Condors in the Pacific Northwest. Corvallis: Oregon State University Press.

Markandya, A., T. Taylor, A. Longo, M.N. Murty, S. Murty, and K. Dhavala. 2008. Counting the cost of vulture decline—an appraisal of the human health and other benefits of vultures in India. Ecol. Econ. 67:194–204.

Neruda, P. 1996. Arte De Pajaros/Art of Birds. Barcelona: Lynx Ediciones.

Newton, I. 1990. Human impacts on raptors. *In* Birds of Prey. (ed., I. Newton.) New York: Facts on File.

Oaks, J.L., M. Gilbert, M.Z. Virani, R.T. Watson, C.U. Meteyer, B.A. Rideout, H.L.

Ogada, D.L., F. Keesing, and M.Z. Virani. 2012. Dropping dead: Causes and consequences of vulture declines worldwide. Ann. N.Y. Acad. Sci. 1249:57–71.

Shivaprasad, S. Ahmed, M.J.I. Chaudhry, M. Arshad, S. Mahmood, A. Ali, and A.A. Khan. 2004. Diclofenac residues as the cause of vulture population decline in Pakistan. Nature 427:630–633.

Zalles, J.I., and K.L. Bildstein. 2000. Raptor watch: A global directory of raptor migration sites. Cambridge: BirdLife International; Kempton: Hawk Mountain Sanctuary.

INDEX

Note: Page numbers followed by a *T* refer to tables; page numbers followed by an *F* refer to illustrations